I0819616

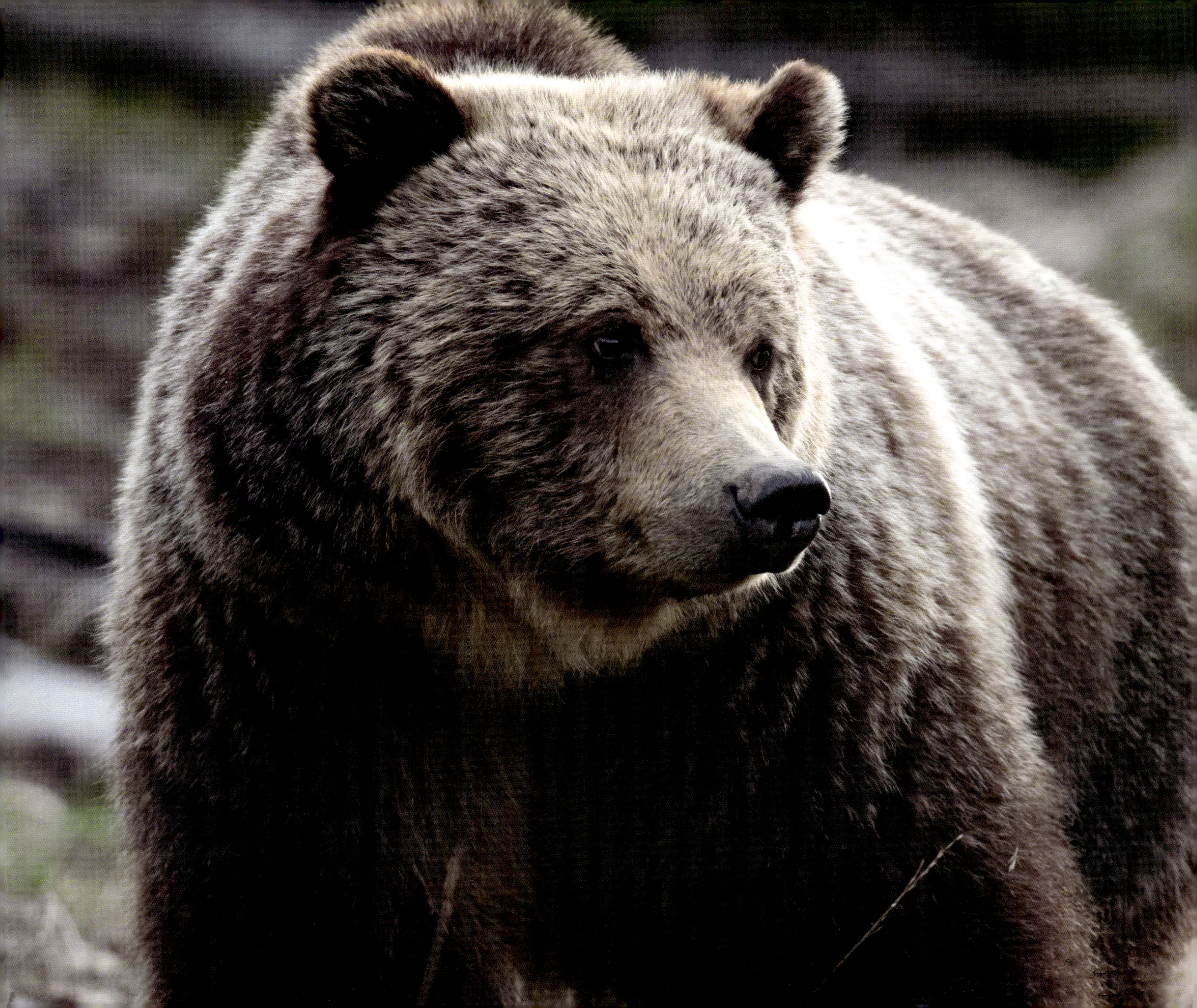

BEARS

THE MIGHTY GRIZZLIES *of the* WEST

JULIE ARGYLE

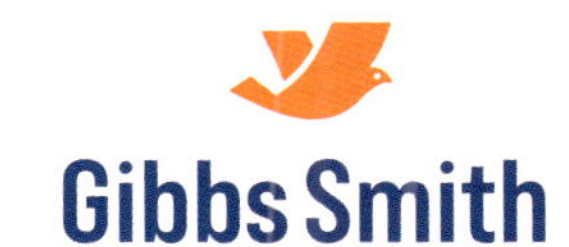

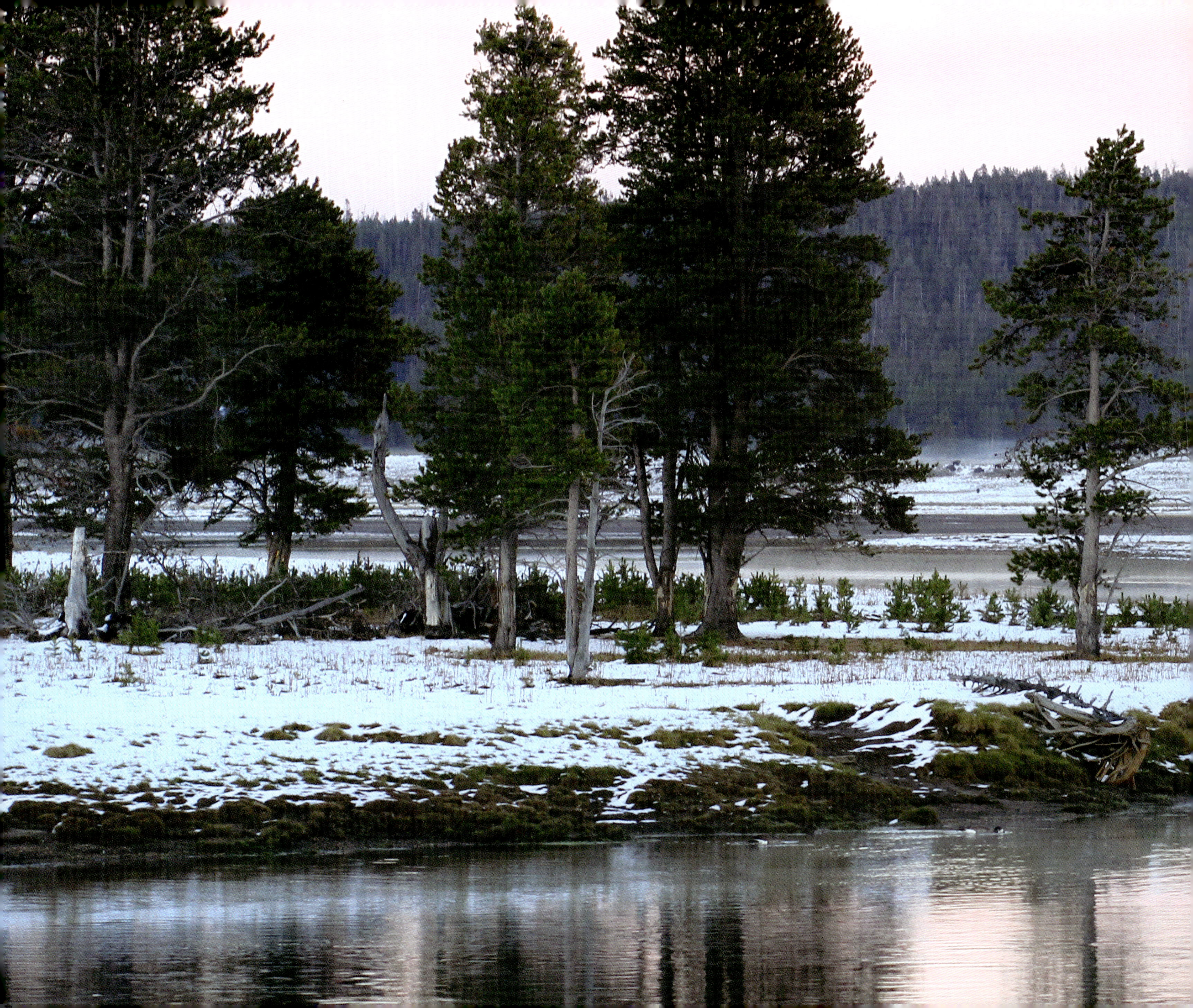

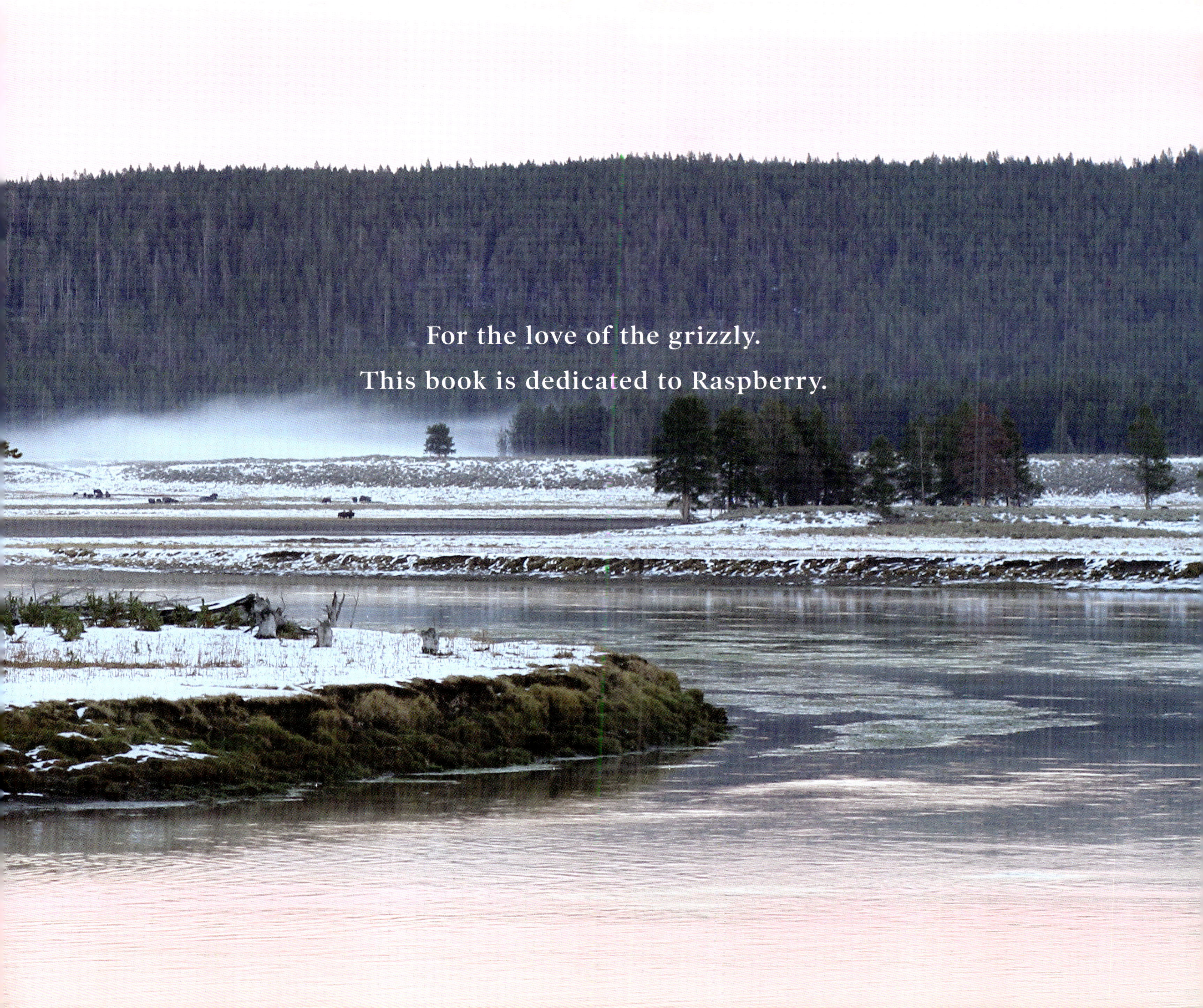

For the love of the grizzly.

This book is dedicated to Raspberry.

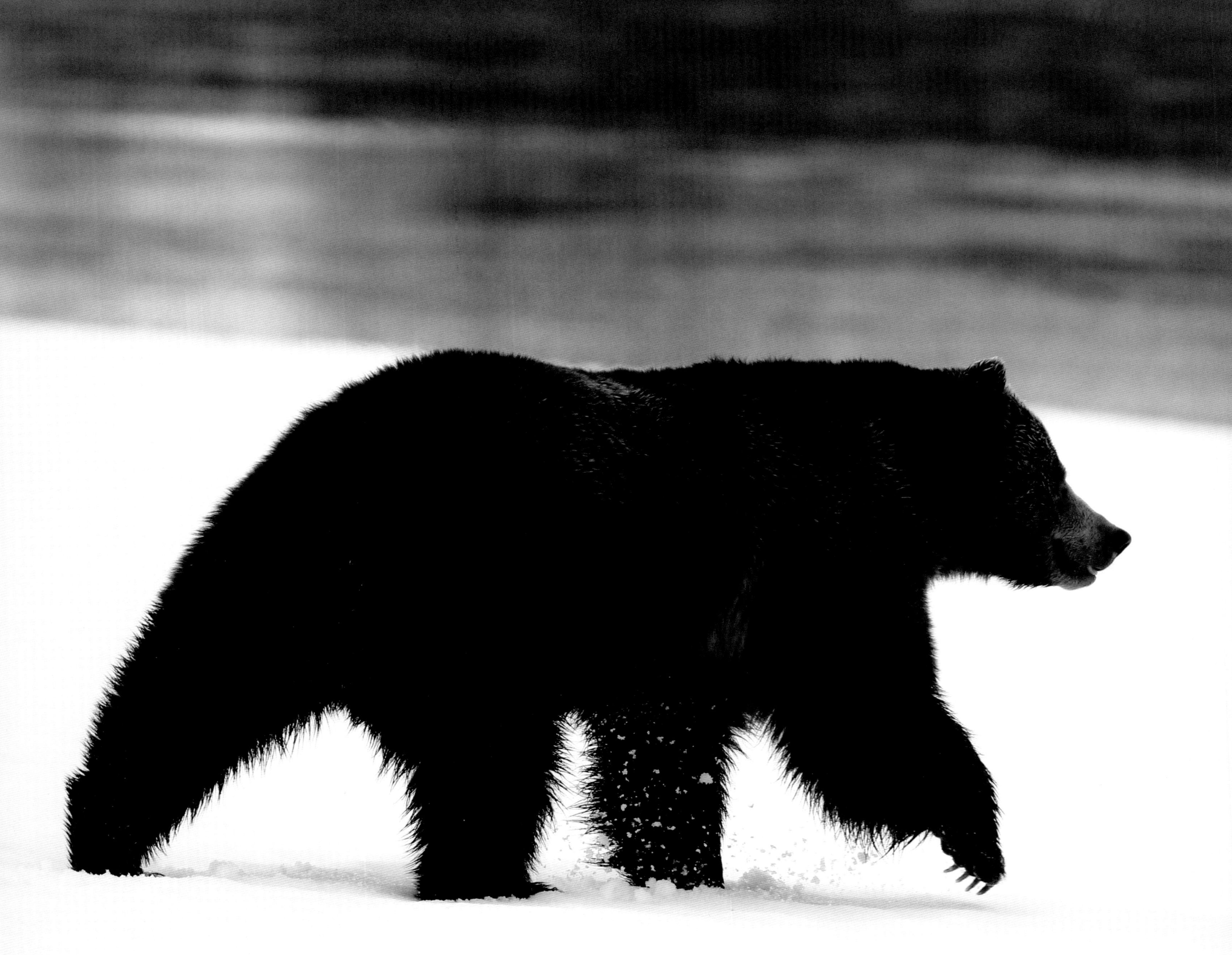

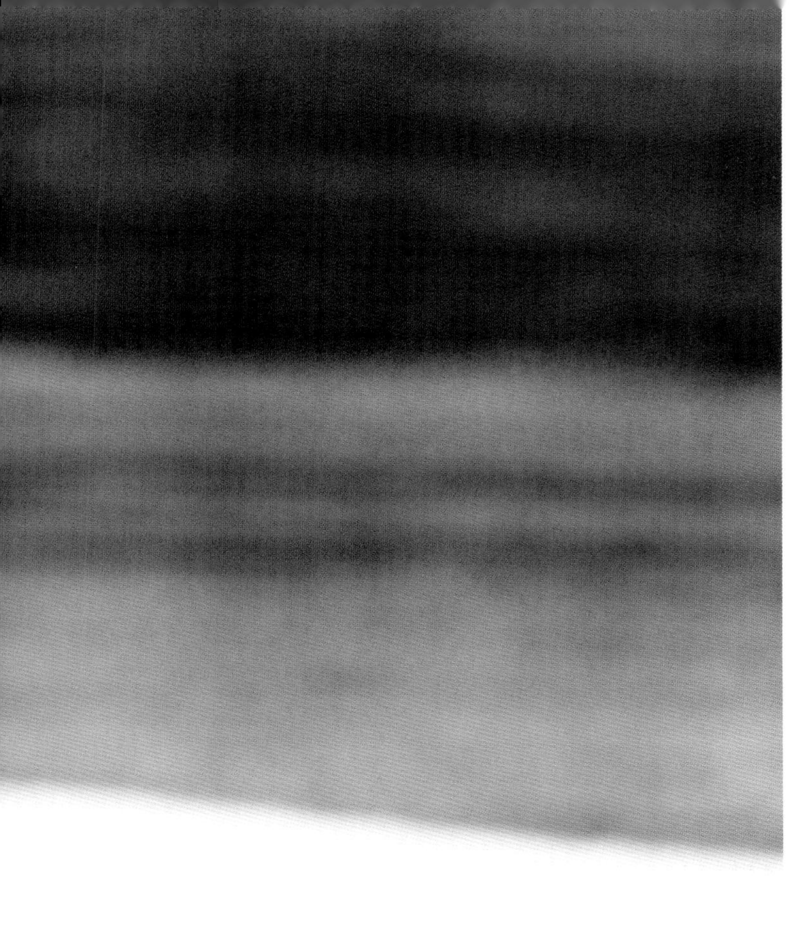

CONTENTS

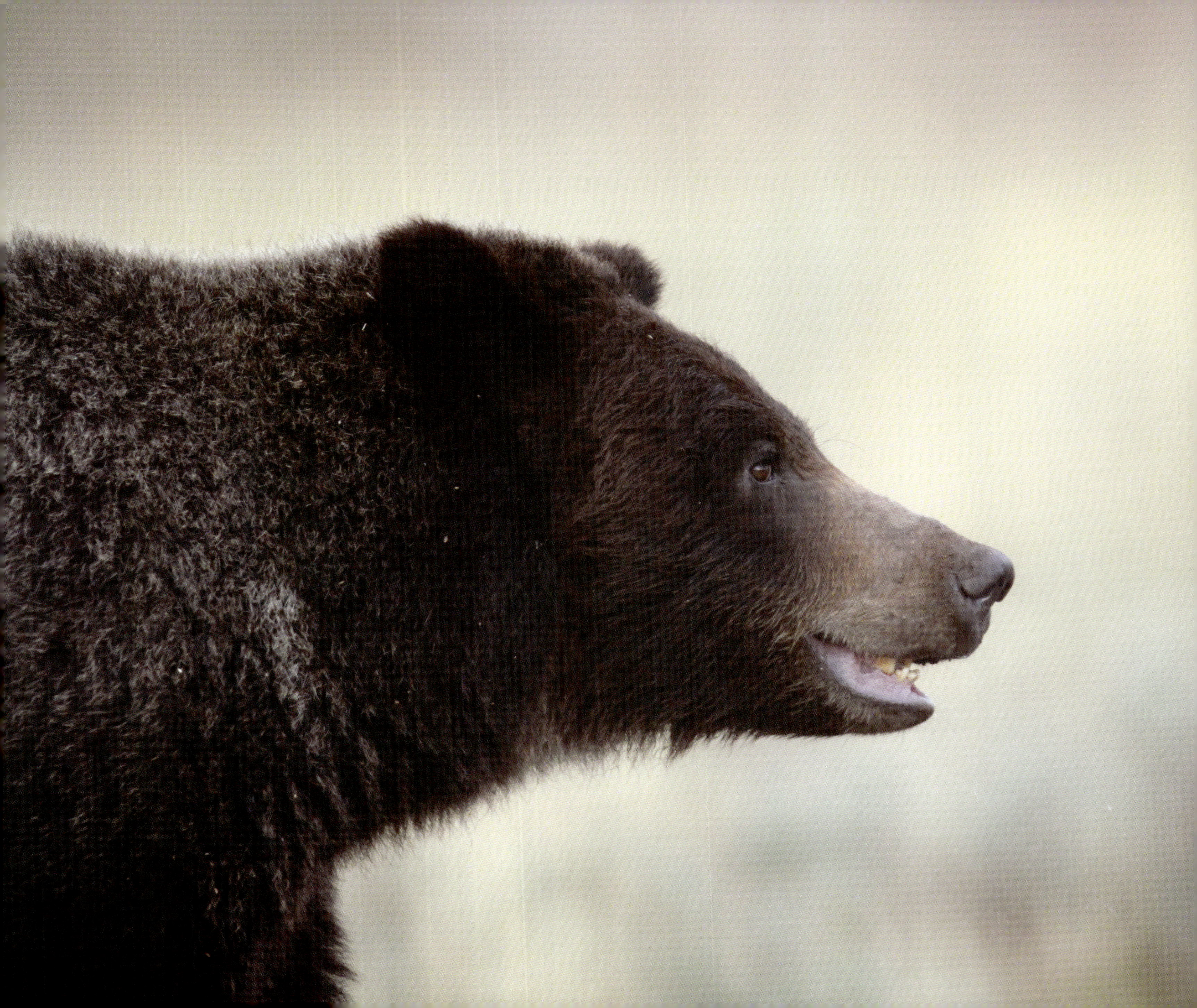

Bears are not companions of men,
but children of God, and His
charity is broad enough for both.

—*John Muir*

FOREWORD

Vóhpåhtse-náhkohe, or "white-mouthed bear," is what my Native American friend calls them, not grizzly bears. The fact his people named this bear for the saliva produced around its mouth when agitated, and not for the silvery, grizzled tips of its outer guard hairs as seen from a distance, speaks to their intimate interactions with the great bear. The Cheyenne, Crow, Lakota, Blackfeet, and First Nation's peoples around the world have not only coexisted in reverence with the grizzly bear, but sought them out as wise grandfathers and keepers of a very sacred, healing medicine—some of the most powerful among shamans.

So when I went to work on a grizzly bear film with fellow advocate and friend, Doug Peacock, one of the first people we reached out to was our Native friend who works with tribes across North America. I asked him if there was any opportunity to film, and/or sit in on a grizzly bear ceremony, somewhere within a day's drive of my home in Montana.

He told me that once the grizzly bear was harassed, murdered, and eventually driven to the most remote and desolate regions of the American West, not unlike indigenous cultures, the sacred relationship between the two was disrupted. That when the grizzly bear began to disappear off of the physical landscape, and holy men with grizzly bear medicine moved on to the next world, there was no one to pass the eons old knowledge to. He told me that sadly, once the grizzly was exterminated, the medicine and ceremonies usually went with it. There was a great heaviness between us.

We have more at stake than just the survival of an animal species. I believe that whatever happens to the grizzly bear will eventually befall humans. The two are that connected. For when we lose the ability to appreciate the intricacies of coexistence, and compassion for another five-toed sentient being trying to find its way in this world too, we lose that luminous fiber that connects us all, and become slightly less human.

I used to tell people that grizzly bears saved my life after immeasurable tragedy, so I felt an obligation to protect them. My friend, and poet, CMarie Fuhrman summed it up a little better. She offered that "grizzly bears didn't save my life, I saved my life, grizzly bears just showed me why." I like that better.

The following stories are told from a place of love and respect for two species I hold dear: humans and grizzlies. Enjoy every word.

For the grizzly inside us all,
— BRAD ORSTED
Wildlife Photographer

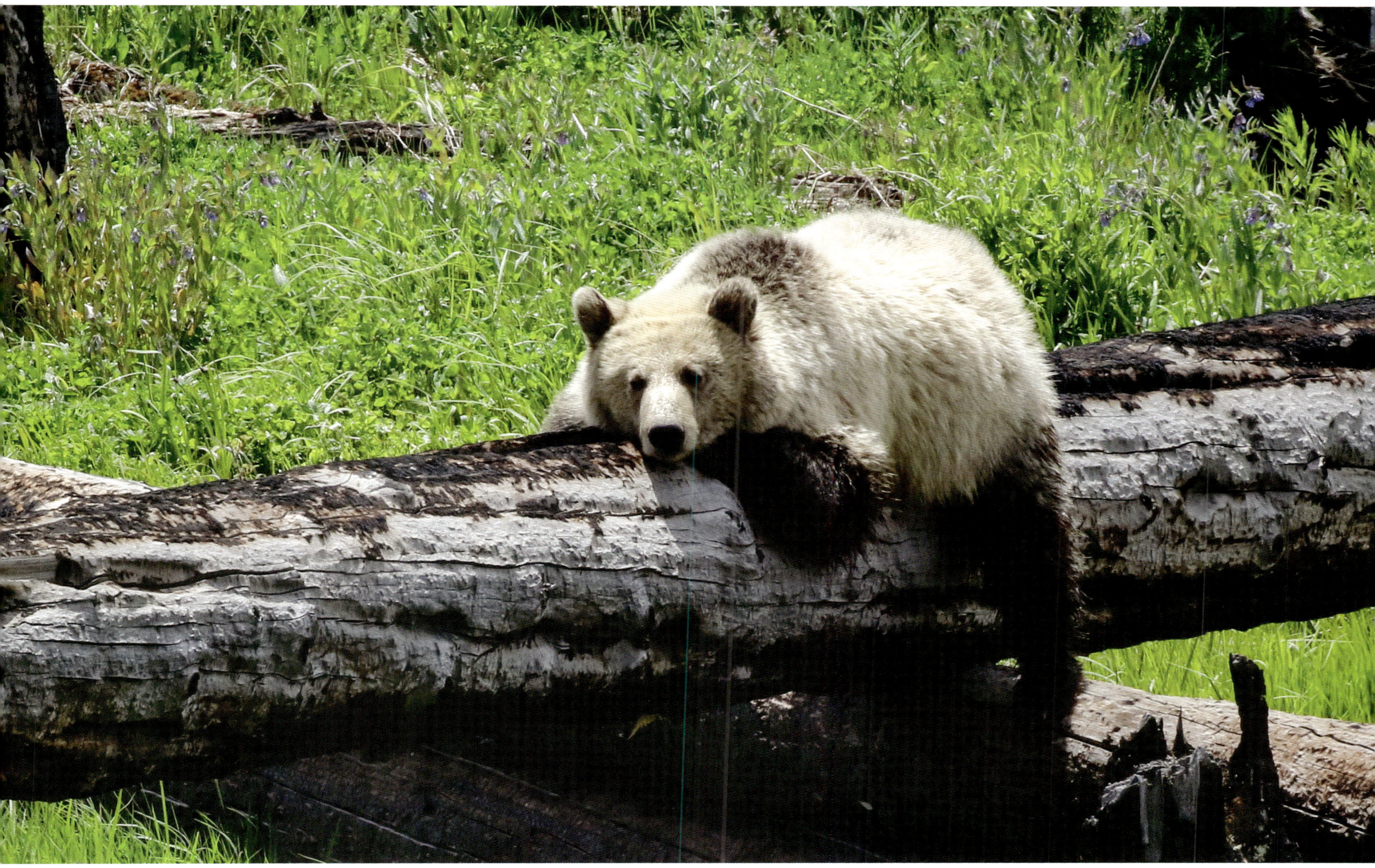

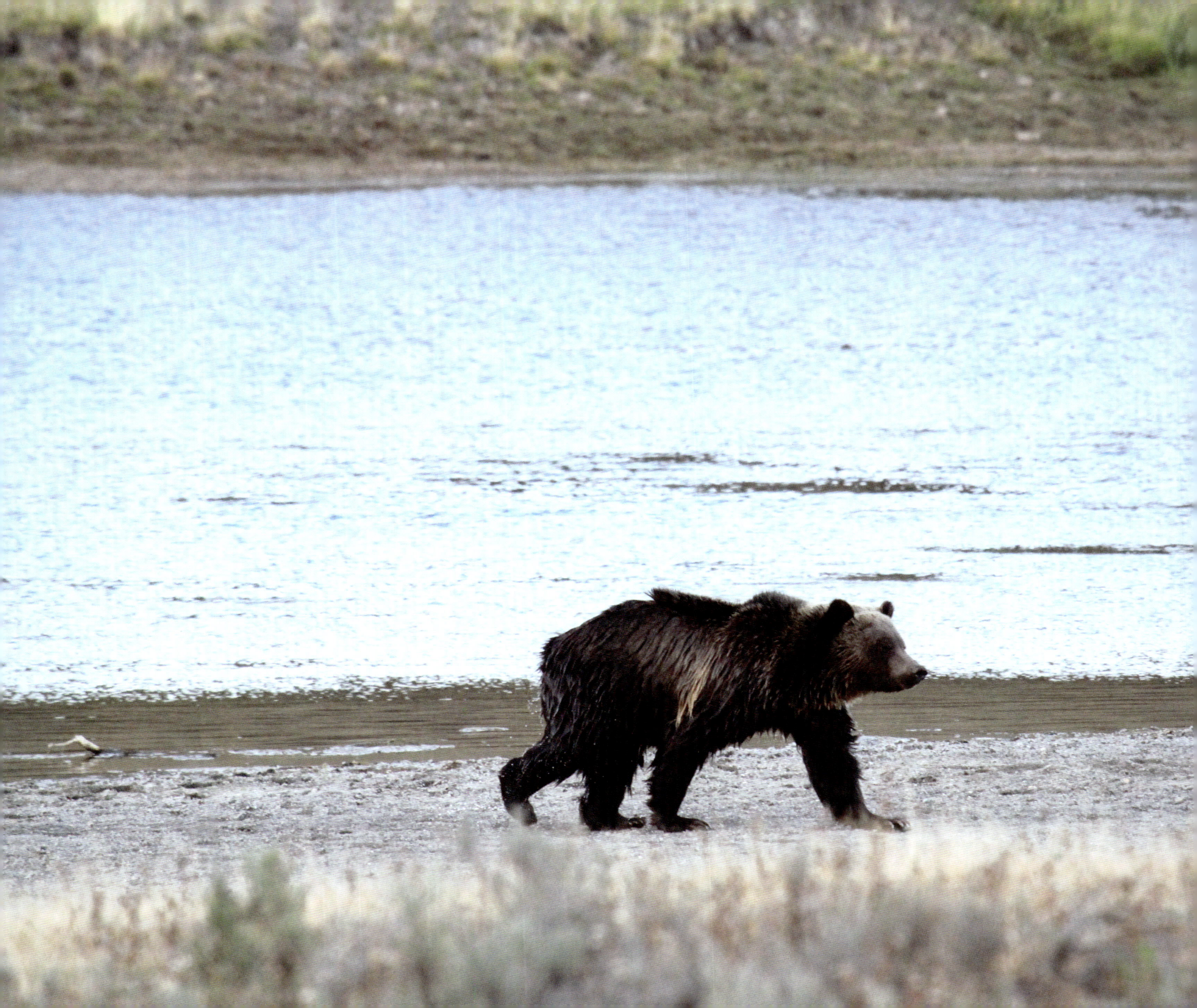

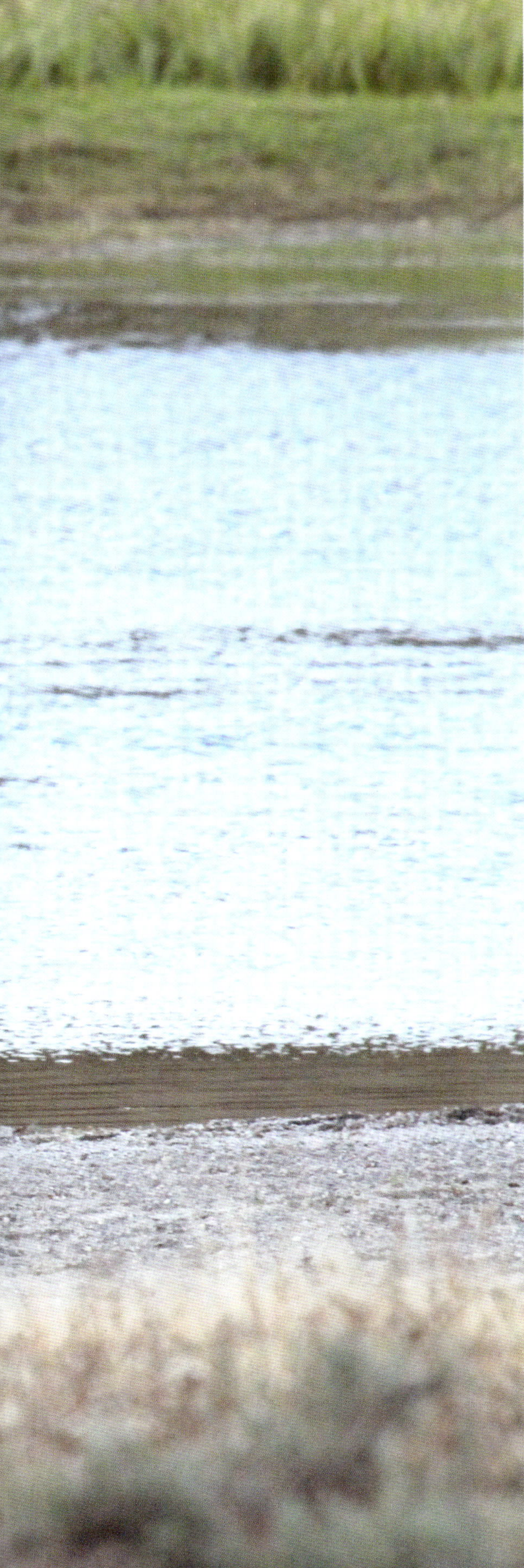

INTRODUCTION

AS THE SNOW STARTS TO MELT AND SPRINGTIME ARRIVES IN THE ROCKIES, many people like me find themselves dreaming not only about the beauty of the wild places you can visit, but also about the wild animals who inhabit those places. It's at this time when I start thinking about the mighty grizzly, and wondering what kind of year I am going to have when I'm out with my camera and shooting photos.

As a young girl, I remember traveling to Yellowstone with my family in hopes of seeing one of these magical bears. Back then we considered ourselves lucky if in a time frame of ten days we were able to see just one, and when we did finally come across one, it was the highlight of our trip. We would then spend as much time as we could watching it. It didn't matter if the bear was hundreds of yards away and looked like an ant on the horizon, we were always in awe and stayed until it disappeared out of sight.

As the years went by, my love for this incredible species grew and I found myself drawn in their direction more than ever. I started visiting Yellowstone more often, hoping to catch a glance and watch them in their natural habitat. I wanted to see sows and cubs as they interacted, and to watch as the big boars made their presence known. I wanted to know what it meant to be a grizzly living in the wild.

The memories I have made over the years are priceless. I remember when the Fishing Bridge campground was located across from the Fishing Bridge store rather than where it is now. It always seemed like a very eerie place to stay. Whether it was daytime or nighttime, I remember thinking that every noise was a bear coming to carry me away. Eventually that campground was closed and the new one was built. That area was given back to the grizzlies because it was considered their passage as they traveled to and from Yellowstone Lake.

At one point in time, there was a grizzly that learned how to get into the bear-proof garbage cans. It would use one paw and stick its claws underneath the handle and open it up and then use the other paw to reach down in and get the garbage out. I'm not sure whatever happened to that bear. I remember them setting a trap for it, but never heard the outcome.

I also remember a small ranger station or warming hut that was located right by the entrance of that first Fishing Bridge campground. We would often stop in there and ask if there were any recent bear sightings. If there were, we hoped that we could make it to the destination before the bear disappeared into the landscape.

Speaking of landscapes, Yellowstone is such a beautiful place for a grizzly to wander. The amazing sunrises and sunsets, not to mention the mountains and valleys, the rivers, the wildflowers, the pines, and all of the smells that go along with them—what an incredible place to call home.

I've watched grizzlies swim and fish in the lake, take down another animal for food, fight another bear, chase wolves, play on a patch of snow, and sleep on a log. I've watched sows and boars mate, cubs hug their mom, and many other wonderful sights. There is no better way to learn about this type of nature than to experience it firsthand. Thankfully I have been able to observe many of the things I always dreamed about.

I hope you enjoy this book as much as I have enjoyed spending so much time in the wilderness.

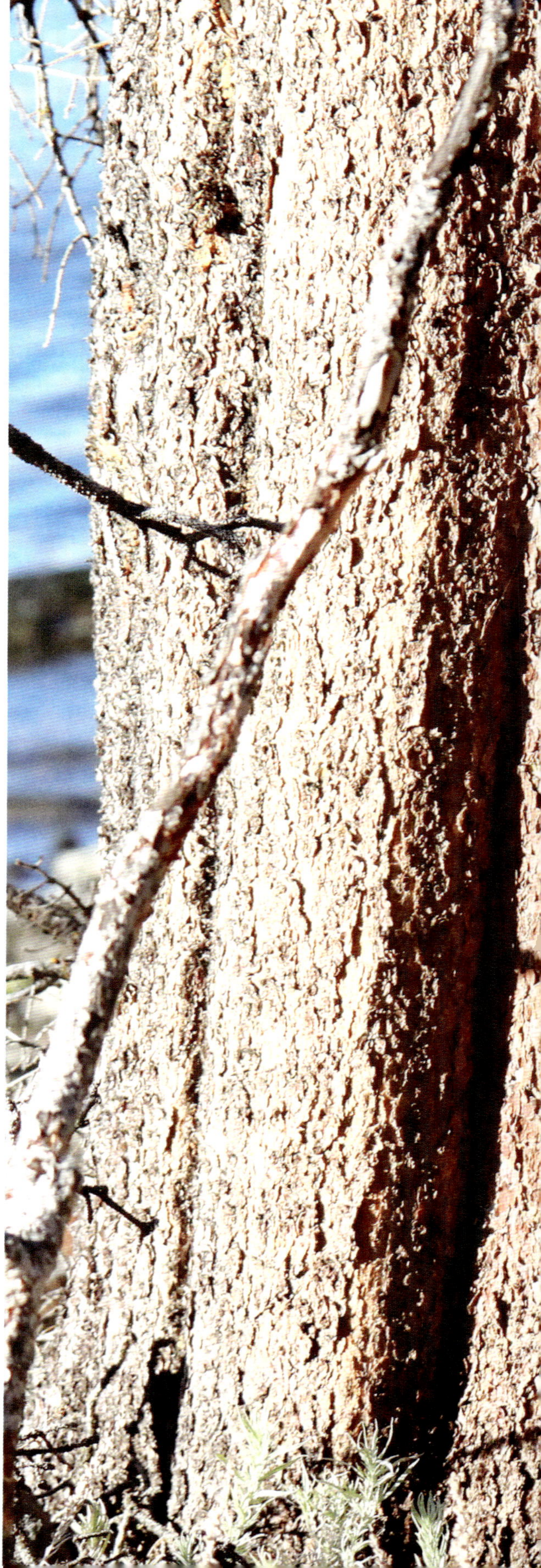

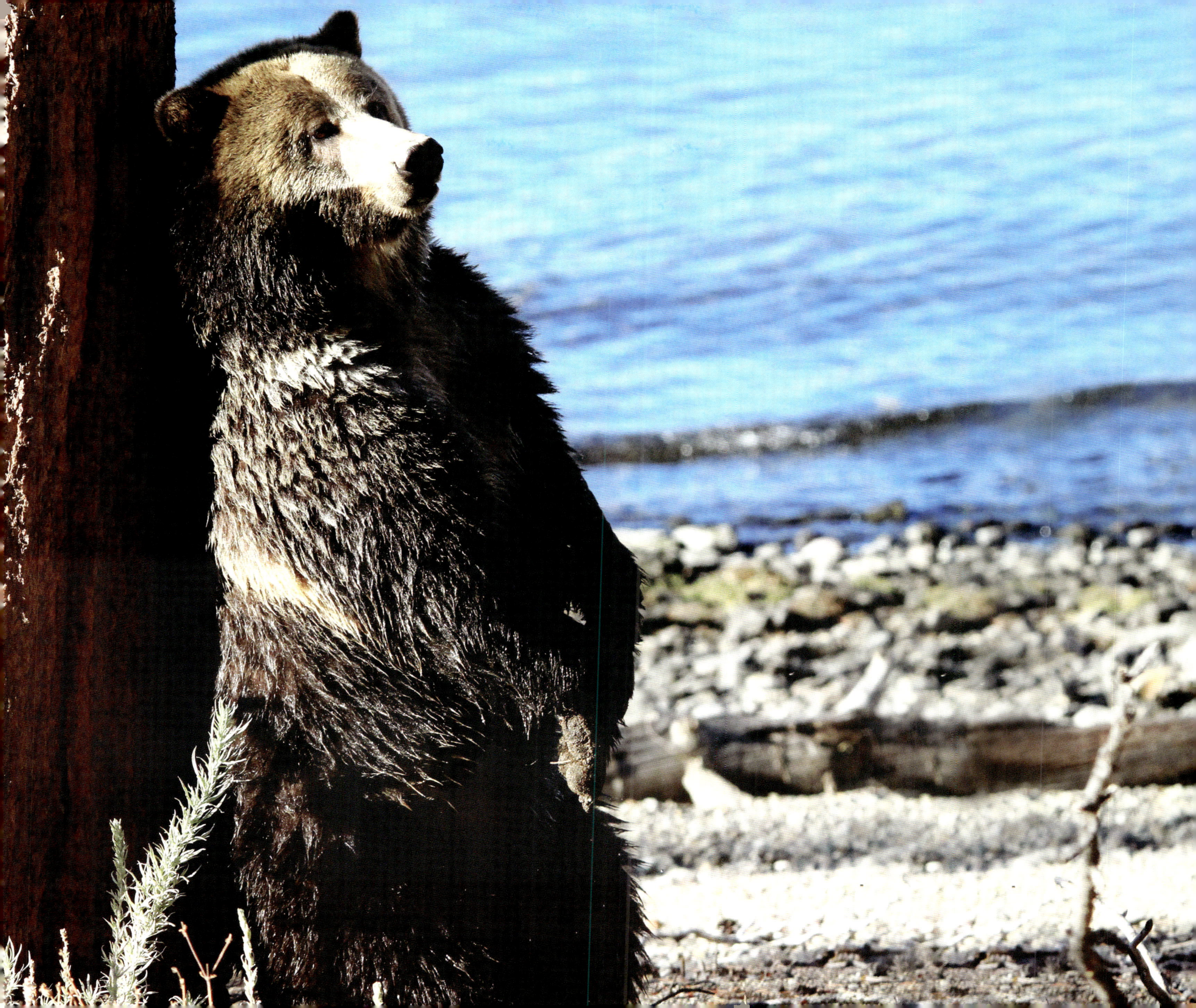

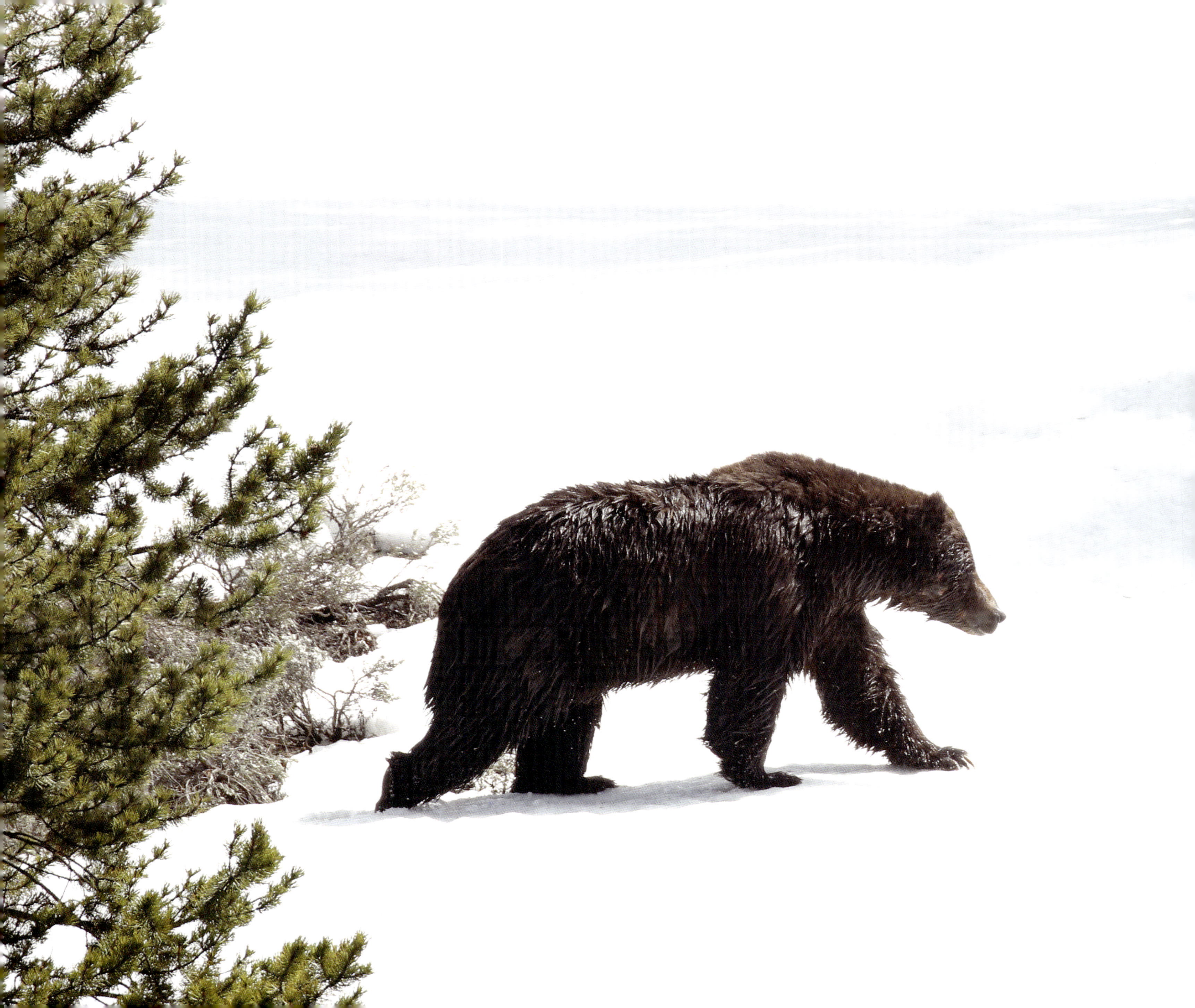

RELATIONSHIP *with* NATIVE AMERICANS

MANY NATIVE AMERICANS CONSIDER THE GRIZZLY TO BE A SACRED ANIMAL. They think of them as brothers and believe that they give power to their people. James "Jimmy" St. Goddard, a member of the Blackfeet Nation, once said, "The grizzly bear is not only equal to, but also far superior to us pitiful humans. To kill the grizzly is to kill our own kind."

Most Native American cultures consider the grizzly a great and powerful spirit, and believe the grizzly is a gift to Mother Earth and her people. While each tribe's view of the bear varies, most look at the grizzly as a healer and protector representing courage, strength, good medicine, and authority. Bears are also a sign of wisdom. Some tribes have a bear dance, which they believe brings the spirits of their ancestors back.

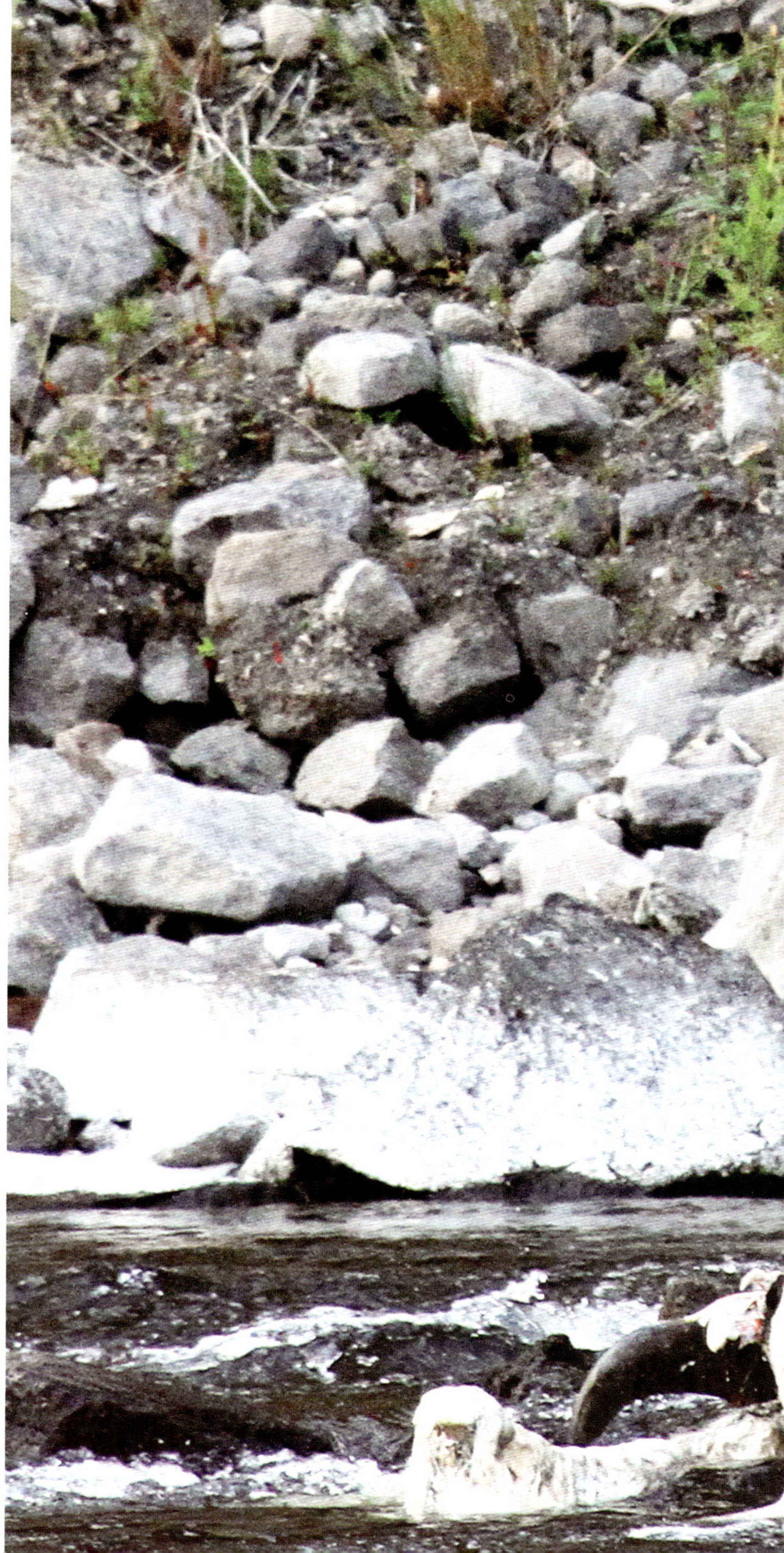

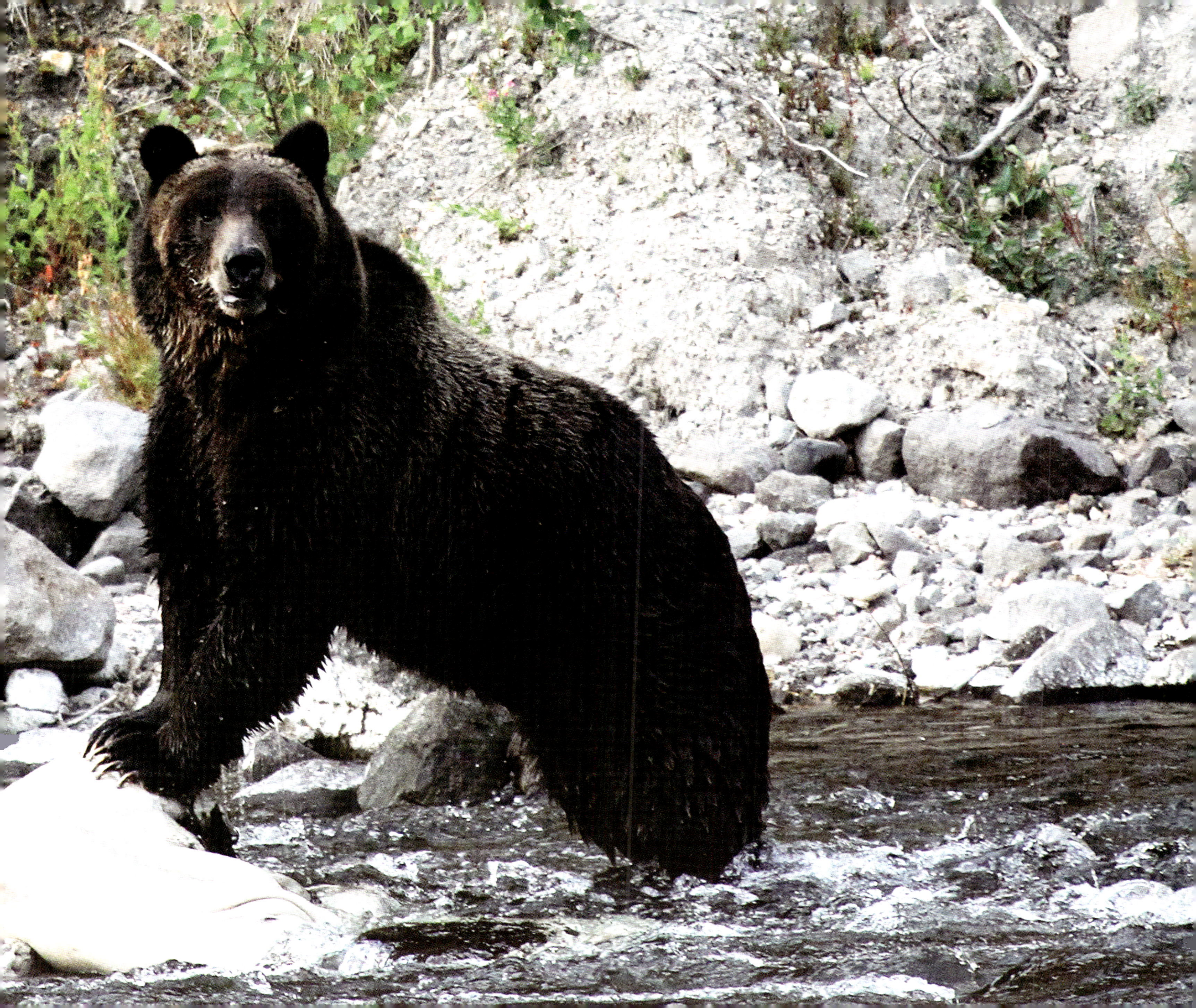

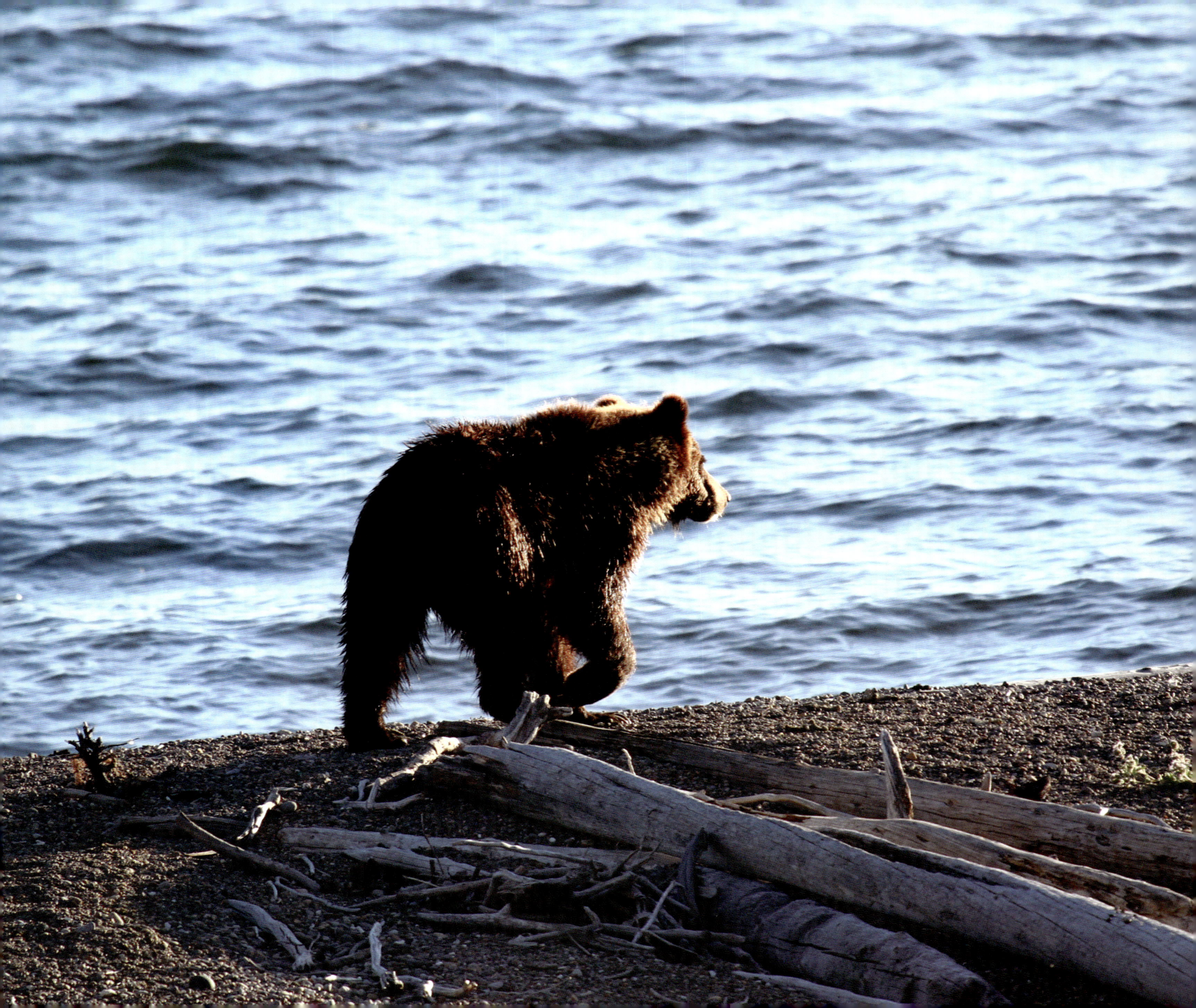

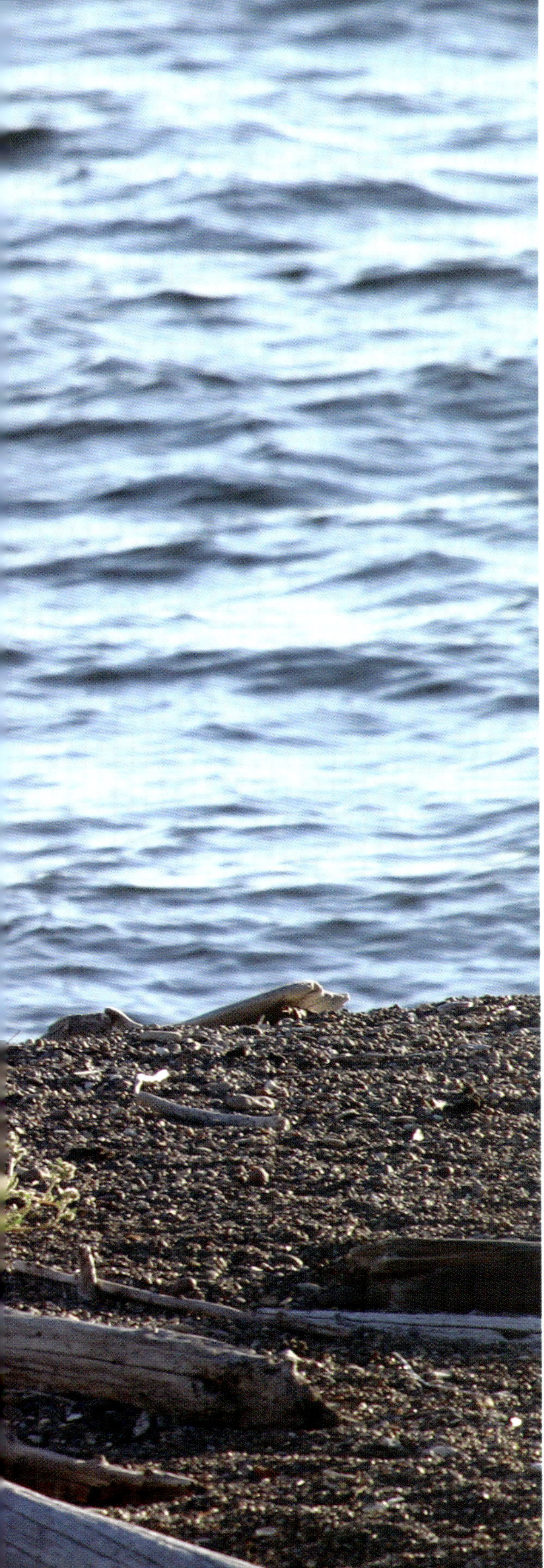

When the grizzly was delisted from the Endangered Species Act, and the hunts were going to start, tribes hosted a prayer ceremony at Glacier National Park in Montana. The Blackfoot Confederacy led the group in prayers for the grizzly. Tribal nations were never consulted in the delisting of the grizzly bear. Don Shoulderblade (Northern Cheyenne), co-founder of the GOAL Tribal Coalition, said, “We will not stand by in the land of our ancestors and watch grizzlies be blown apart by high-powered rifles and mutilated just to satiate the bloodlust of some rich, ‘great white hunters.’”

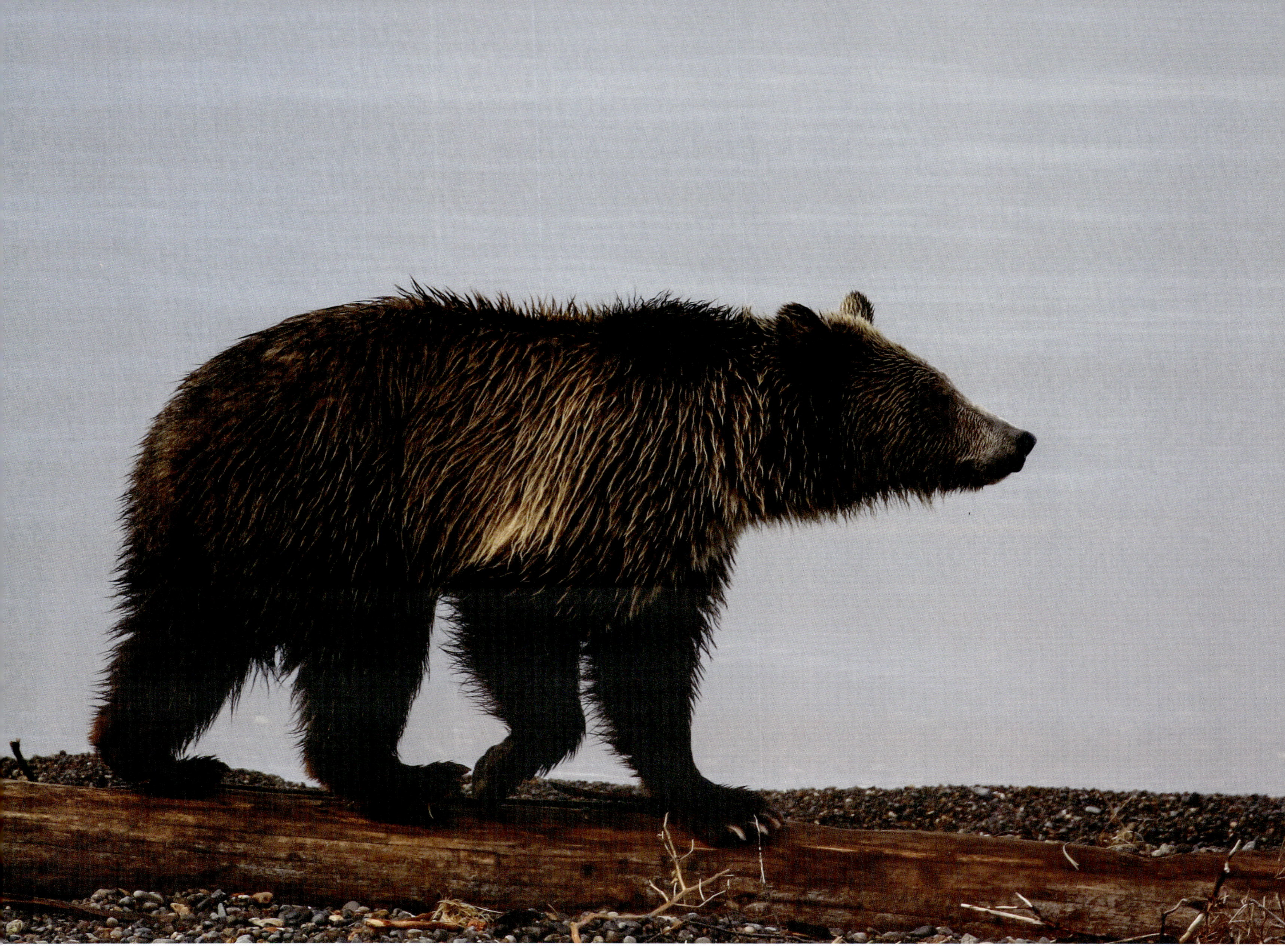

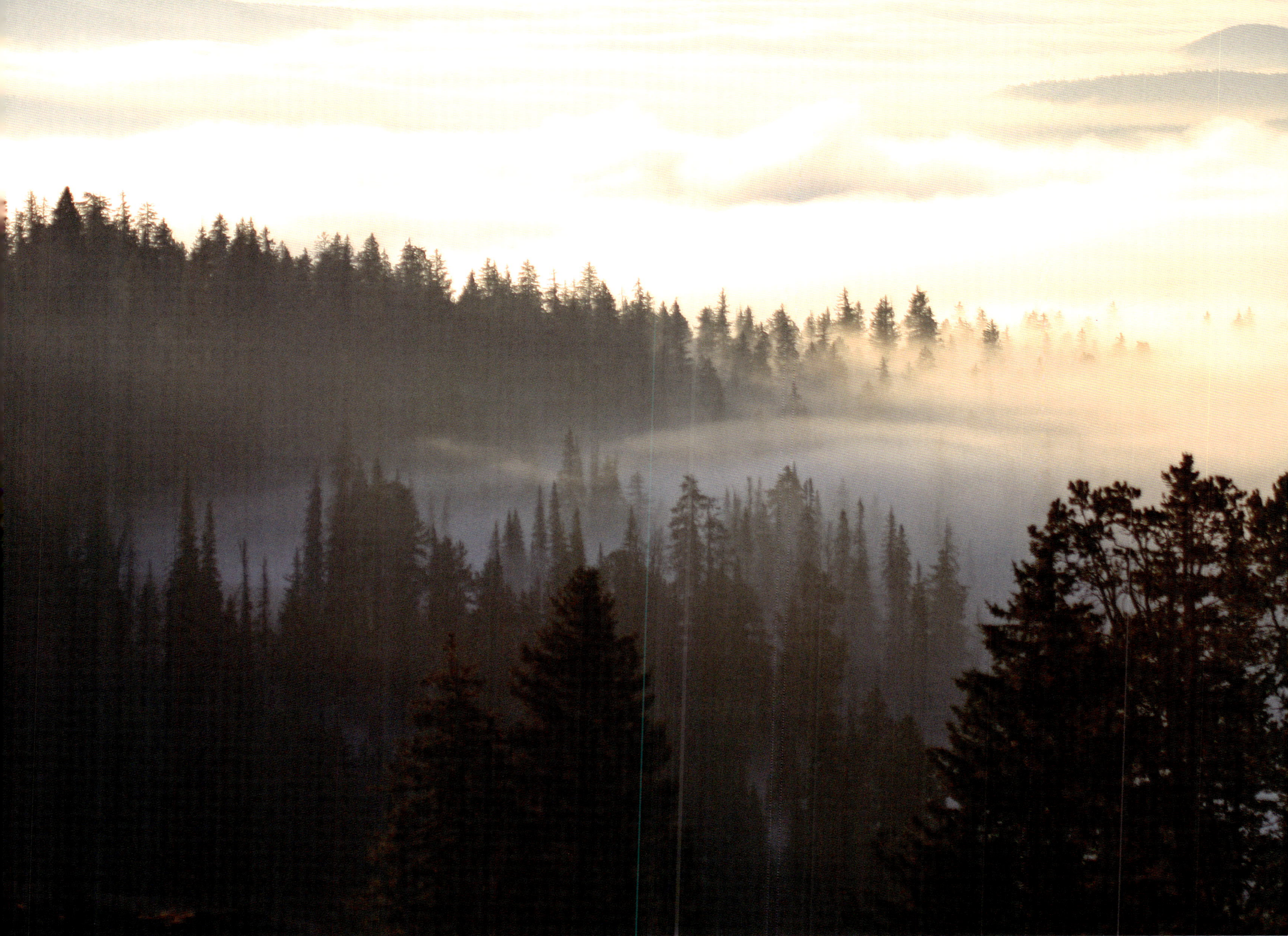

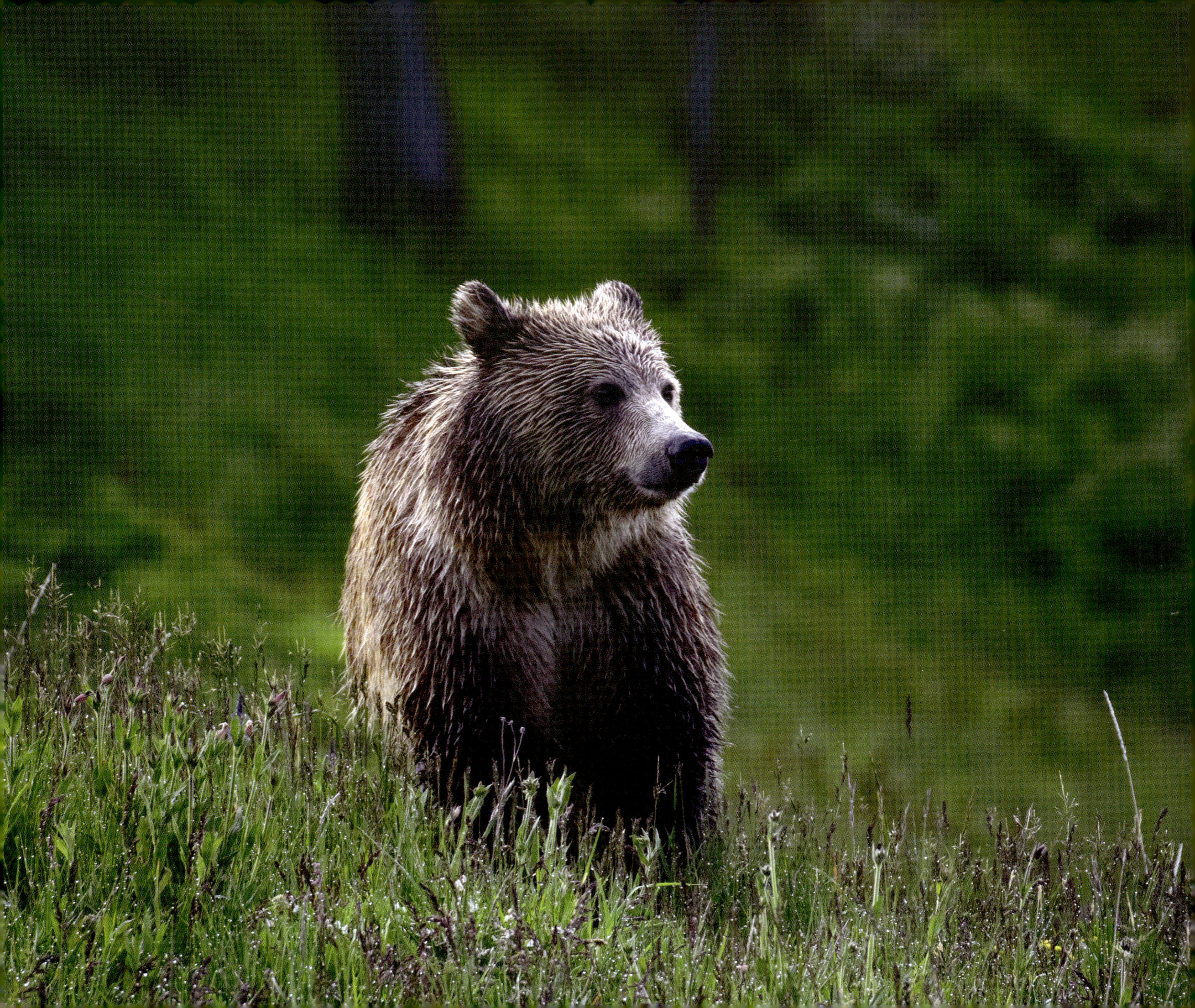

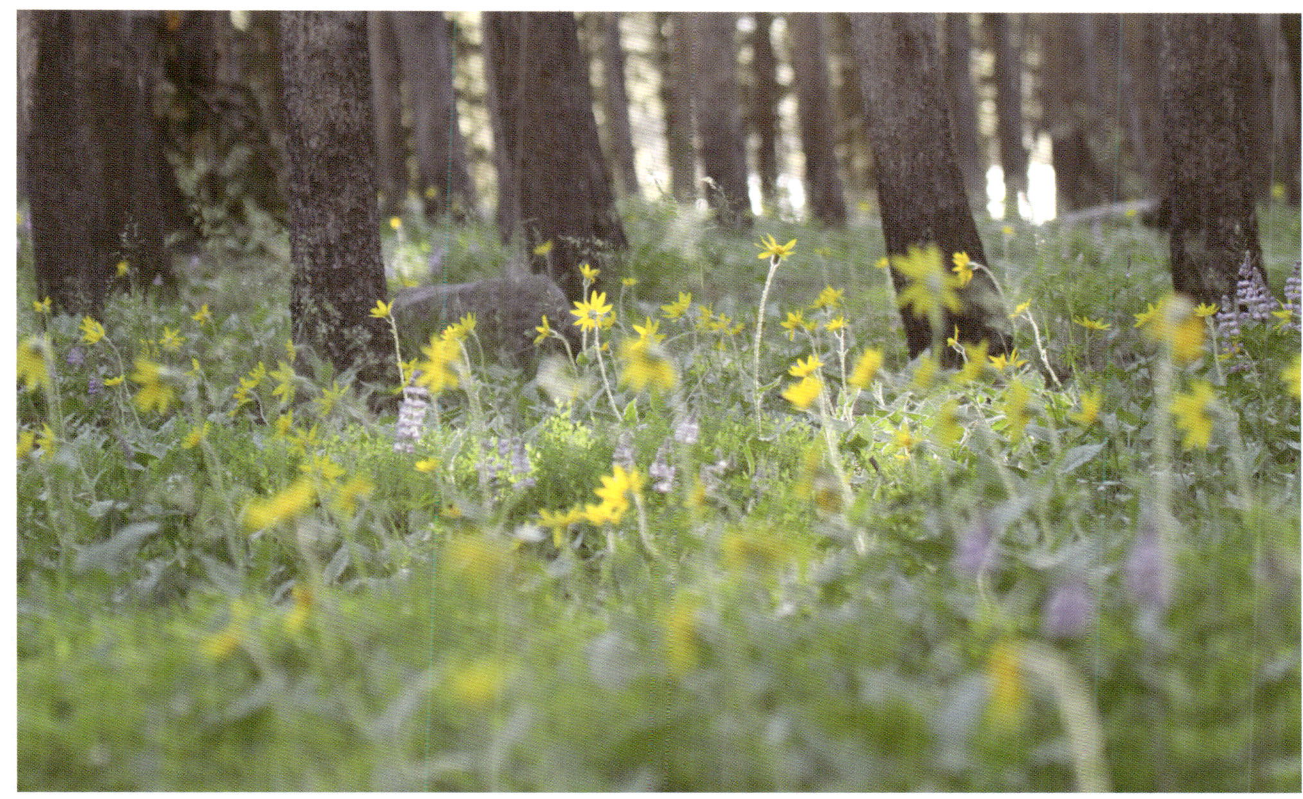

When a pine needle falls in the forest, the eagle sees it,
the deer hears it, and the bear smells it.

—*Unknown*

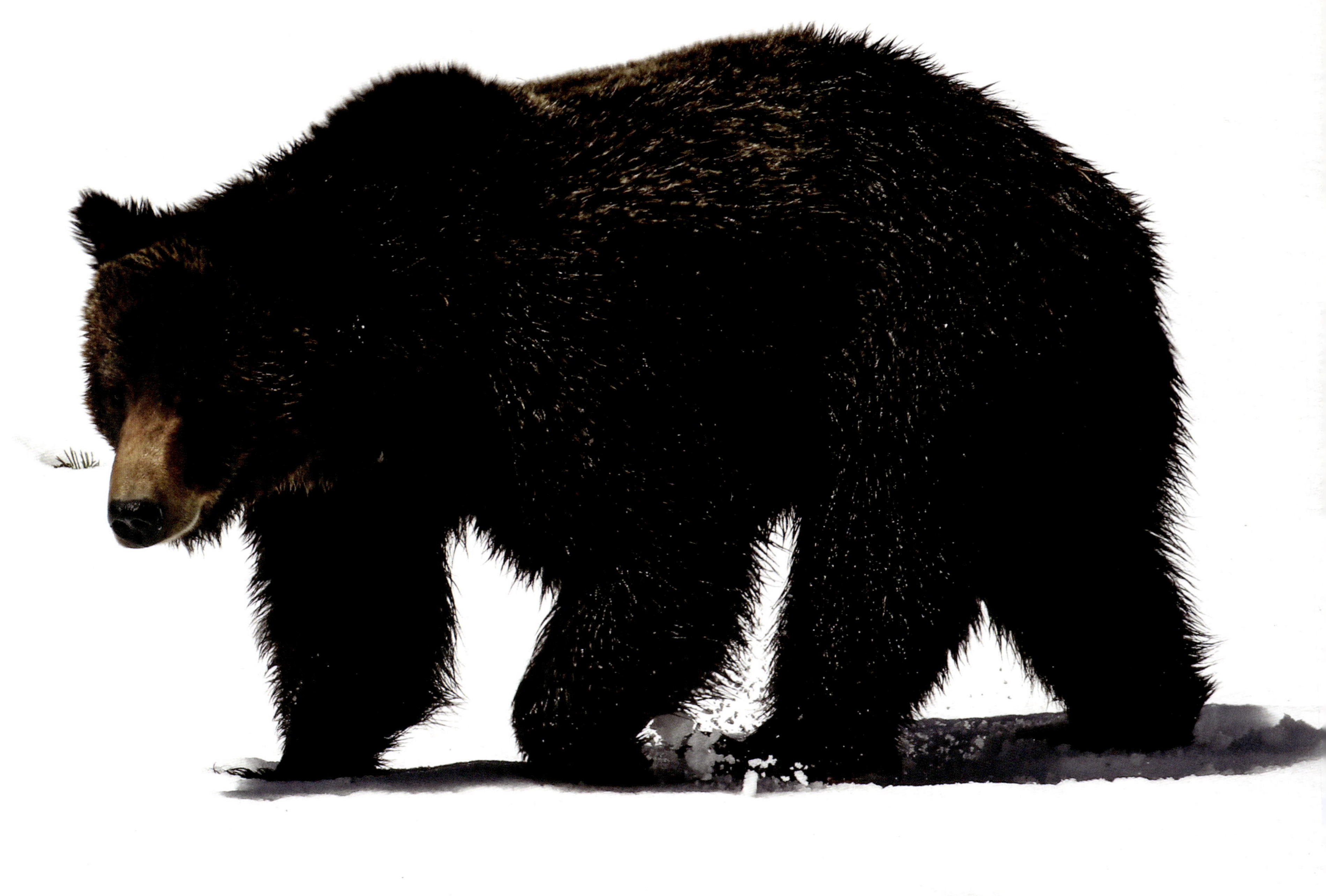

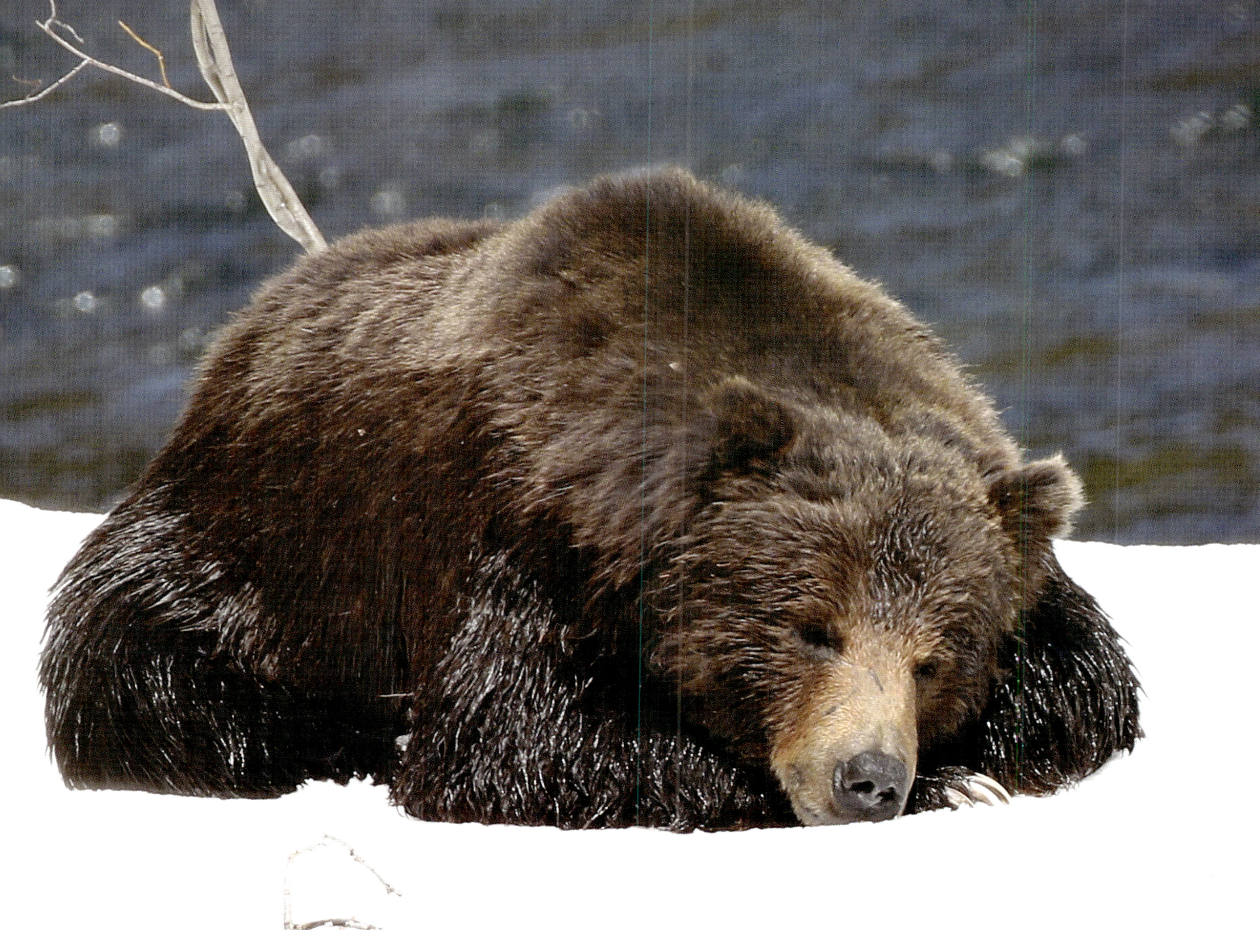

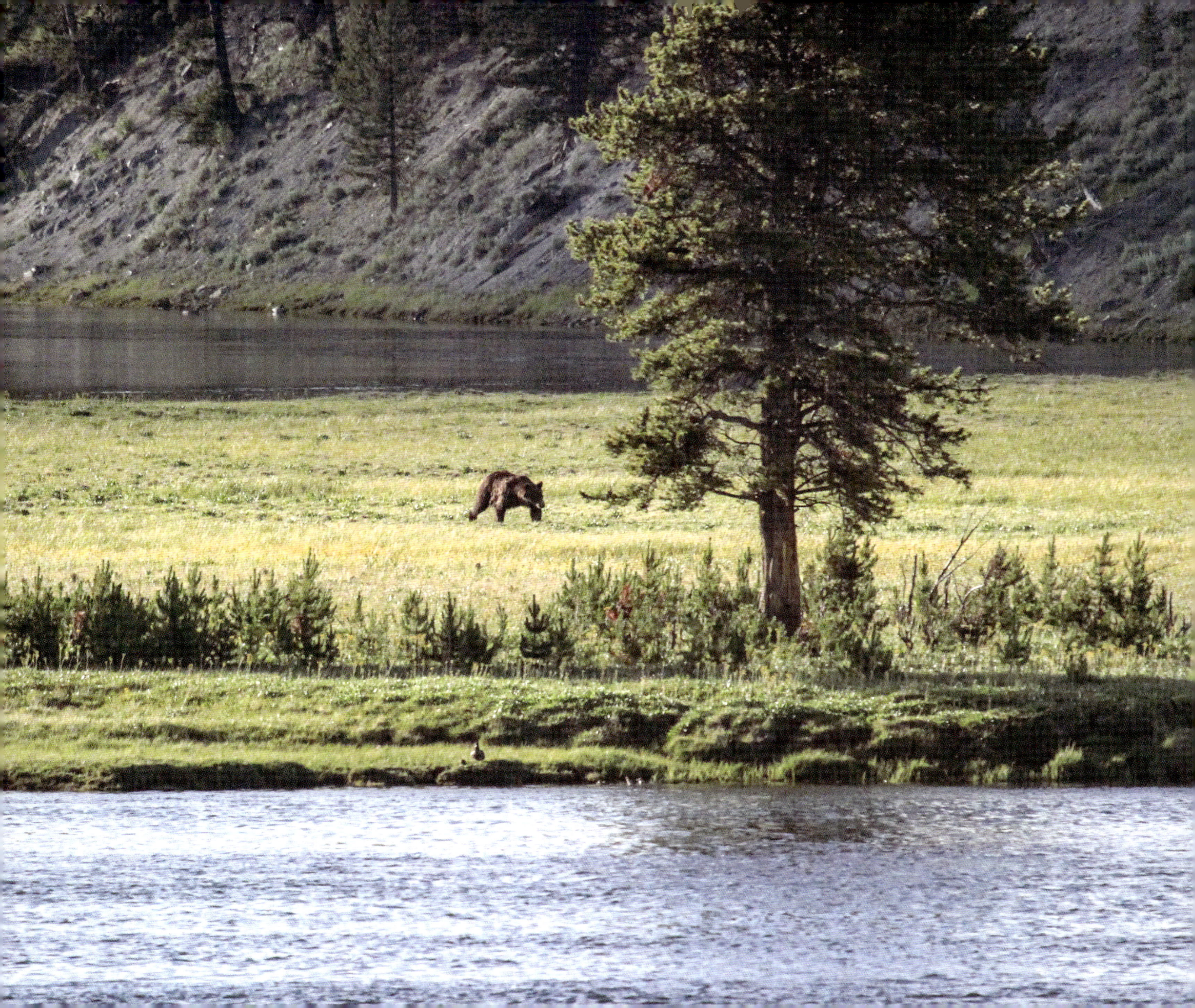

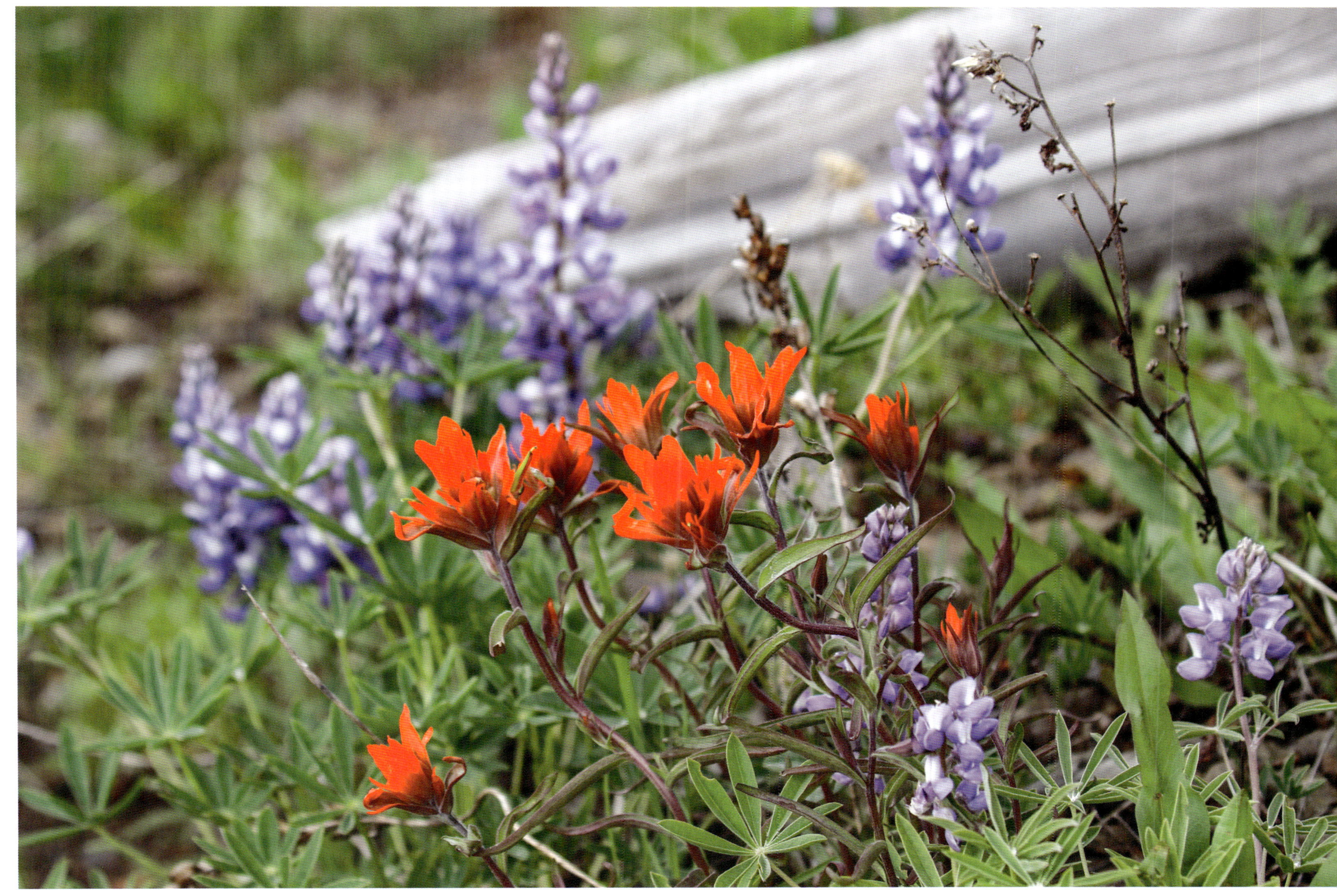

Grizzly bears have made a remarkable recovery and are considered a success story. The population in the Greater Yellowstone Ecosystem has definitely grown, but there are still many factors that influence their survival.

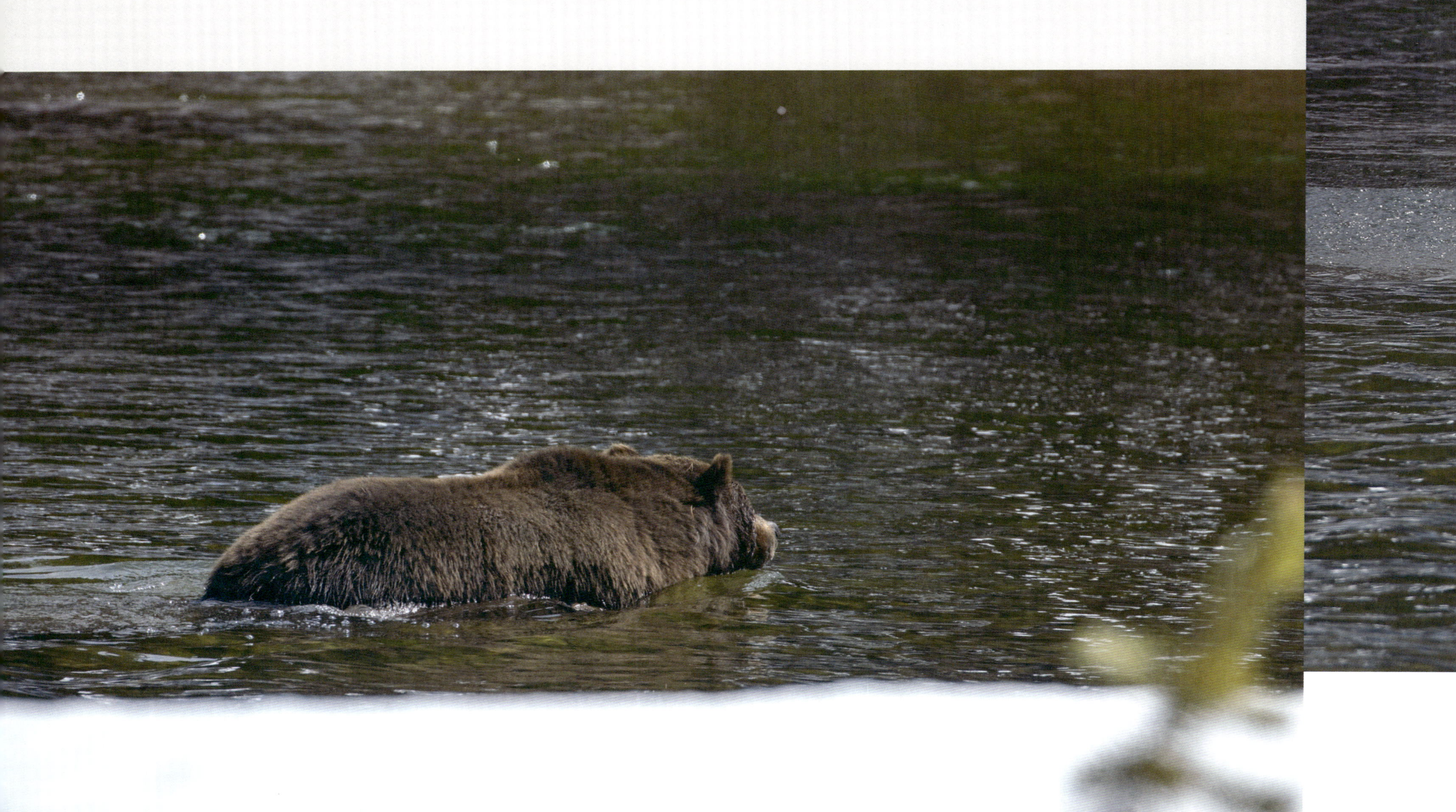

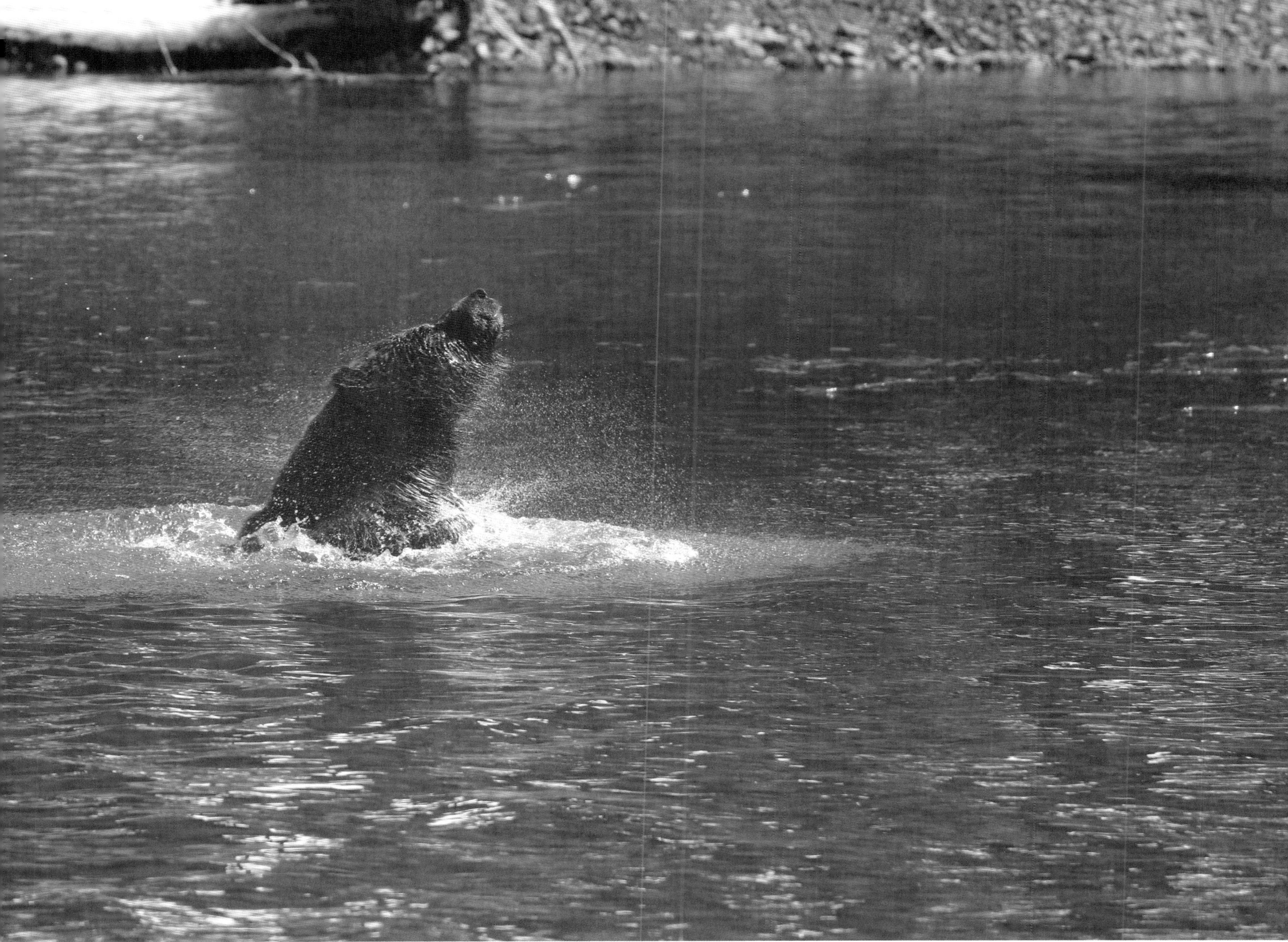

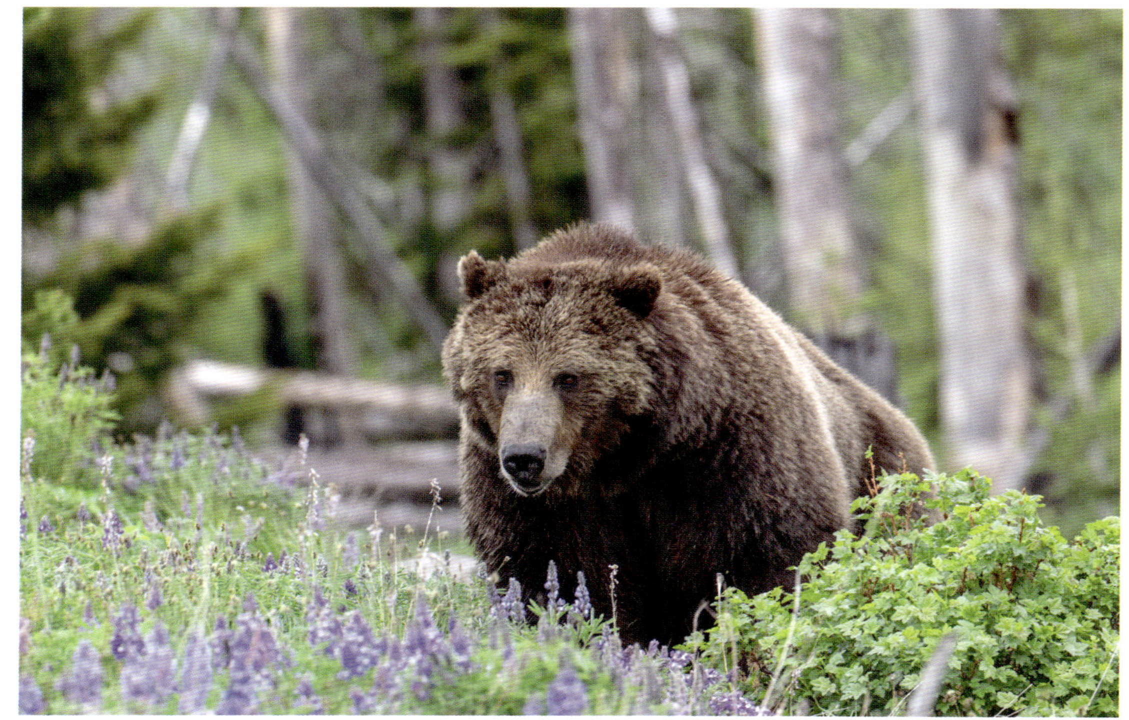

The old Lakota was wise. He knew that man's heart, away from nature, becomes hard; he knew that lack of respect for growing, living things soon led to lack of respect for humans too.

—Luther Standing Bear

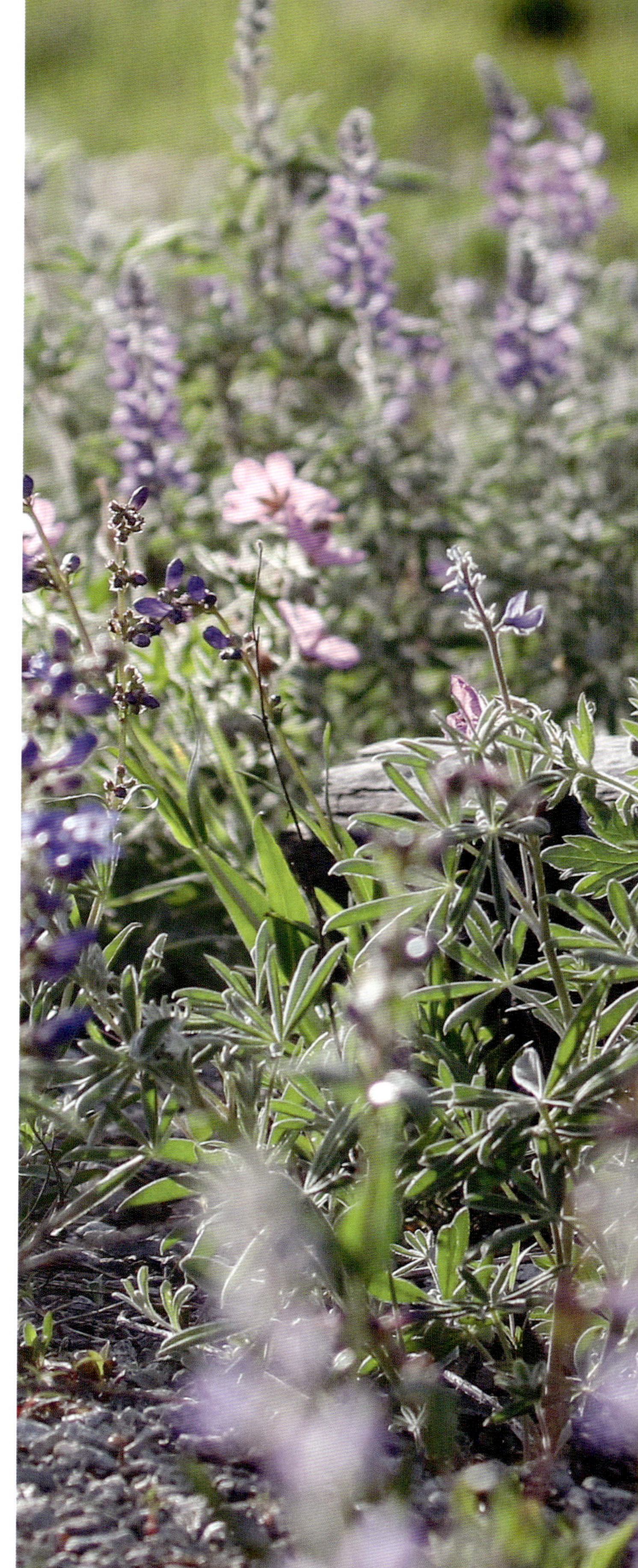

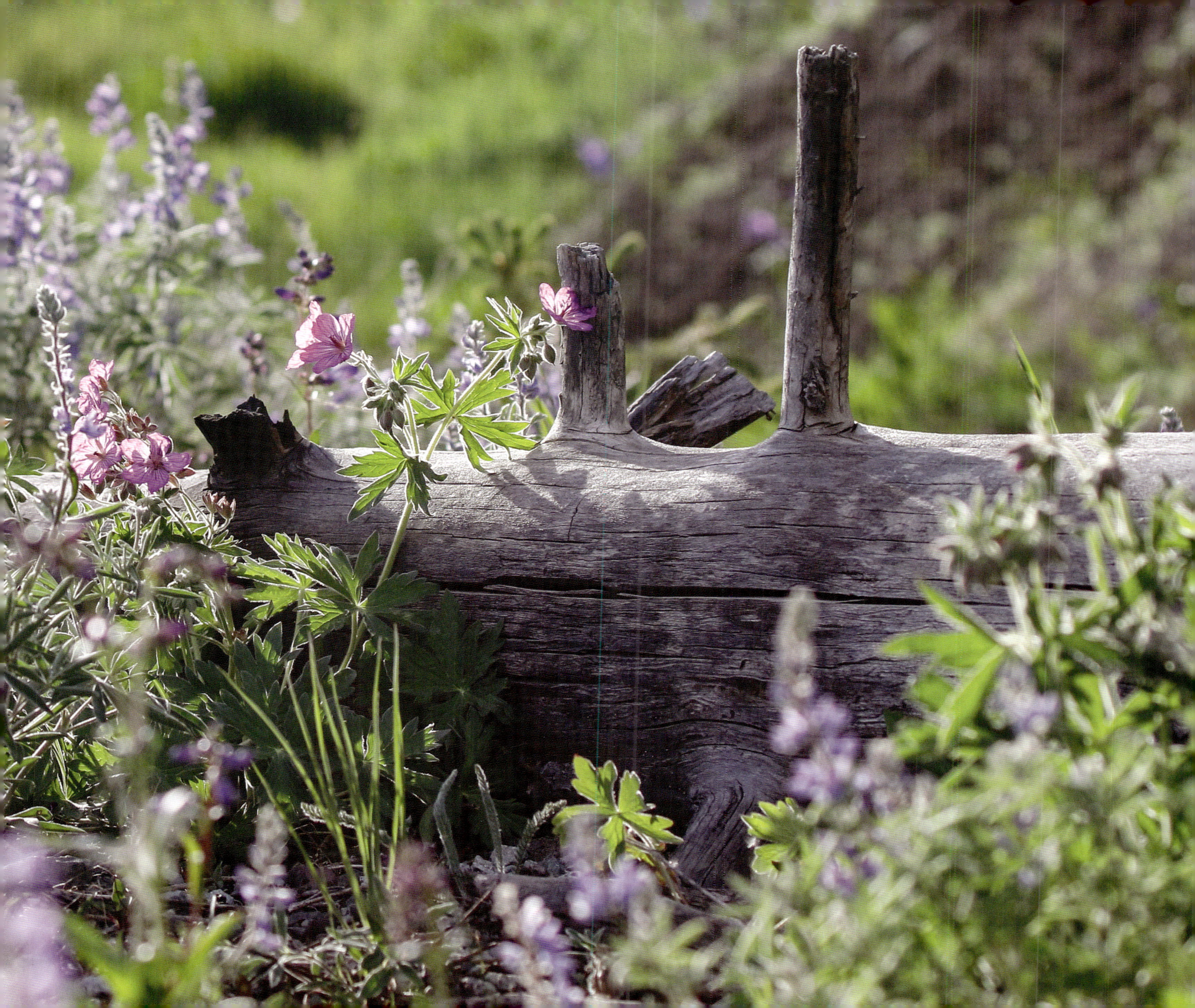

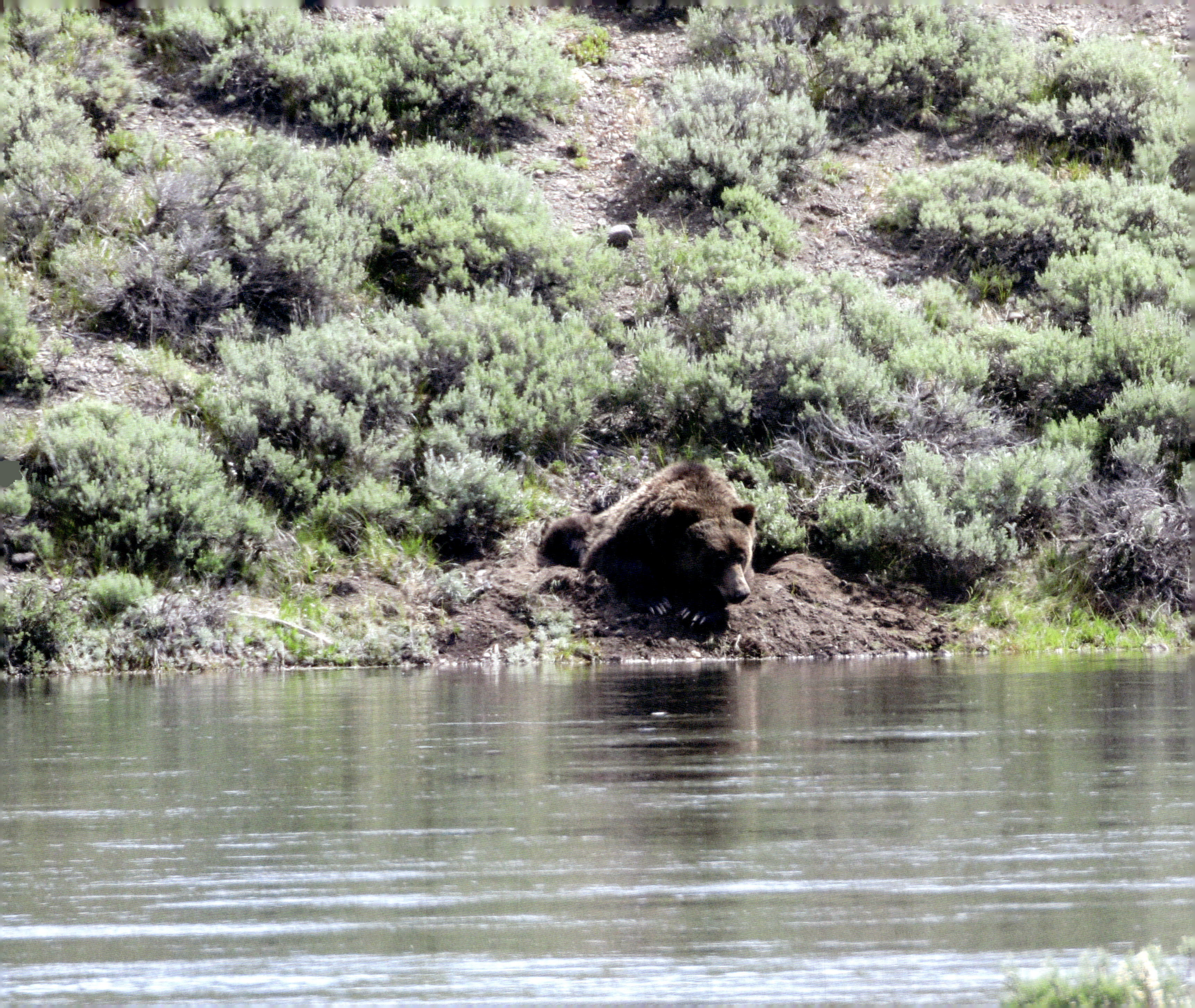

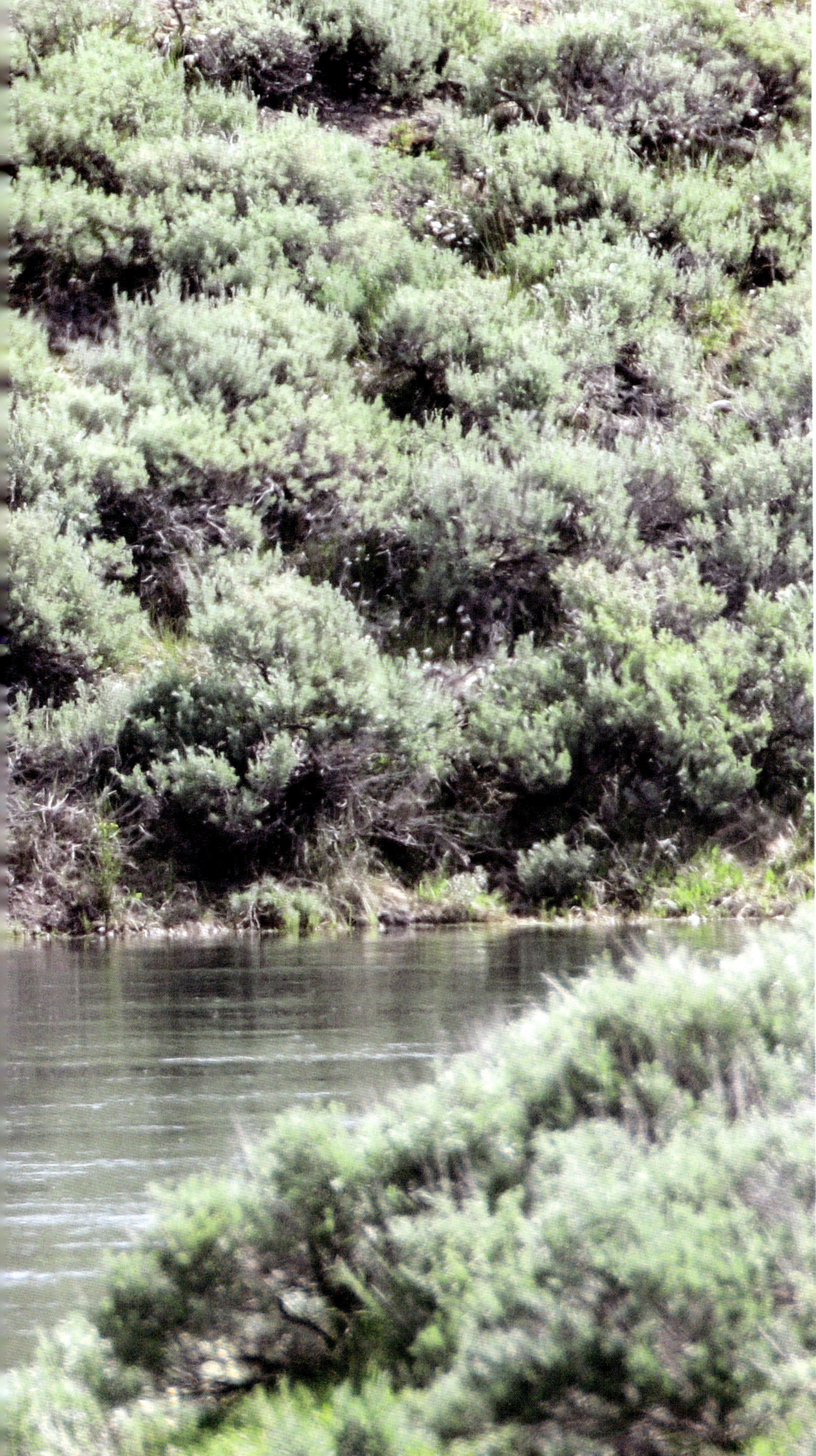

It's amazing. Bears occupy the same spaces we do. They breathe the same air and dust. They dig their dens in some of the most magnificent areas. Here in this ecosystem, people build their cabins and residences with the intent to have great views of the forest and mountains. I can assure you, a bear's den never has a bad view and will give you a run for your money.

—Tyler Brasington, Bear Biologist, College of Natural Resources, University of Wisconsin —Stevens Point, Wildlife Management, and Bear Management Office, Grand Teton National Park

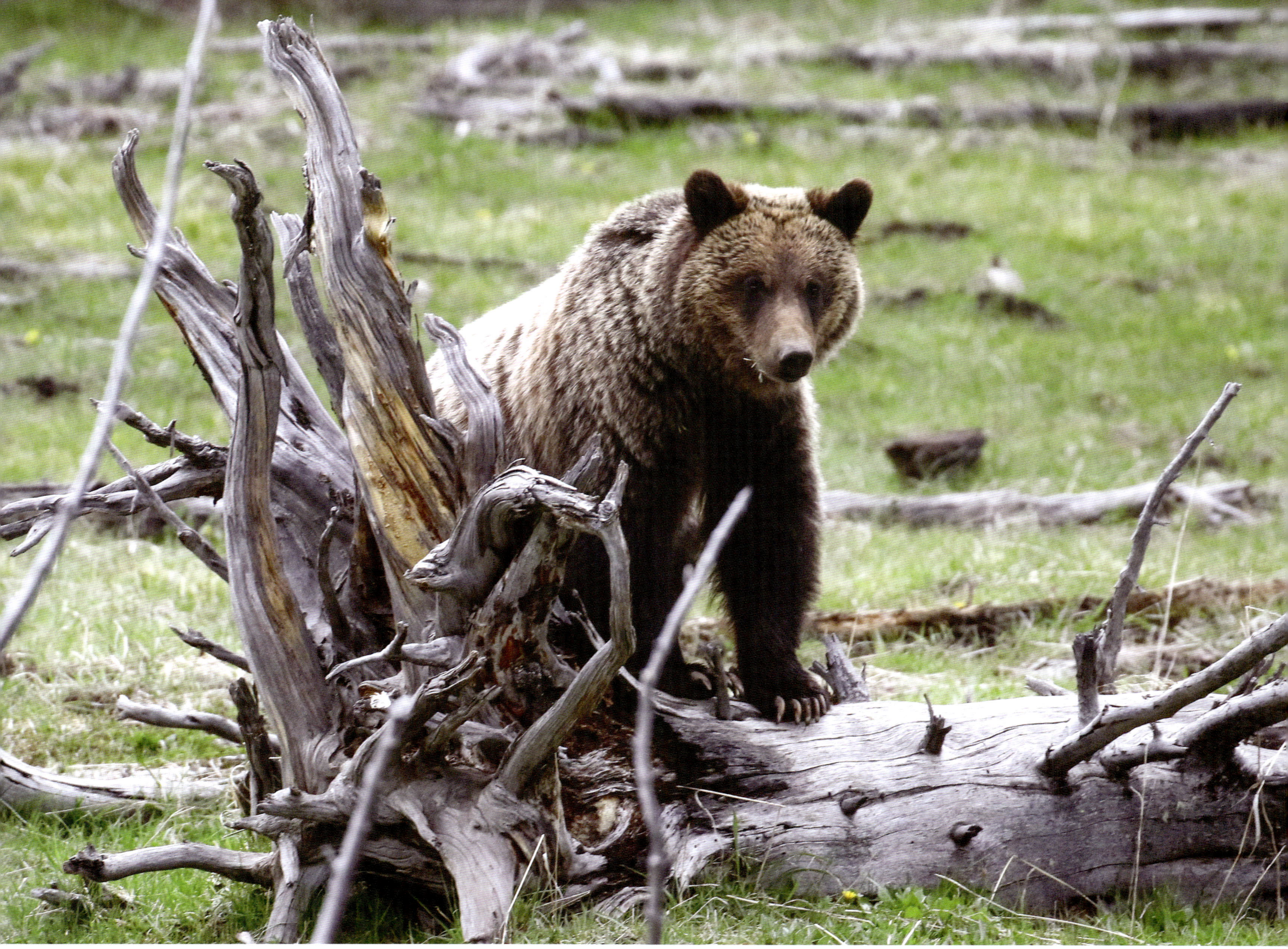

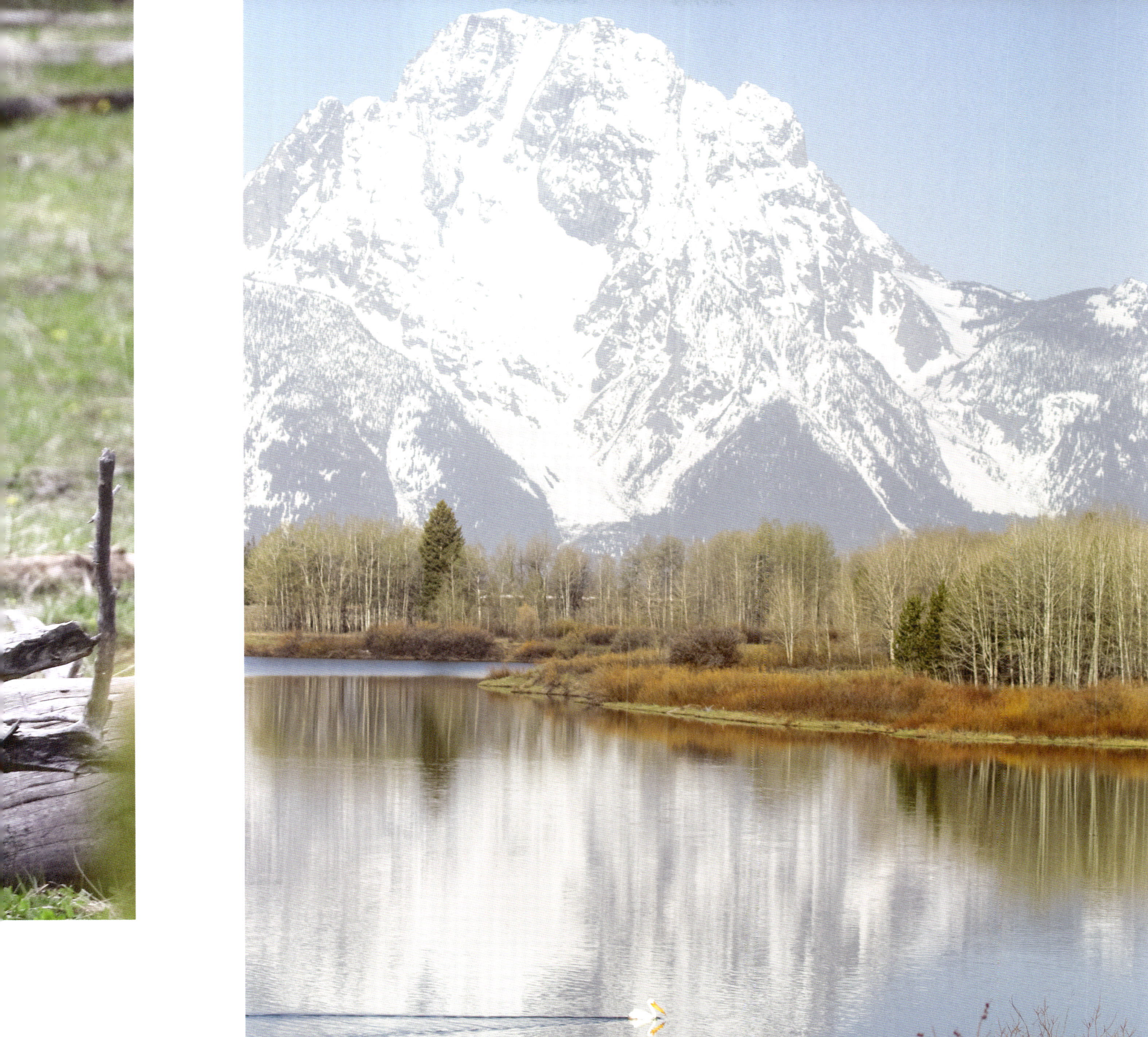

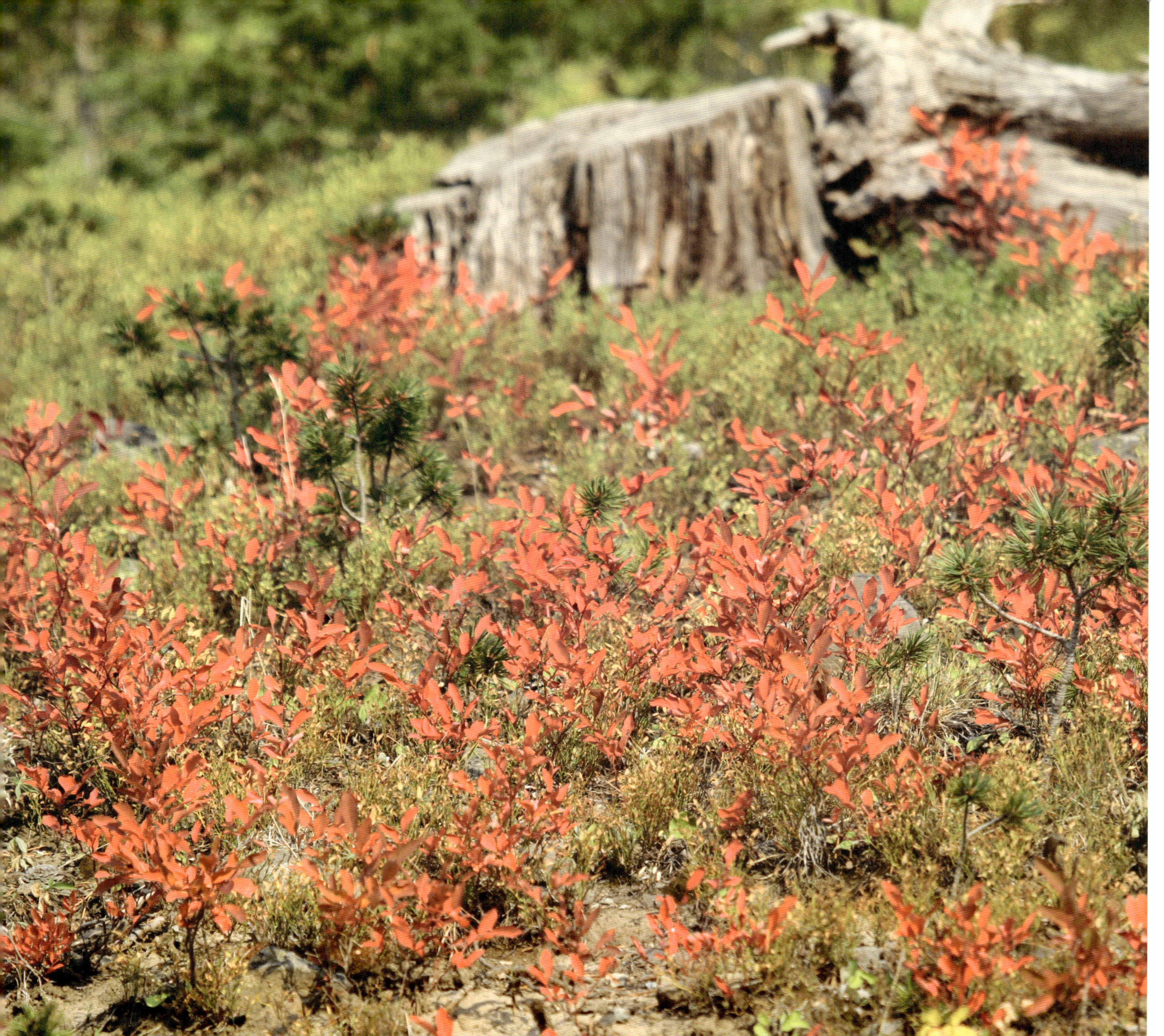

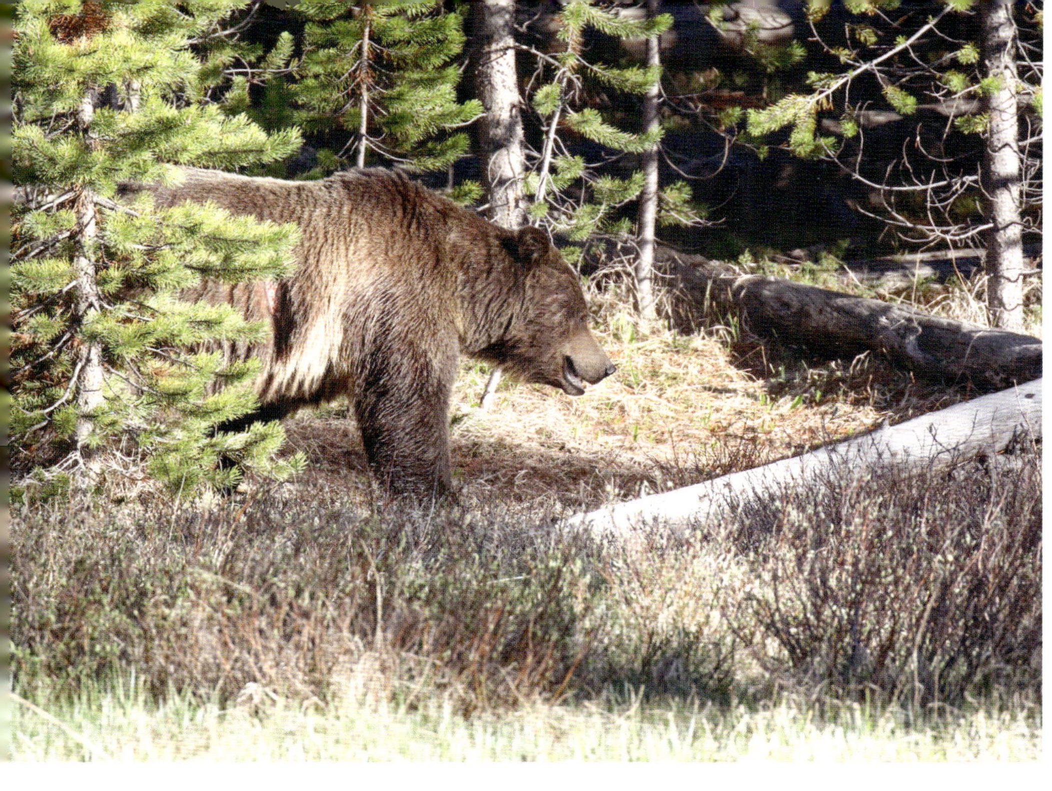
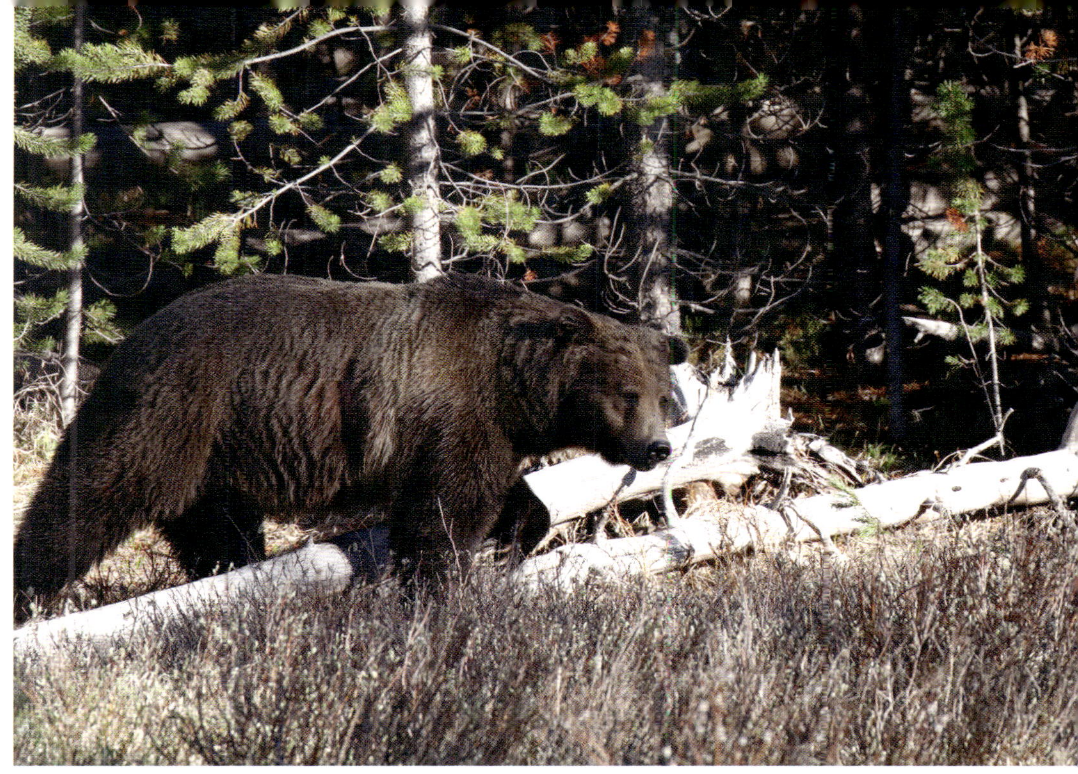
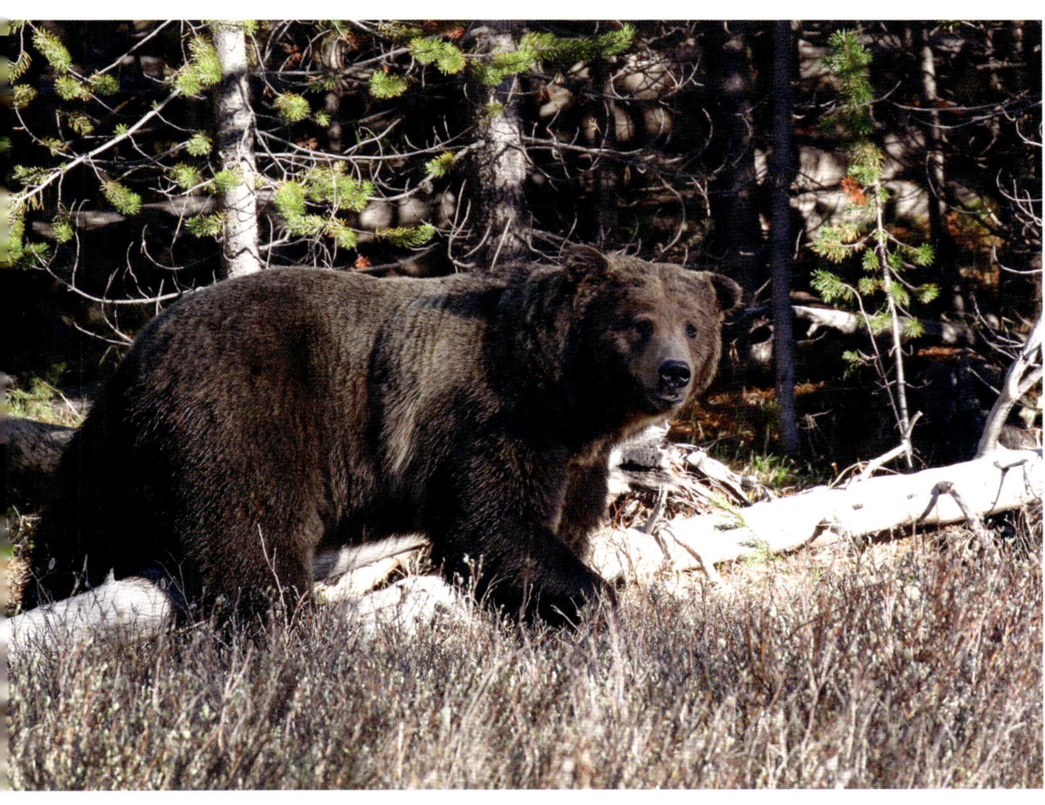
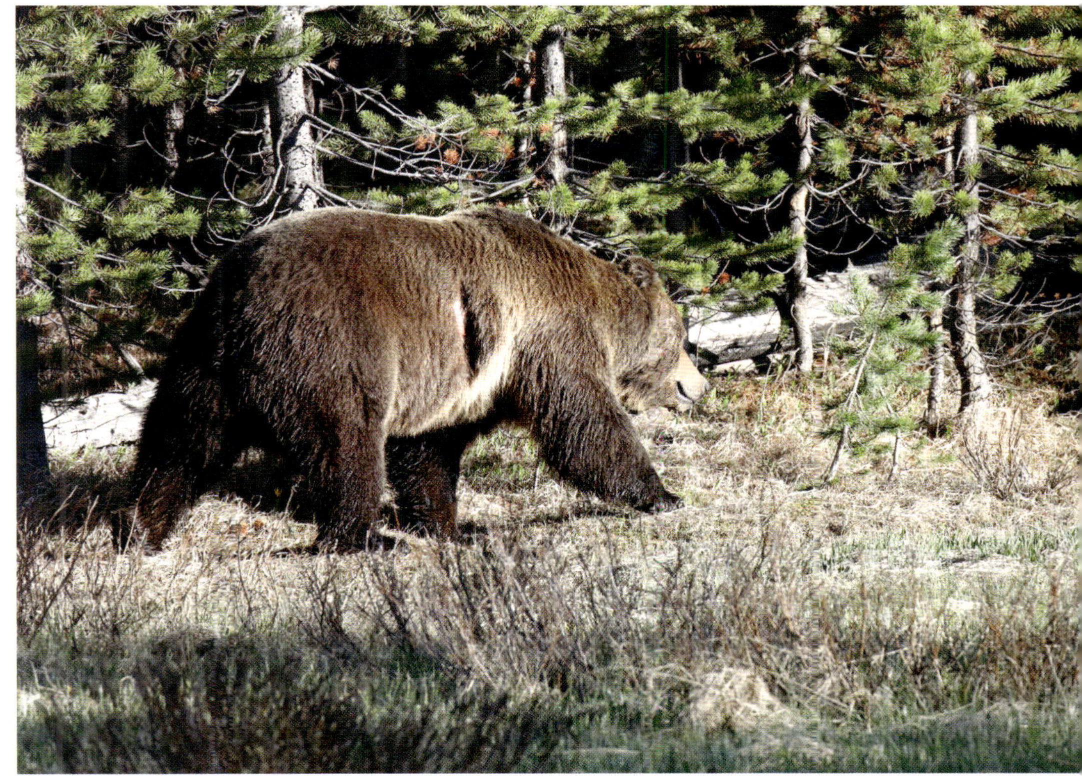

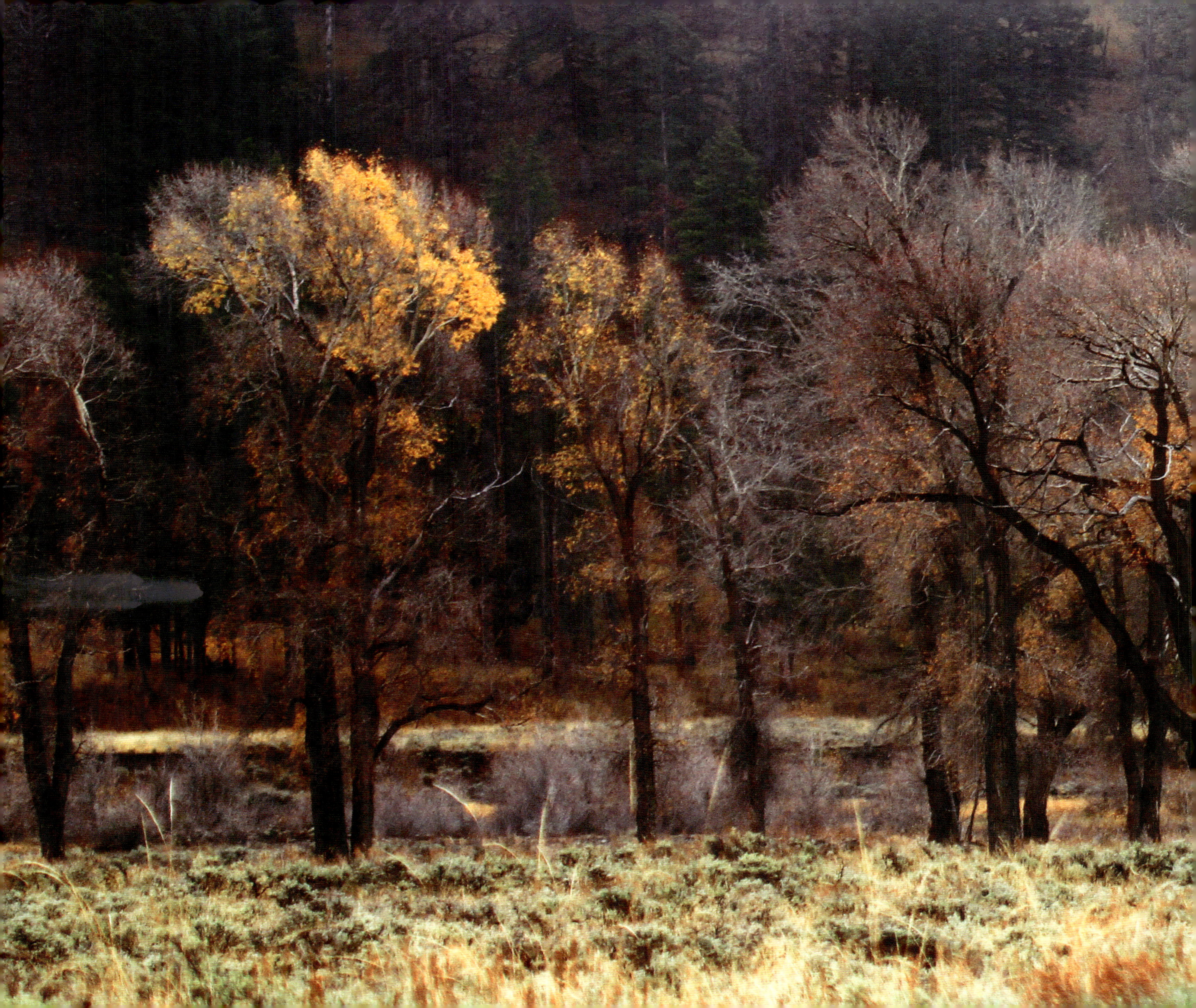

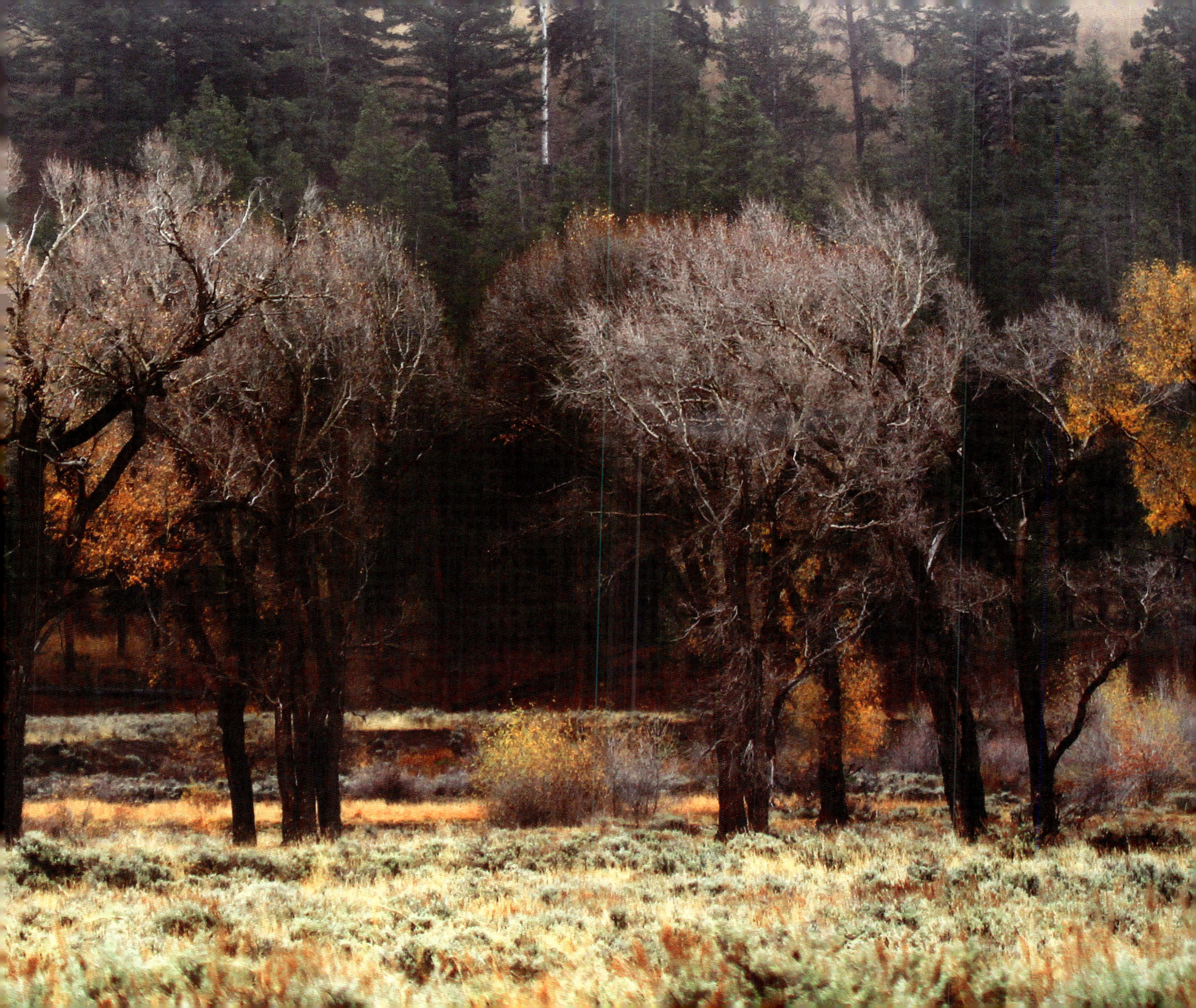

It would be fitting, I think, if among the last man-made tracks on earth would be found the huge footprints of the great brown bear.

—*Earl Fleming*

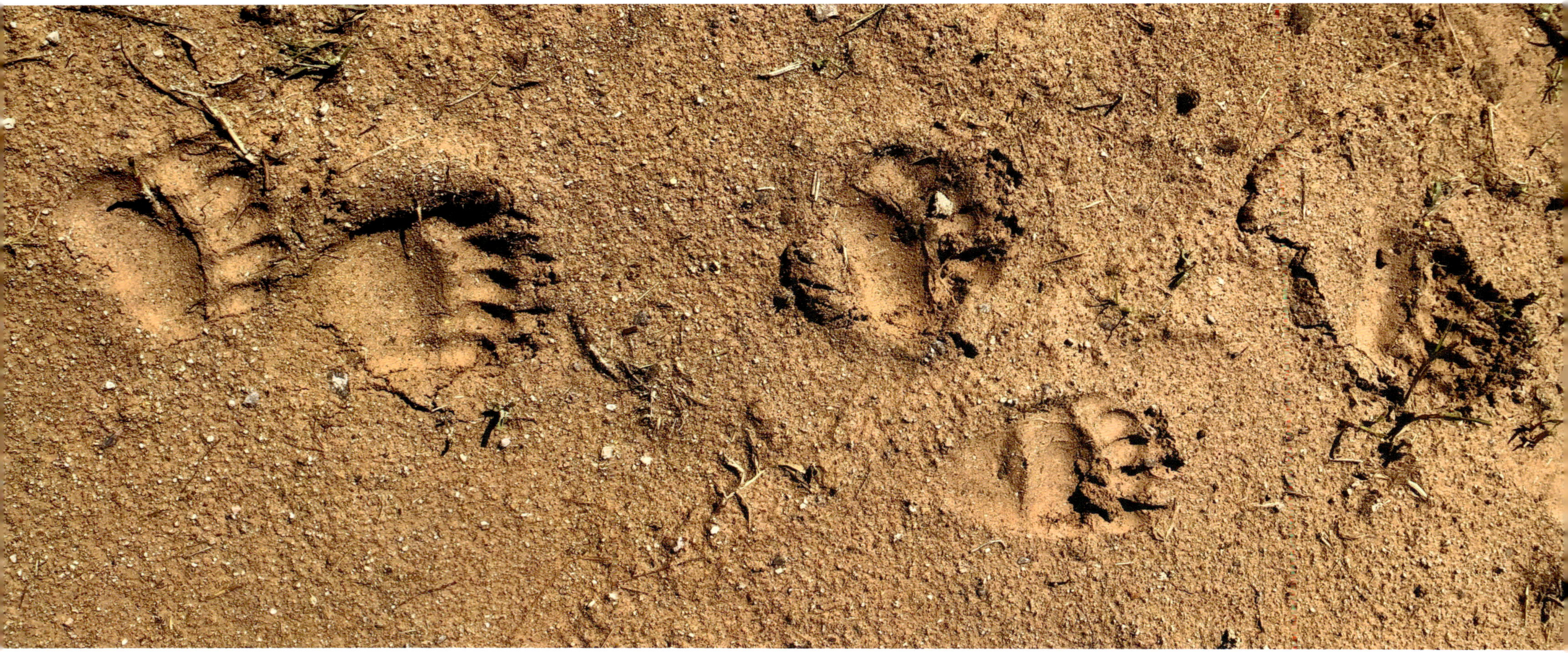

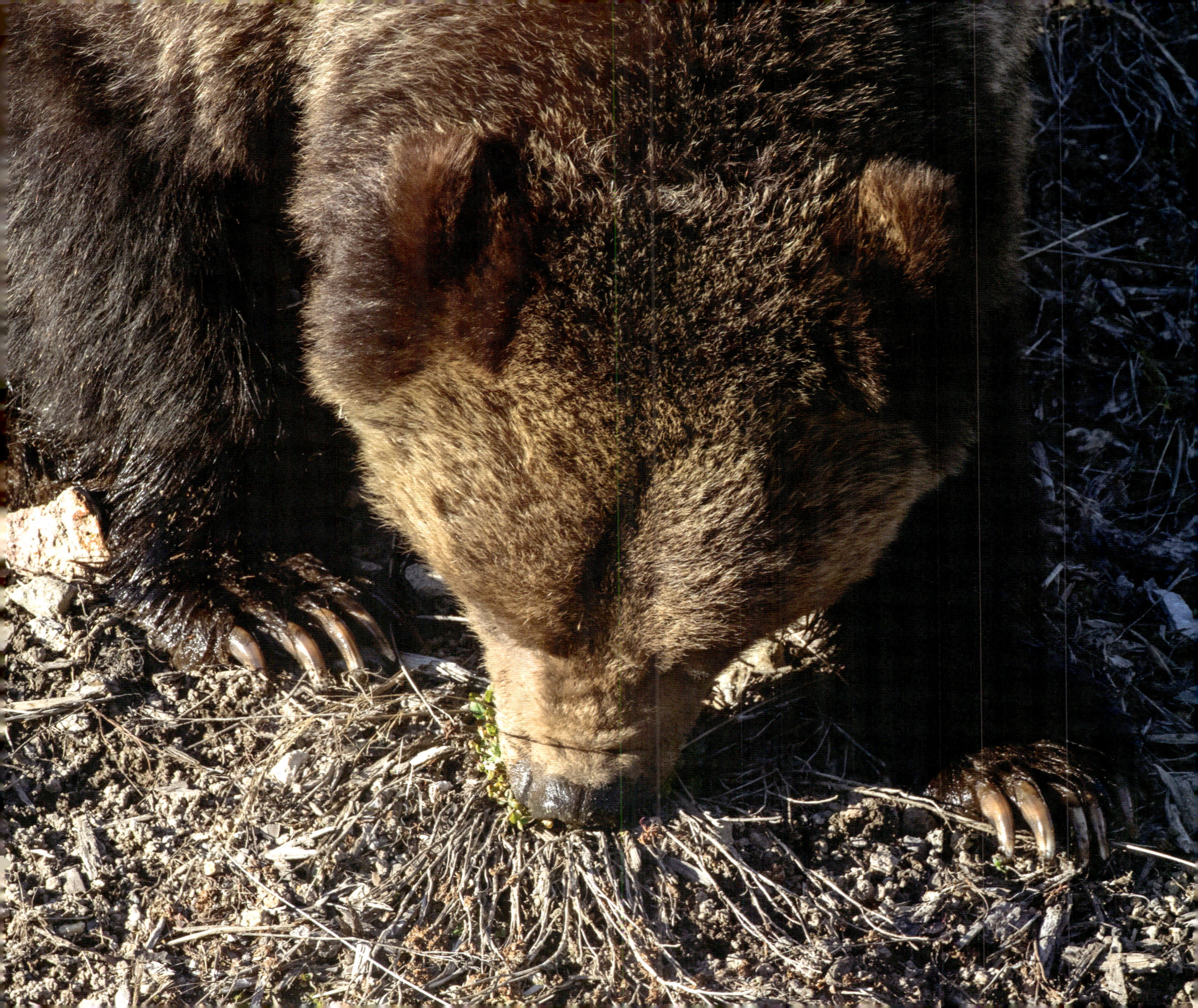

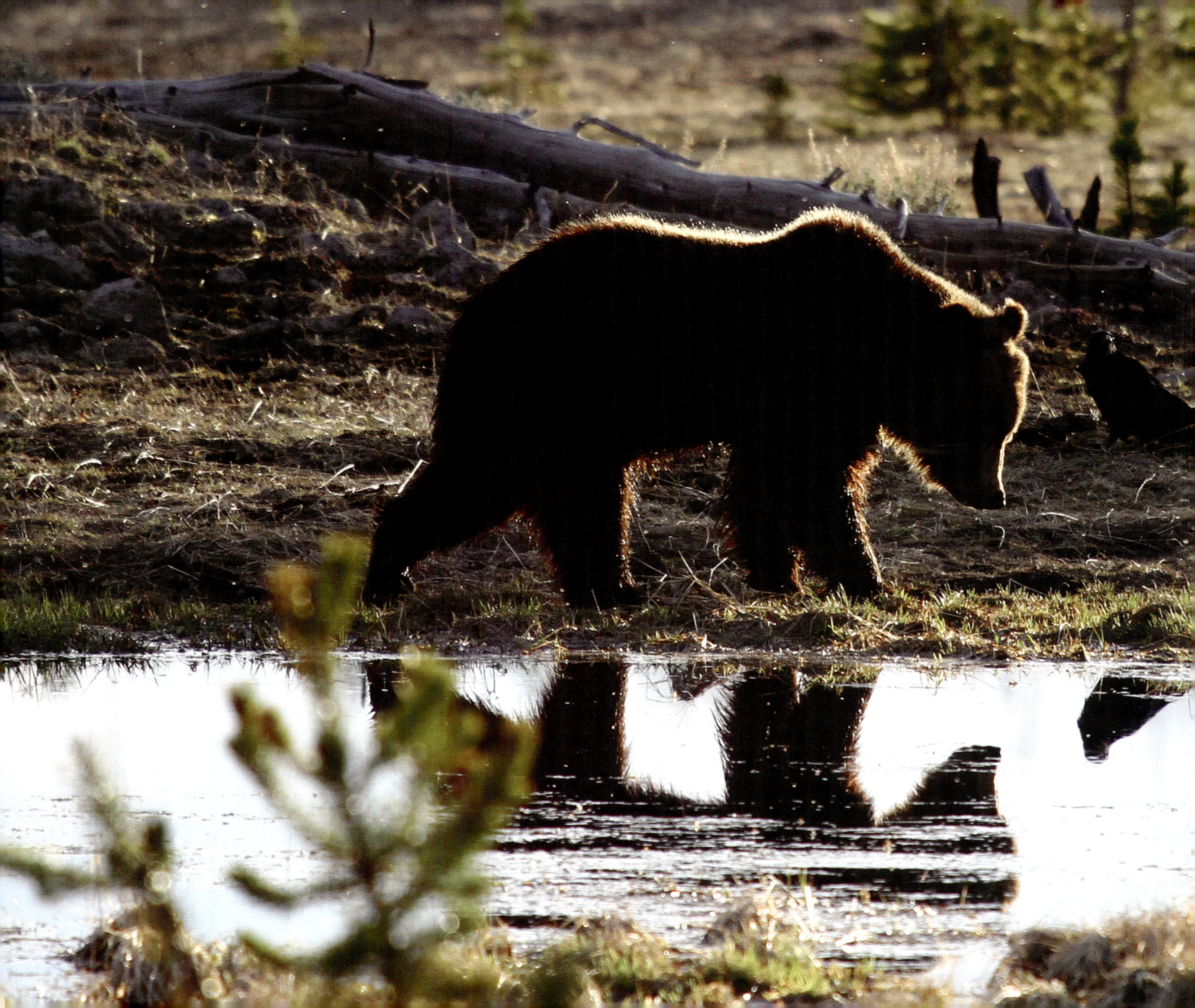

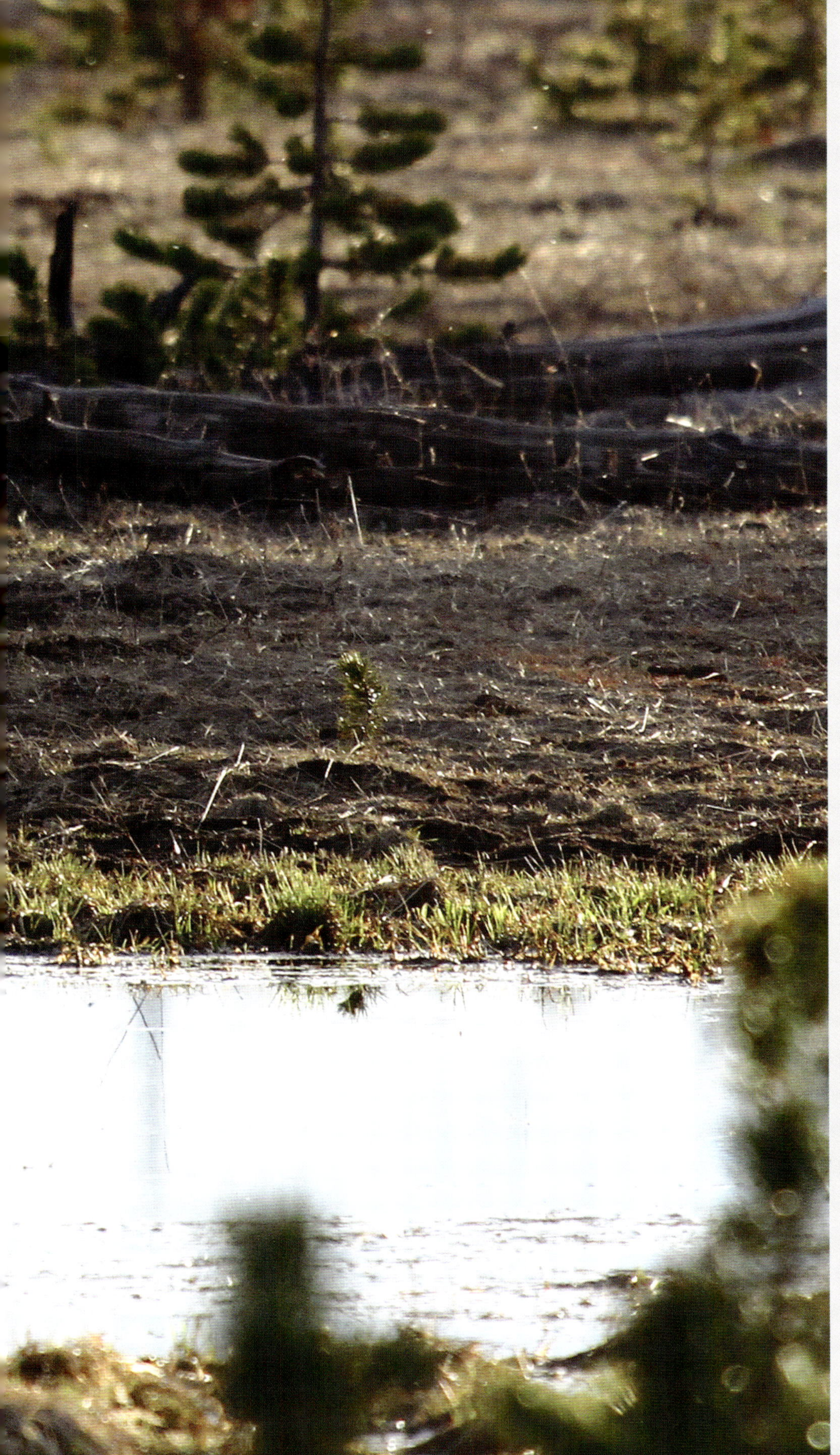

Grizzlies range in color from blonde to black, and in the Greater Yellowstone Ecosystem many have a light brown or blonde girth band. They have a dished face, a distinctive shoulder hump, claws that are slightly curved and range from two to four inches long, and they can run up to forty miles per hour. Their life span is between twenty and thirty years long, with the oldest known in the greater Yellowstone area as being thirty-four years old.

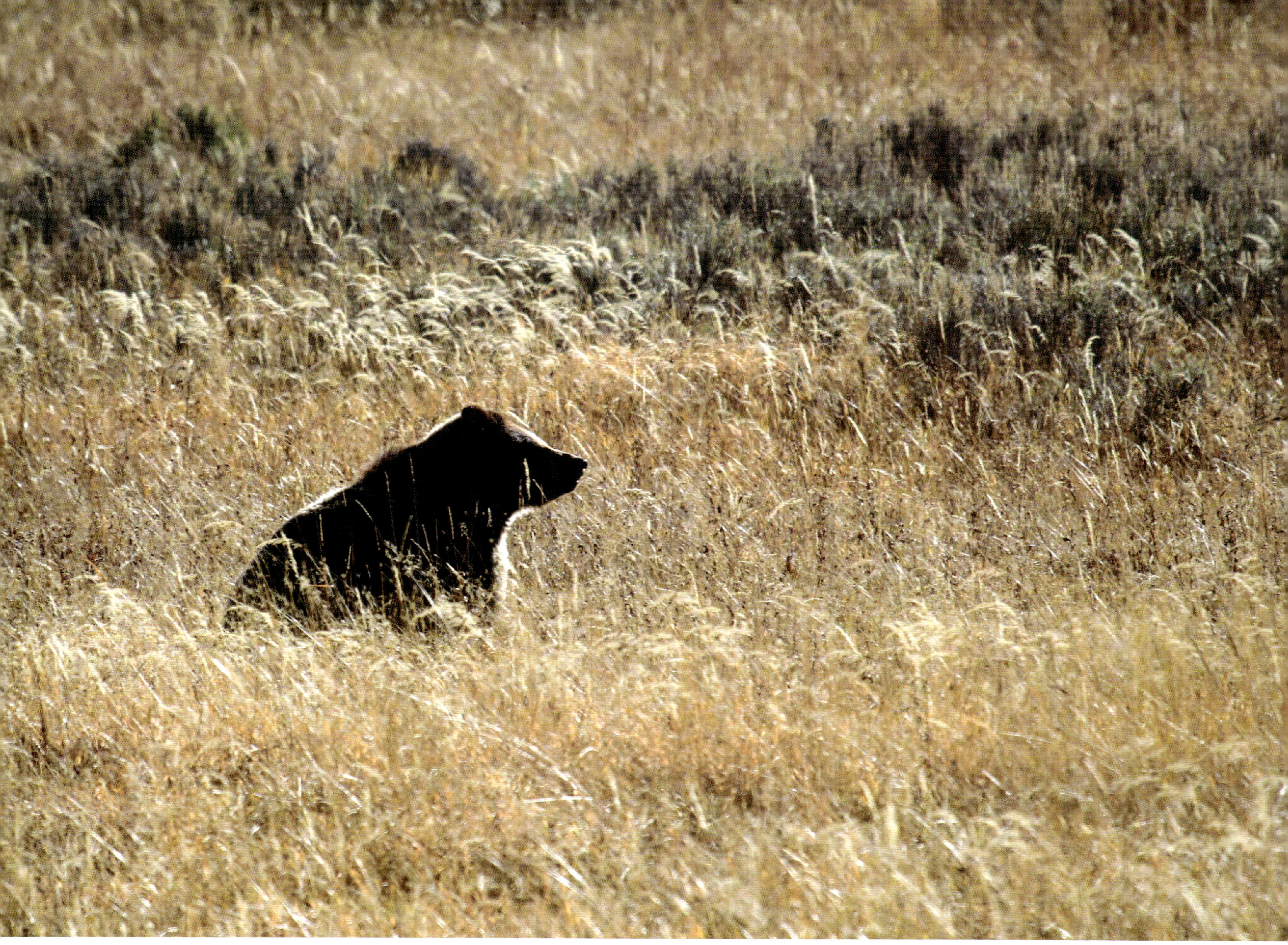

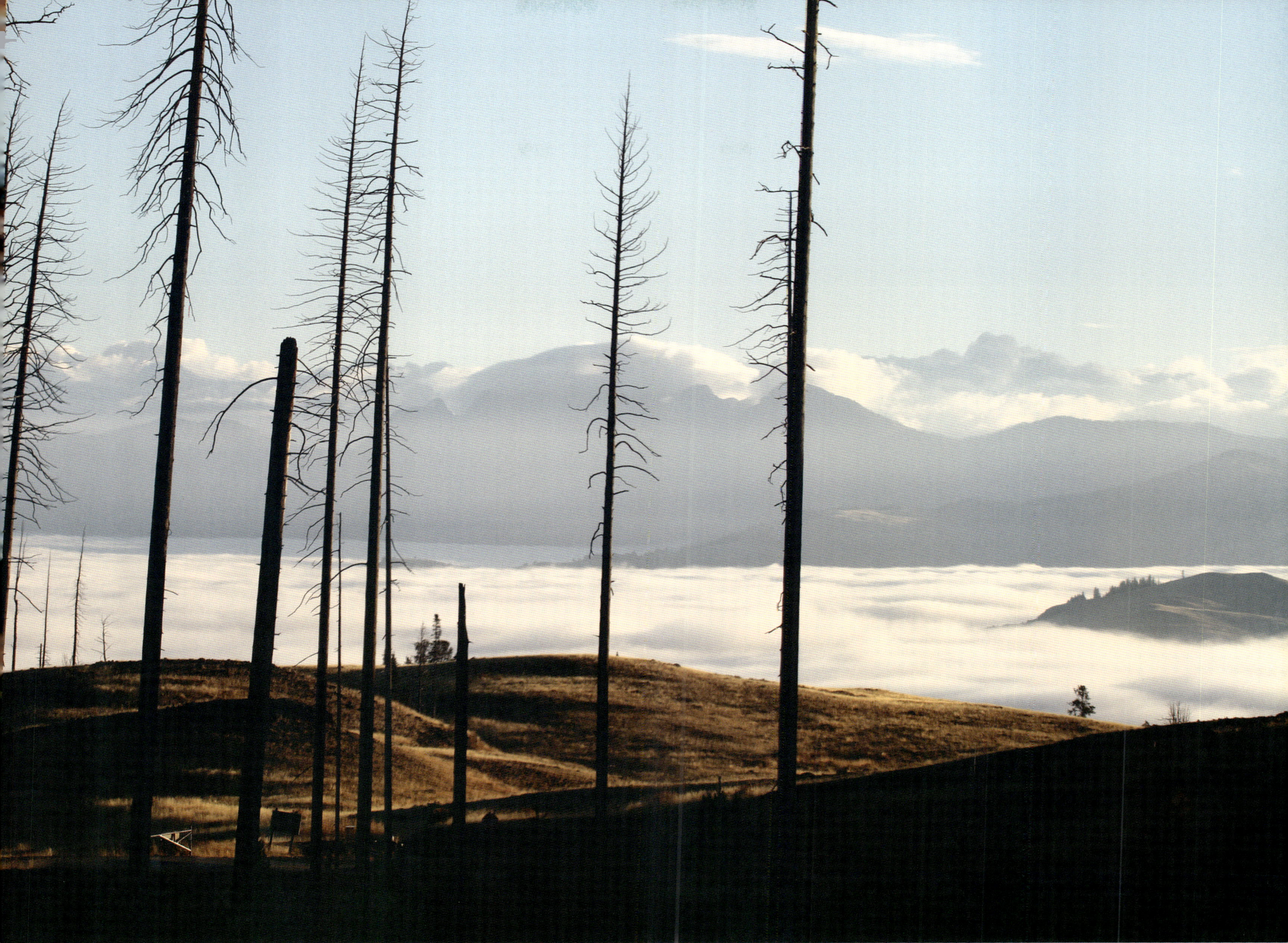

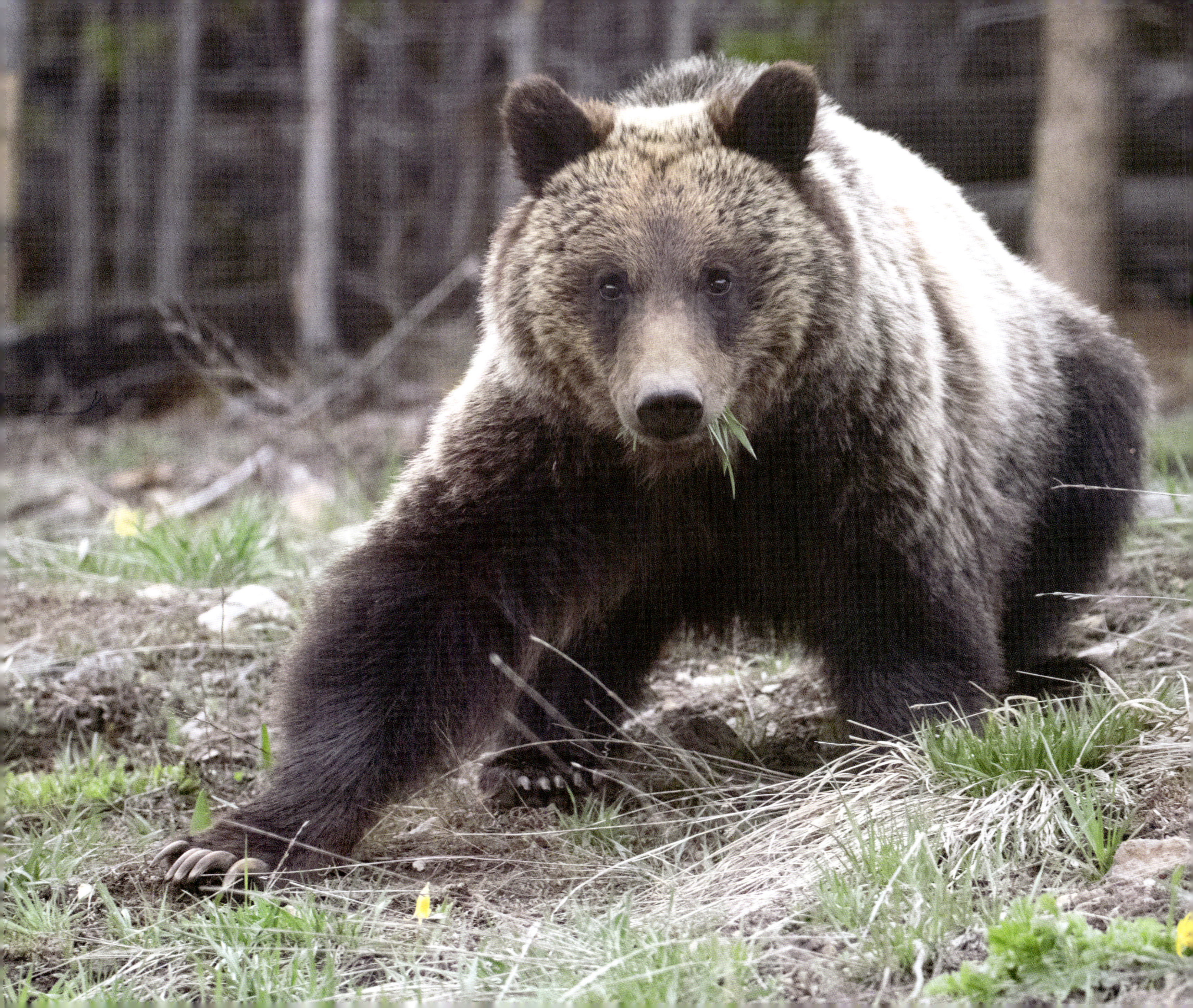

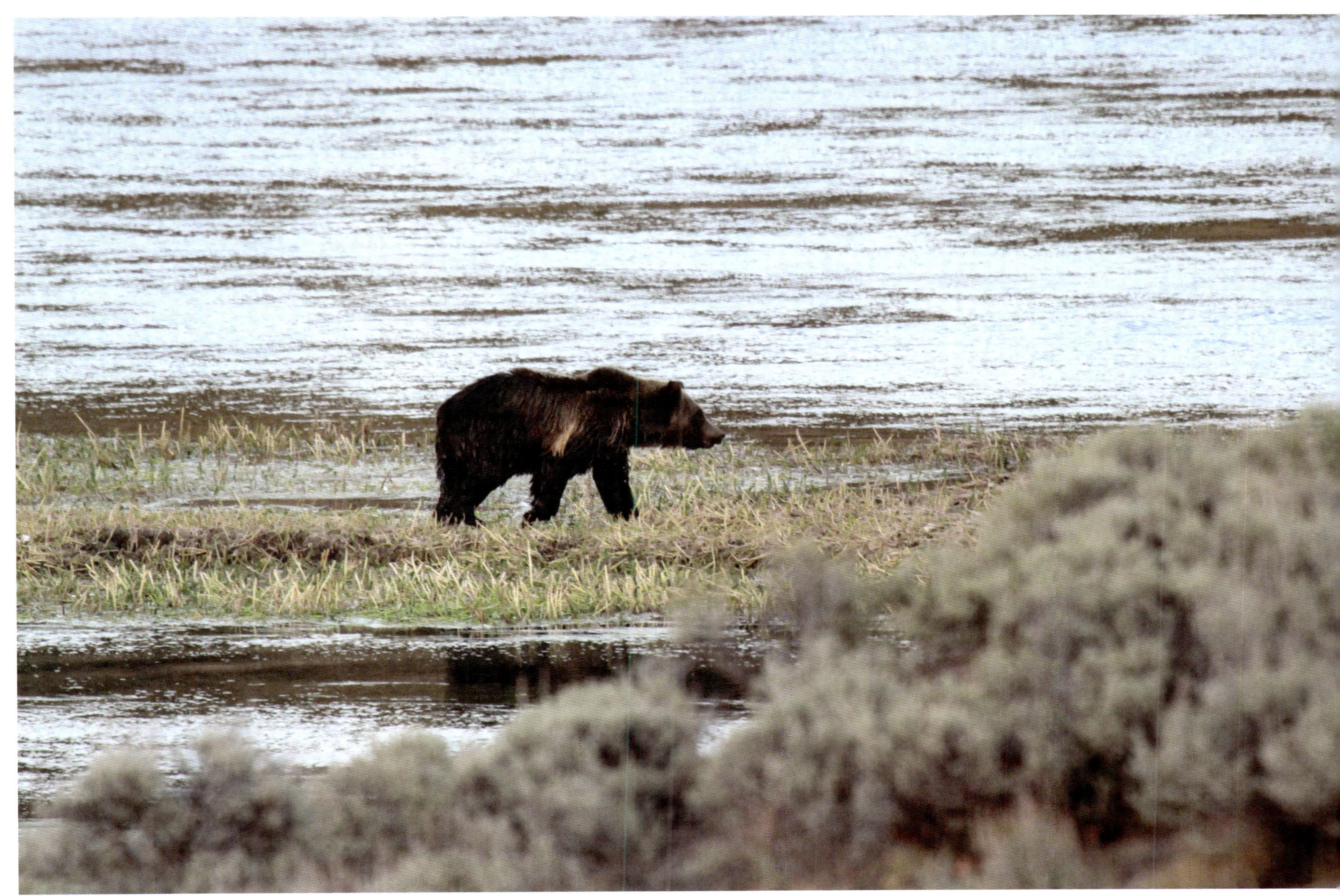

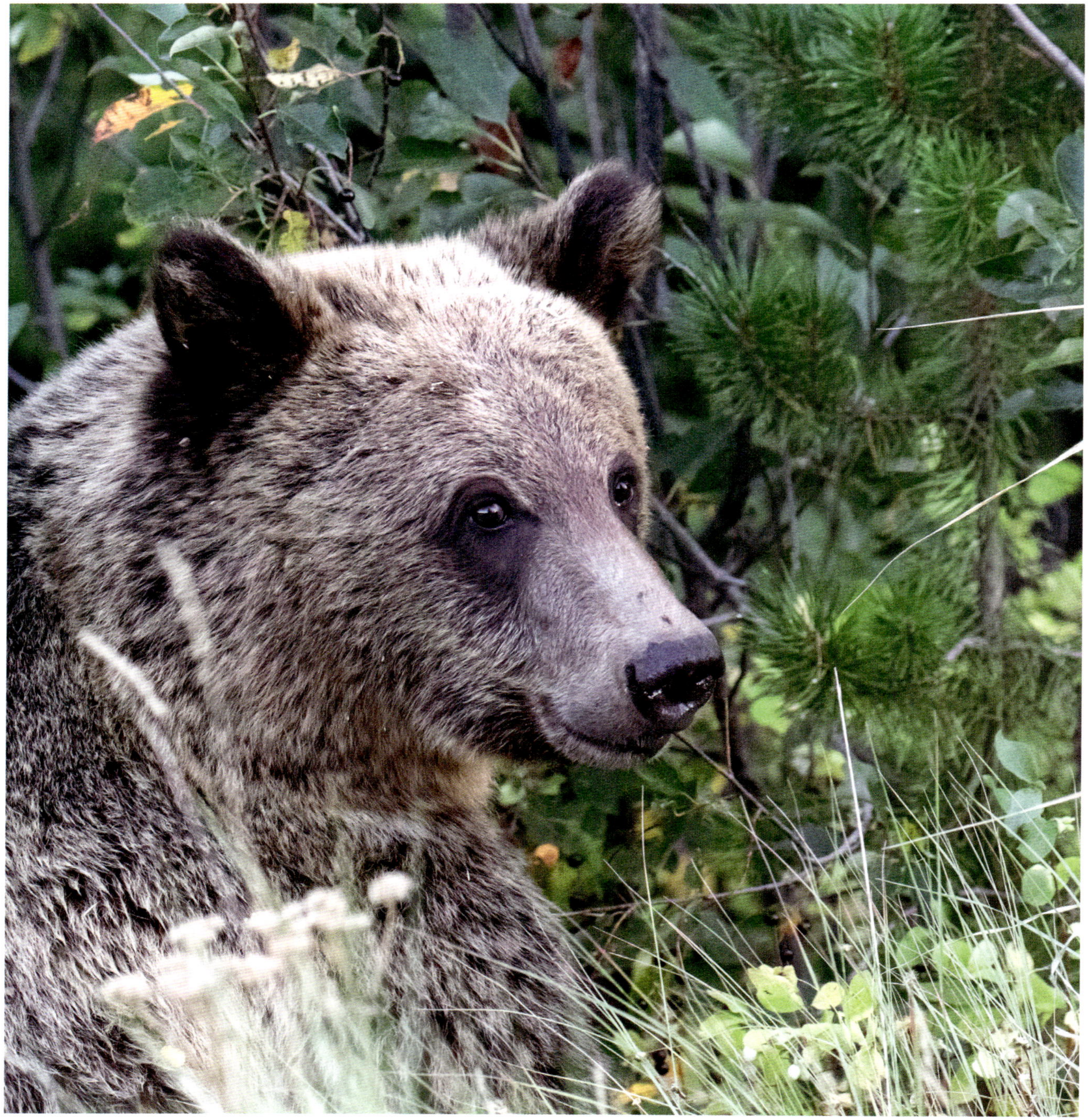

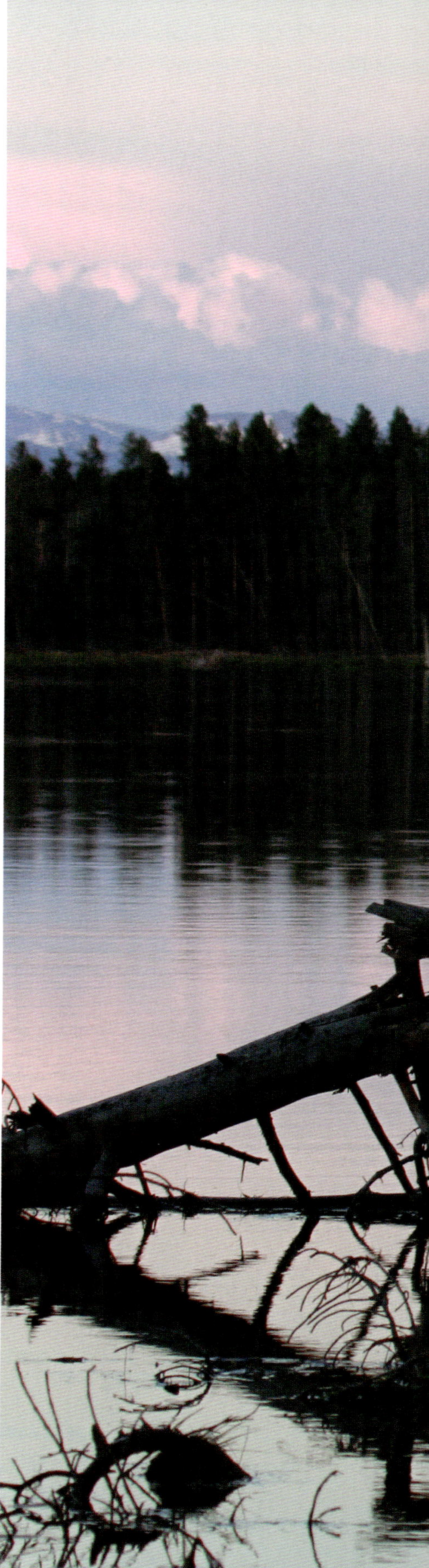

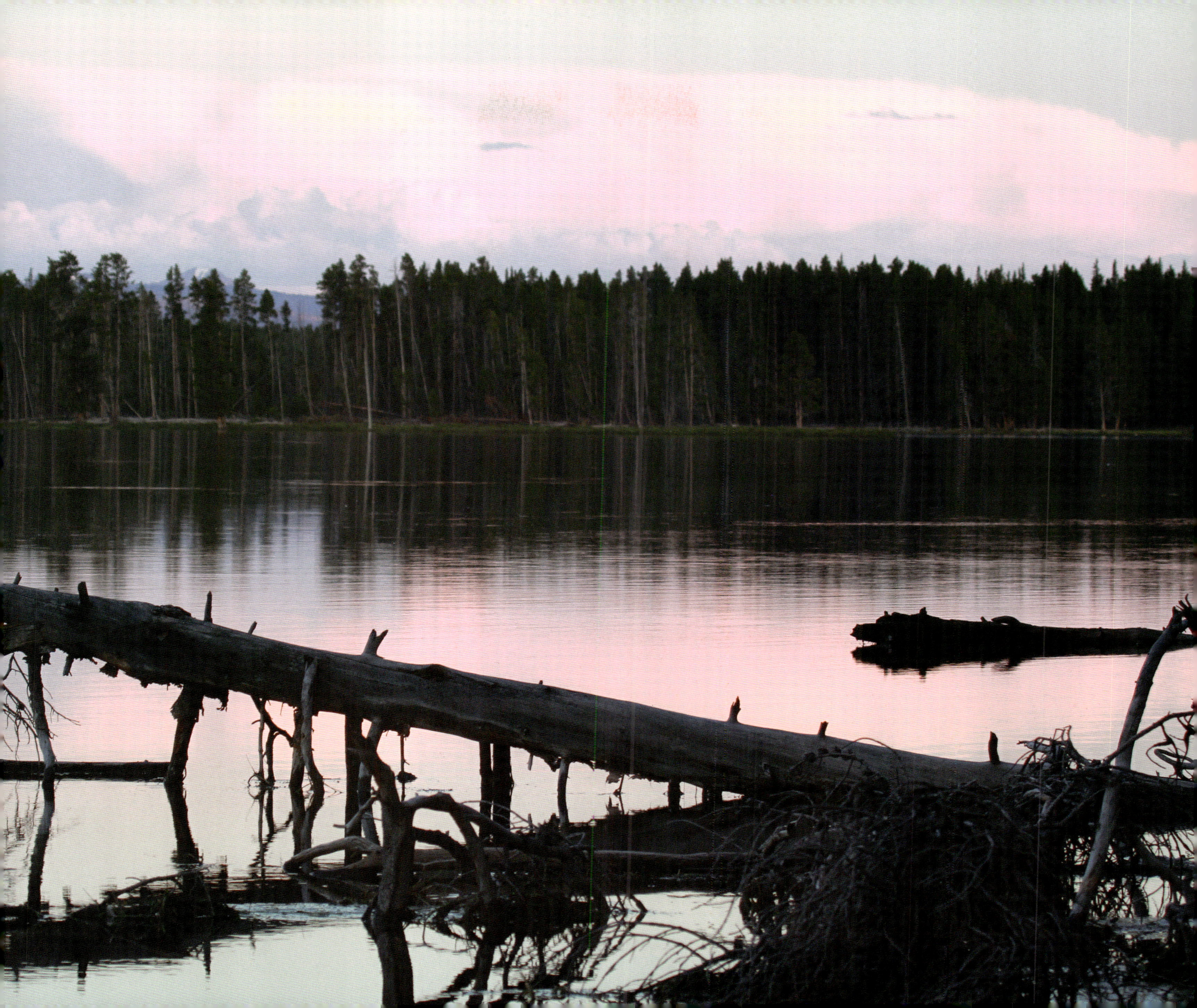

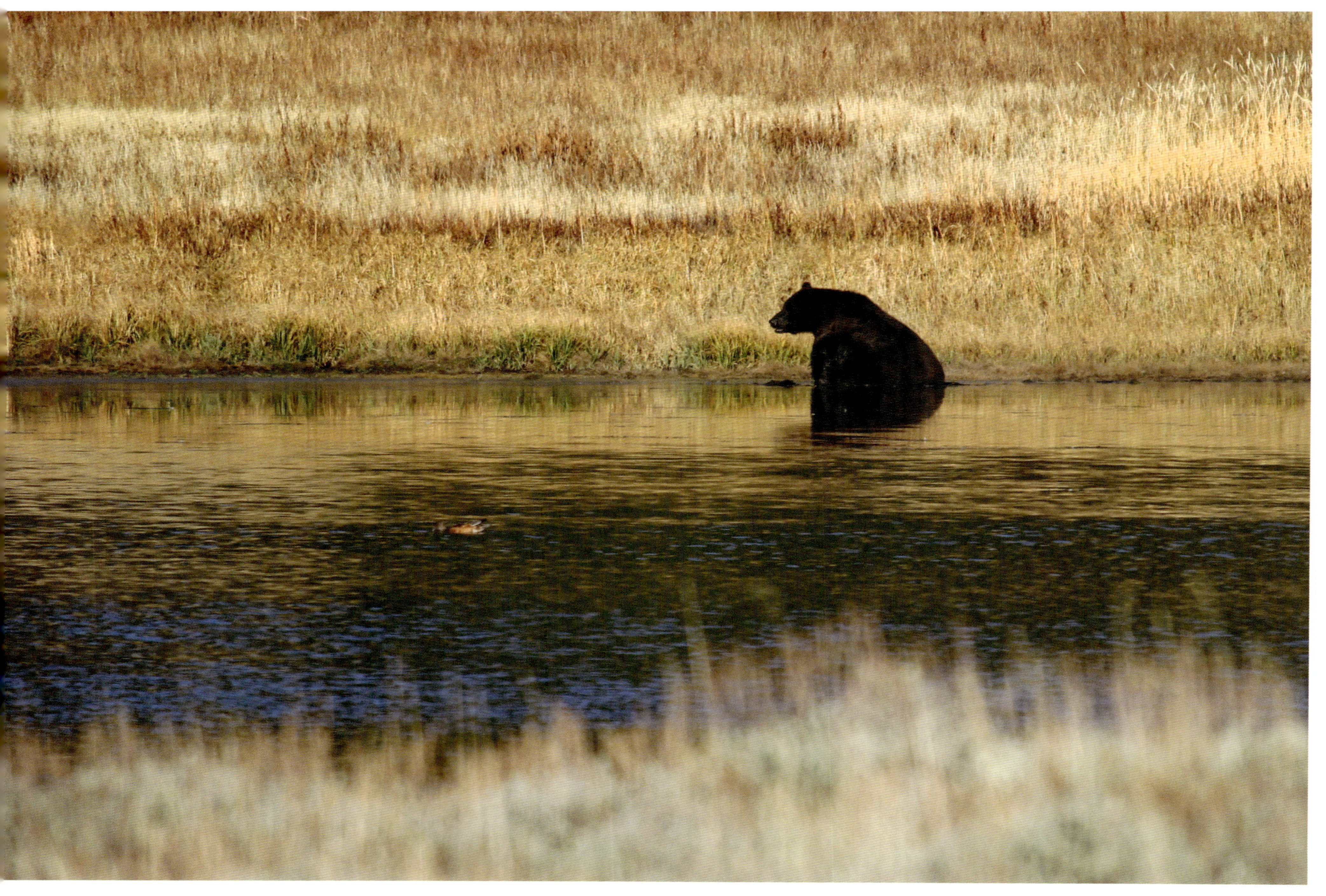

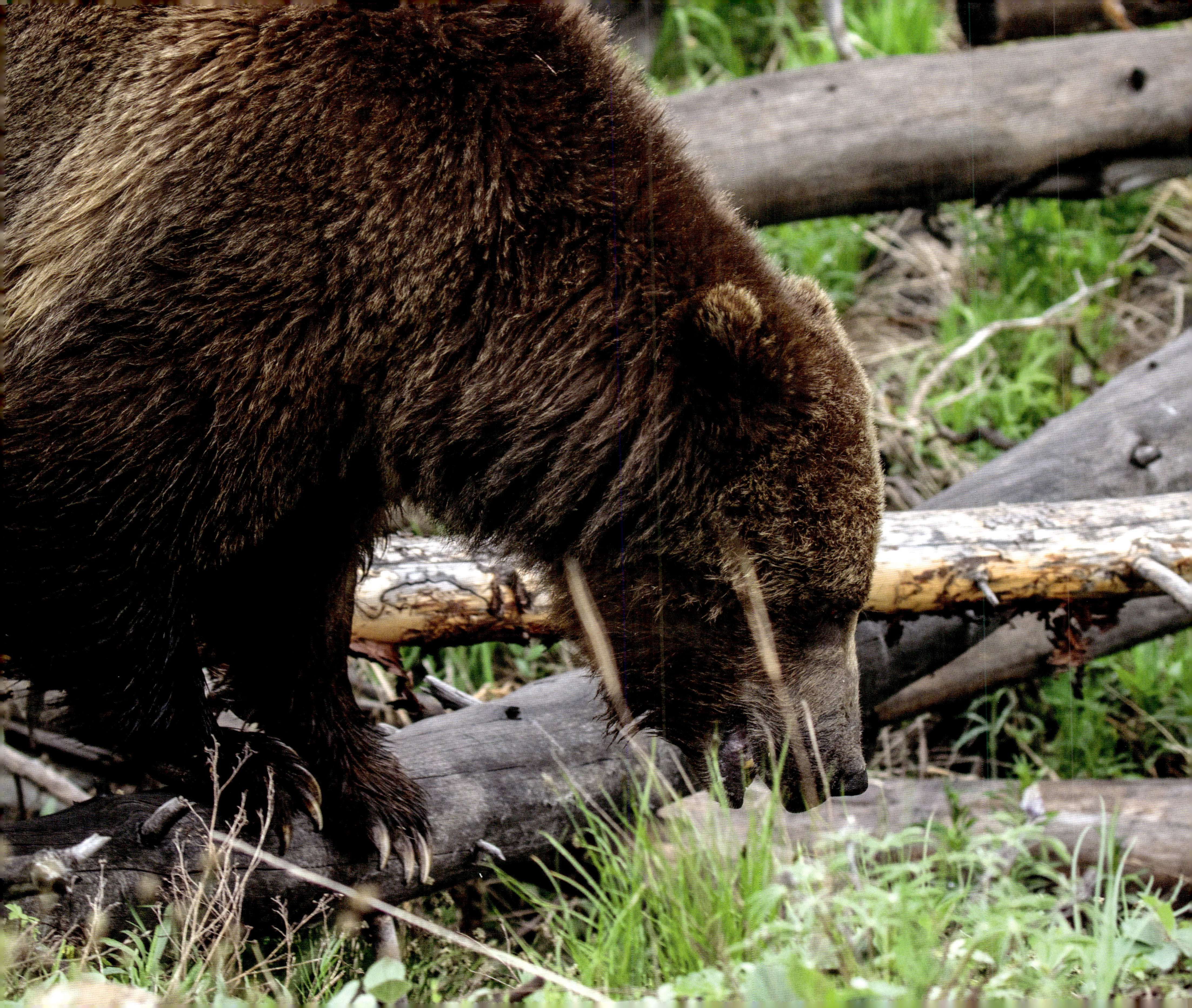

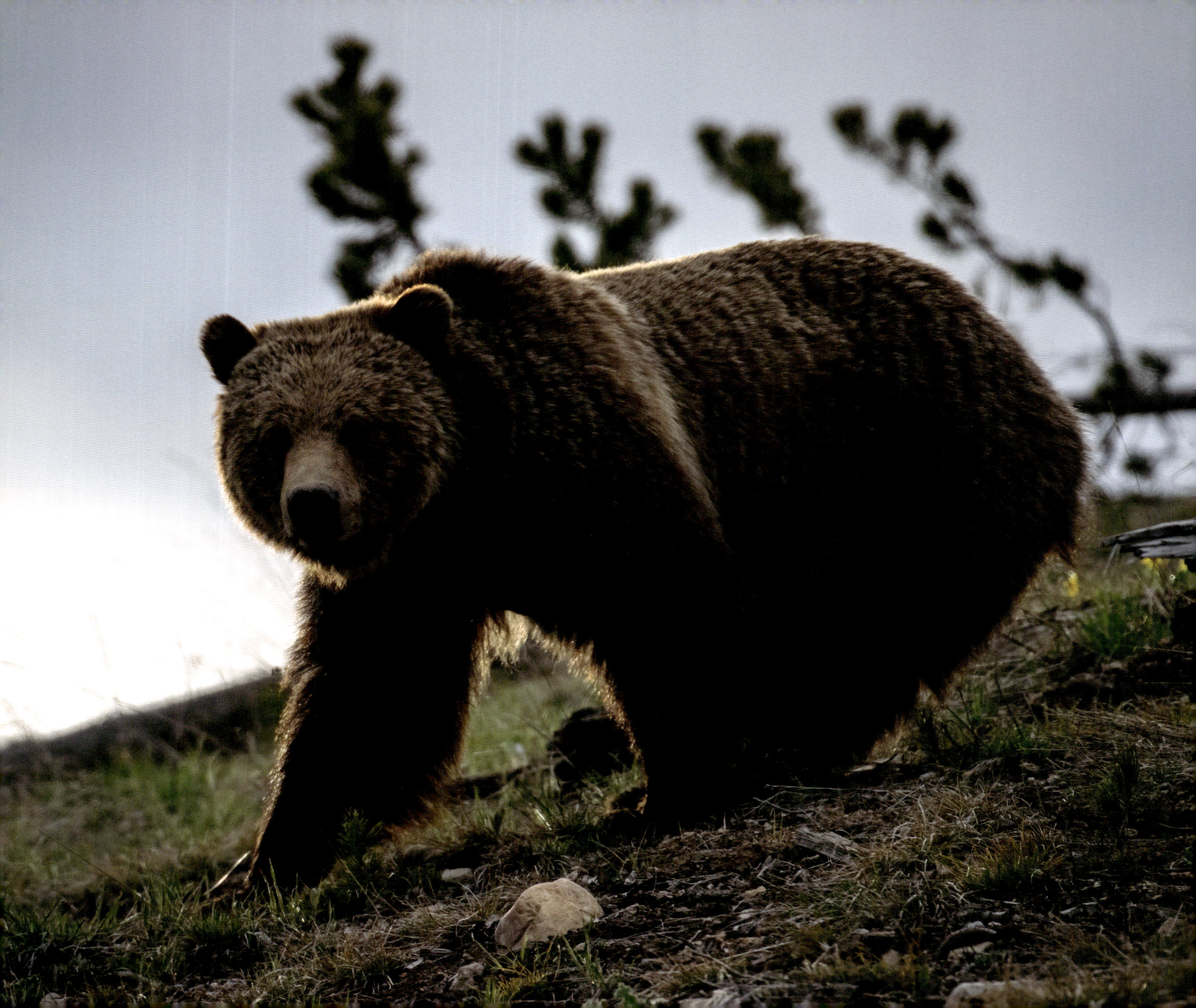

The ENDANGERED SPECIES ACT

IN 1975 GRIZZLY BEARS WERE LISTED UNDER THE ENDANGERED SPECIES ACT. At that time, the population was believed to be between 800 to 1,000 bears, with only 136 bears living in the Greater Yellowstone Ecosystem.

The goal of the Endangered Species Act of 1973 was to recover a species to a viable self-sustaining population where protection is no longer needed. In order to do this, federal and state agencies stopped hunting of grizzly bears in the areas outside the national park boundaries, which are known as the Greater Yellowstone Ecosystem. They also established the Yellowstone Grizzly Bear Recovery Zone, created the Interagency Grizzly Bear Study Team, and established the Interagency Grizzly Bear Committee. Each of these resources has an important role to play in grizzly bear recovery.

In 1993 a recovery plan was implemented with three specific recovery goals that had to be met for six consecutive years that focused on sufficient habitat quality to support a self-sustaining grizzly bear population, an area where management monitoring efforts could be focused, and where habitat standards could be applied as measurable recovery criteria. In 2002 a conservation strategy was approved after a public comment period where 16,794 comments were received. This strategy was to be implemented when the grizzly was removed from the threatened species list. In 2003 the recovery goals were met for the sixth year in a row.

In 2005 the U.S. Fish and Wildlife Service proposed to remove the grizzly from the threatened species list. Following that proposal, in 2007, the Greater Yellowstone Ecosystem distinct population segment was removed from the threatened species list. Several groups then filed a lawsuit challenging the decision. In 2009 a federal district judge overturned that decision, stating the conservancy strategy was unenforceable and that the impact of the potential loss of whitebark pine nuts, a grizzly bear food source, was not adequately considered.

In 2010 the U.S. Fish and Wildlife Service appealed that decision, and in 2011 an appeals court ruled the grizzly bear would remain on the threatened species list.

In 2013 the Yellowstone Ecosystem Subcommittee, the Interagency Grizzly Bear Committee, and the Interagency Grizzly Bear Study Team recommended that grizzly bears be removed from the threatened species list because alternate food sources were available and the reduction of whitebark pine was not having a significant impact at that time. In 2017 the U.S. Fish and Wildlife Service removed grizzly bears from the threatened species list.

At that time, the states of Wyoming and Idaho both scheduled grizzly bear hunts for the fall of 2018, something that hadn't happened for over forty years.

Many tourists, photographers, and wildlife advocates were outraged that trophy sport hunting of these animals was going to be allowed. Thomas Mangelsen, a well-known photographer and strong advocate for grizzlies, put in for one of the tags with the intent of shooting with a camera not a gun, and as the luck of the draw would have it, he drew the number eight spot, thus potentially saving the life of one bear.

In August of that same year, just two days before those hunts were to begin, a federal judge put a fourteen-day hold on them in order to consider the recommendations on which the protections had been lifted. On September 24, 2018, that same judge overturned the previous ruling, stating the U.S. Fish and Wildlife Service had been "arbitrary and capricious" in their 2017 decision, and once again returned the grizzly to the Endangered Species Act list.

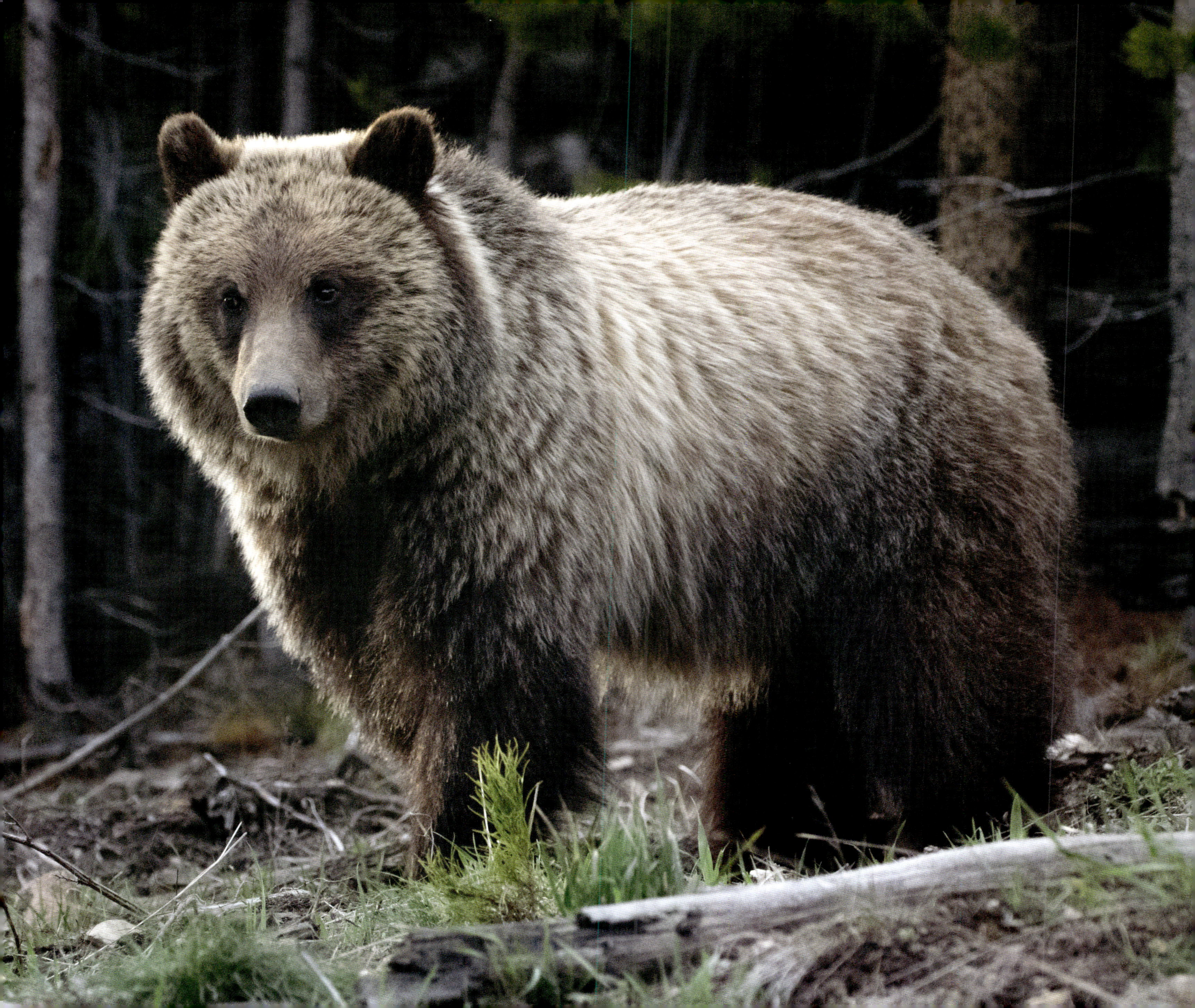

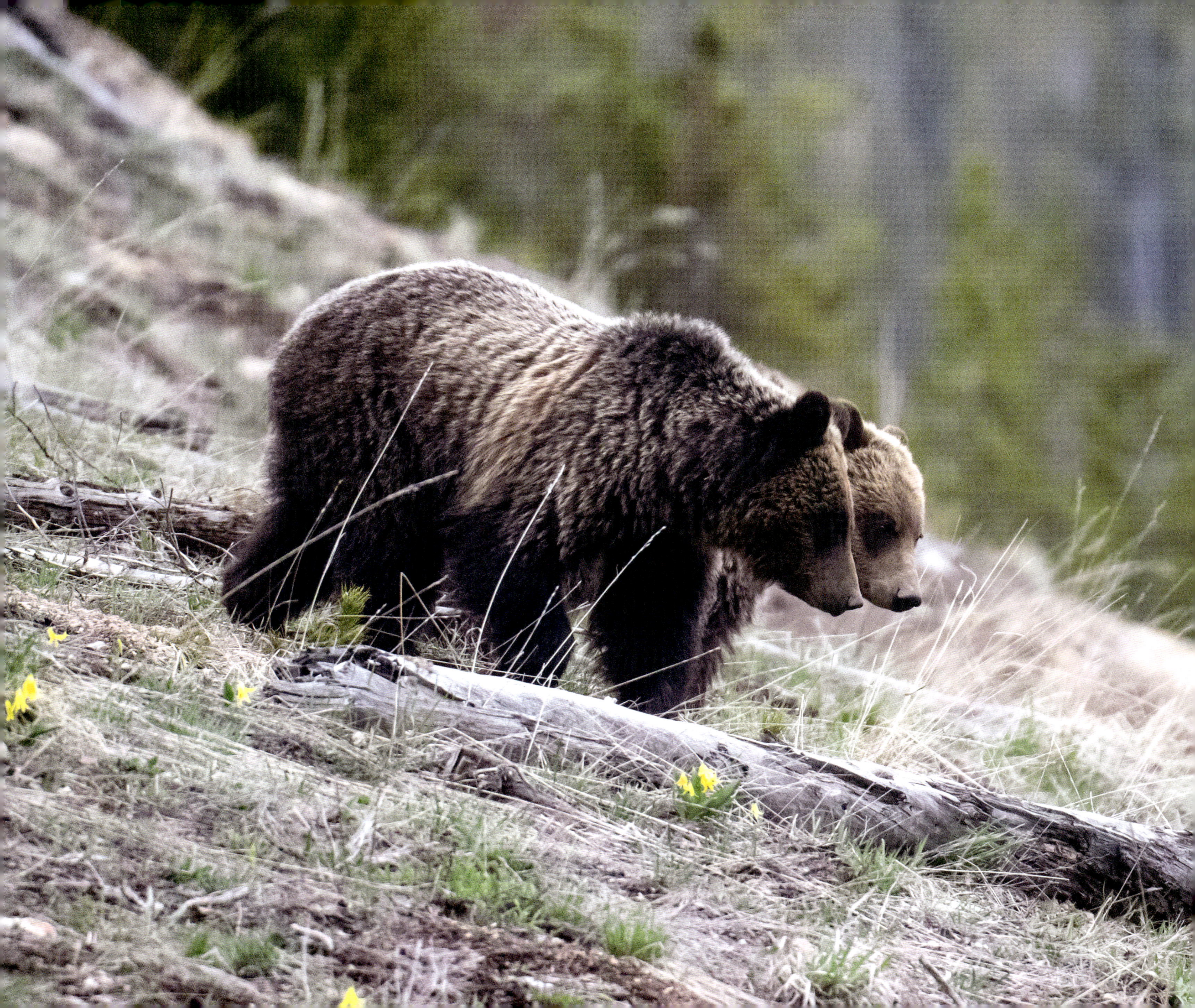

Grizzly bears in the greater Yellowstone are the epitome of wildness. Their recovery serves to represent one of the greatest conservation successes of our lifetime. The tireless work, and effort placed into the conservation of this species the last several decades is not just work, but passion. It's not the talent, or the tremendous masterful knowledge of professionals that have achieved this success—but a combination of passion and enthusiasm for the species. Passion has no bounds—your creativity flourishes like never before; you forget time. It's in this zone where great ideas are born, and where objectives or ideas that may have seemed daunting or impossible become true and take flight. We can thank the countless [wildlife] professionals who have demonstrated this and dedicated their lives and careers over the last fifty years for where grizzly bears are today. Passion still lives in our field, and it's what drives my purpose professionally and personally, to protect these creatures for as long as I live.

—Tyler Brasington, Bear Biologist

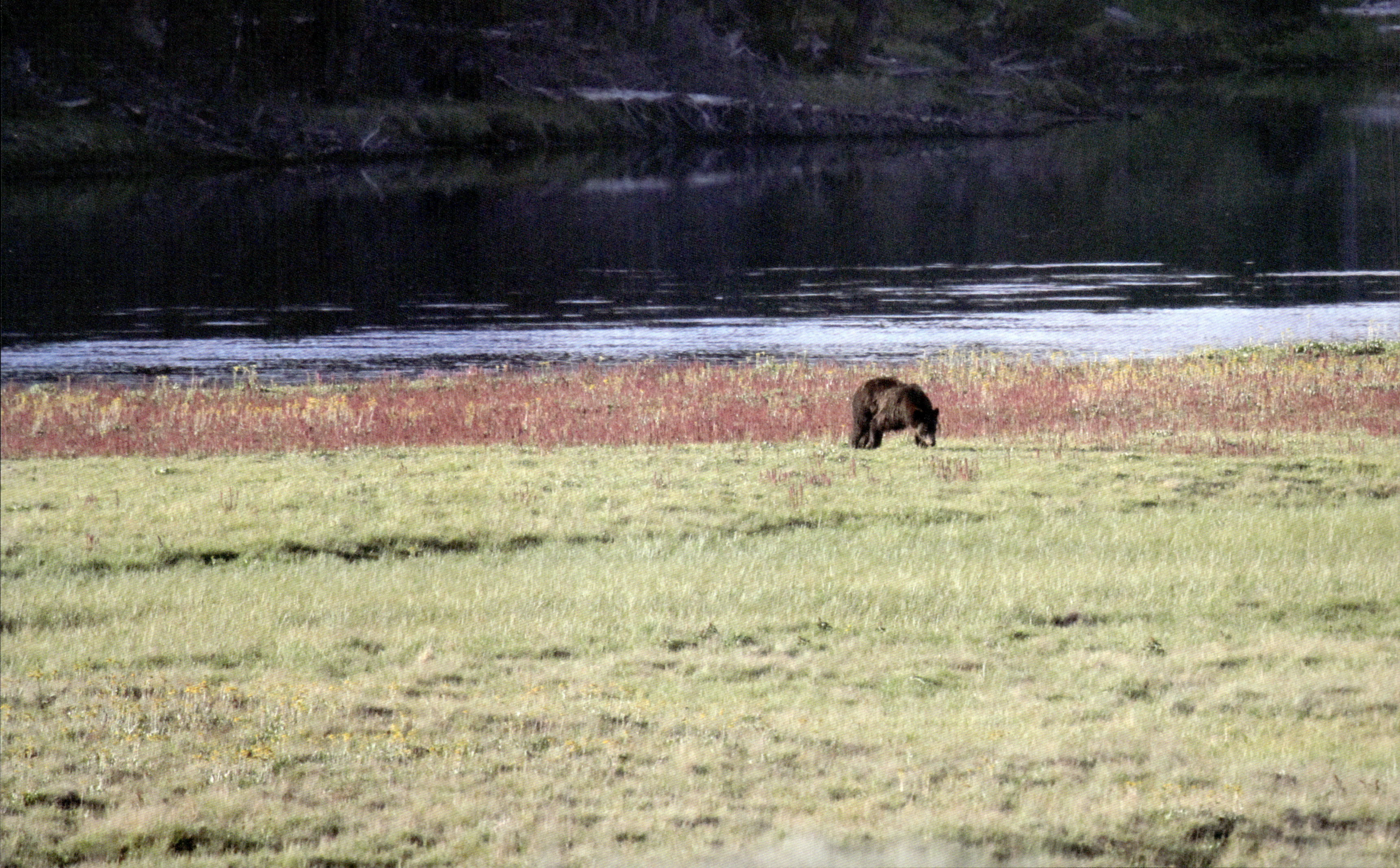

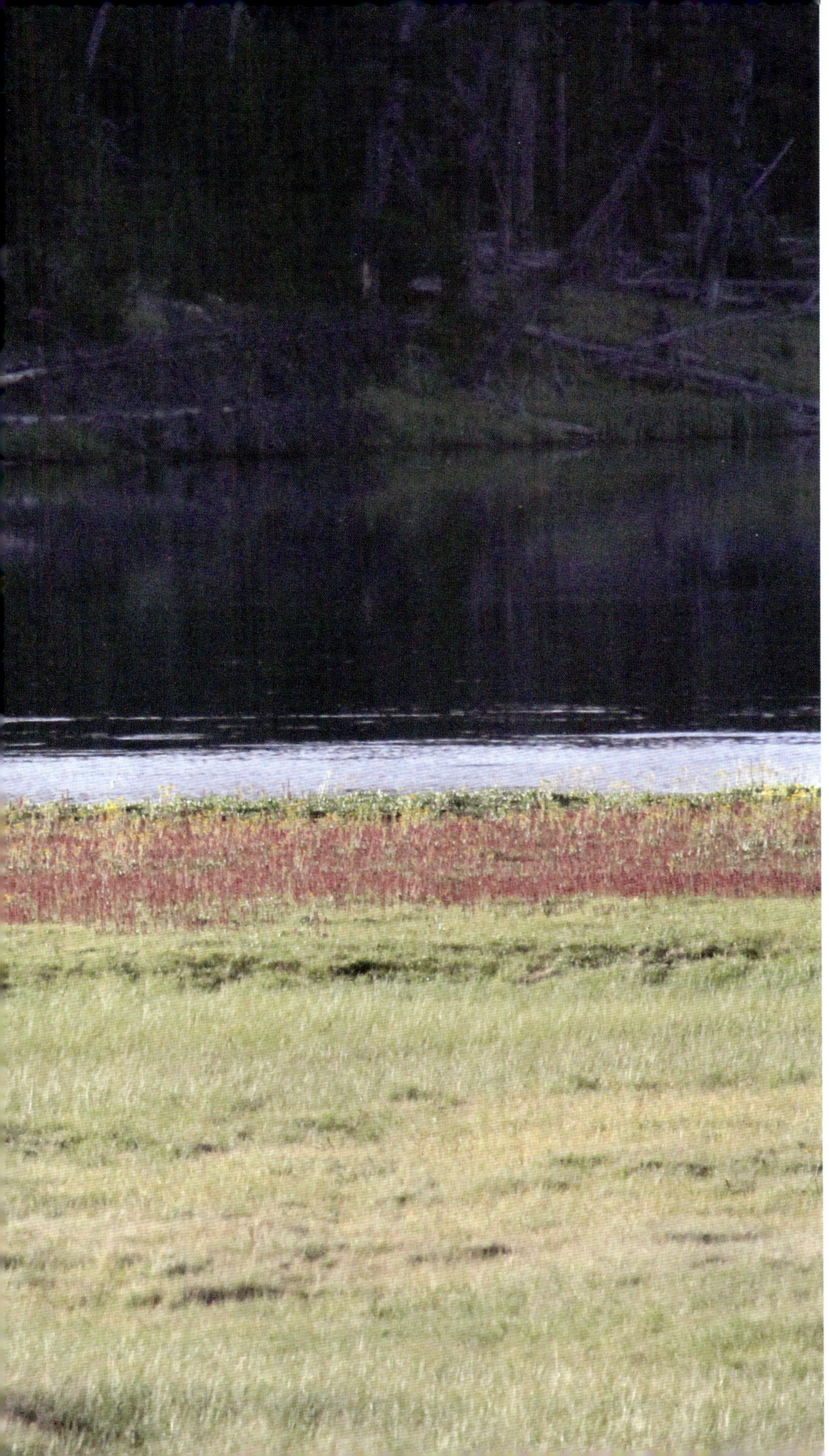

Humility is born in wildness. We are not protecting grizzlies from extinction; they are protecting us from the extinction of experience as we engage with a world beyond ourselves. The very presence of a grizzly returns us to an ecology of awe. We tremble at what appears to be a dream yet stands before us on two legs and roars.

—*Terry Tempest Williams*

The loss of natural food sources due to climate change, drought, and severe winters has a huge effect on reproduction and the survival rate of grizzly bears.

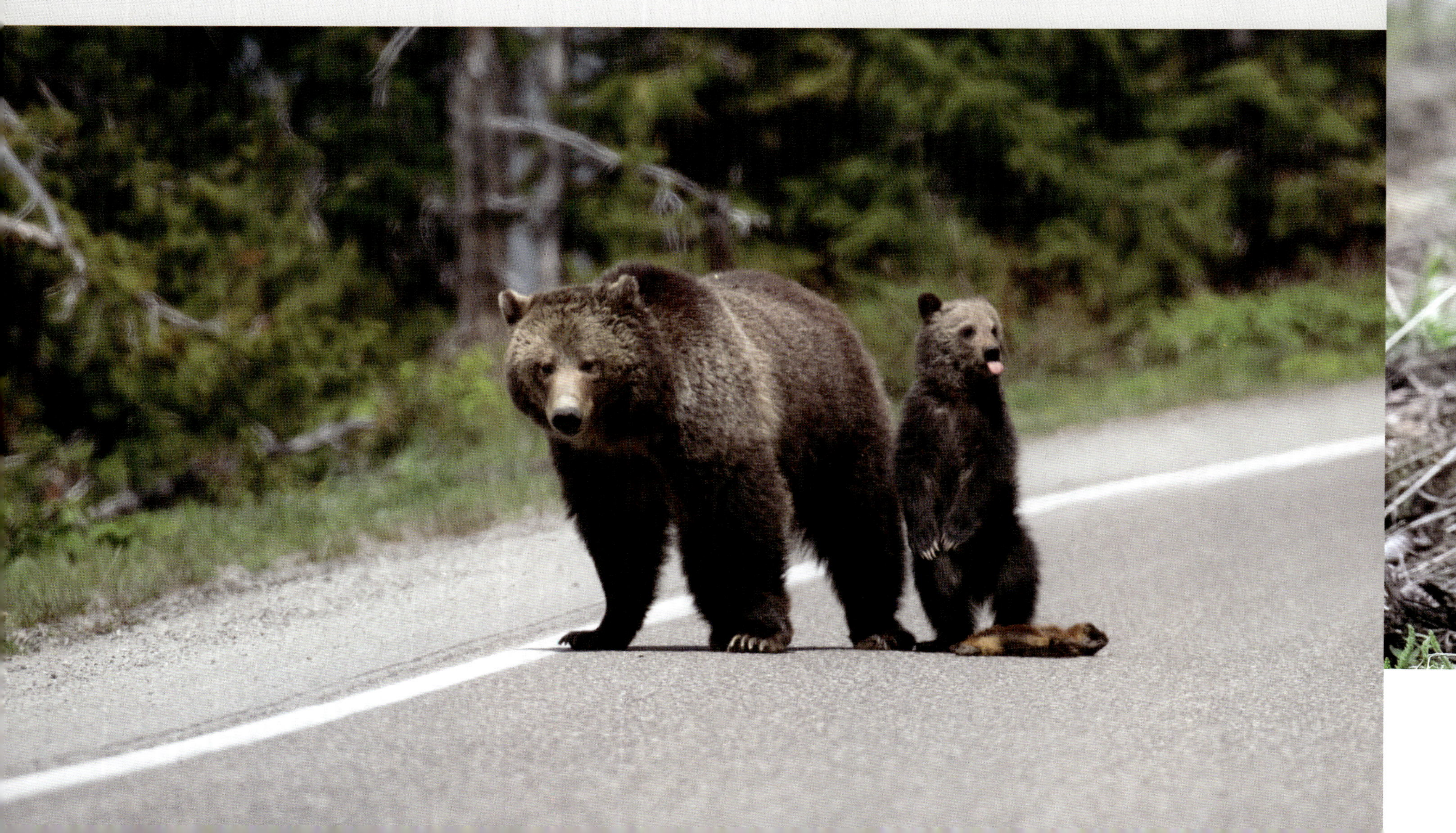

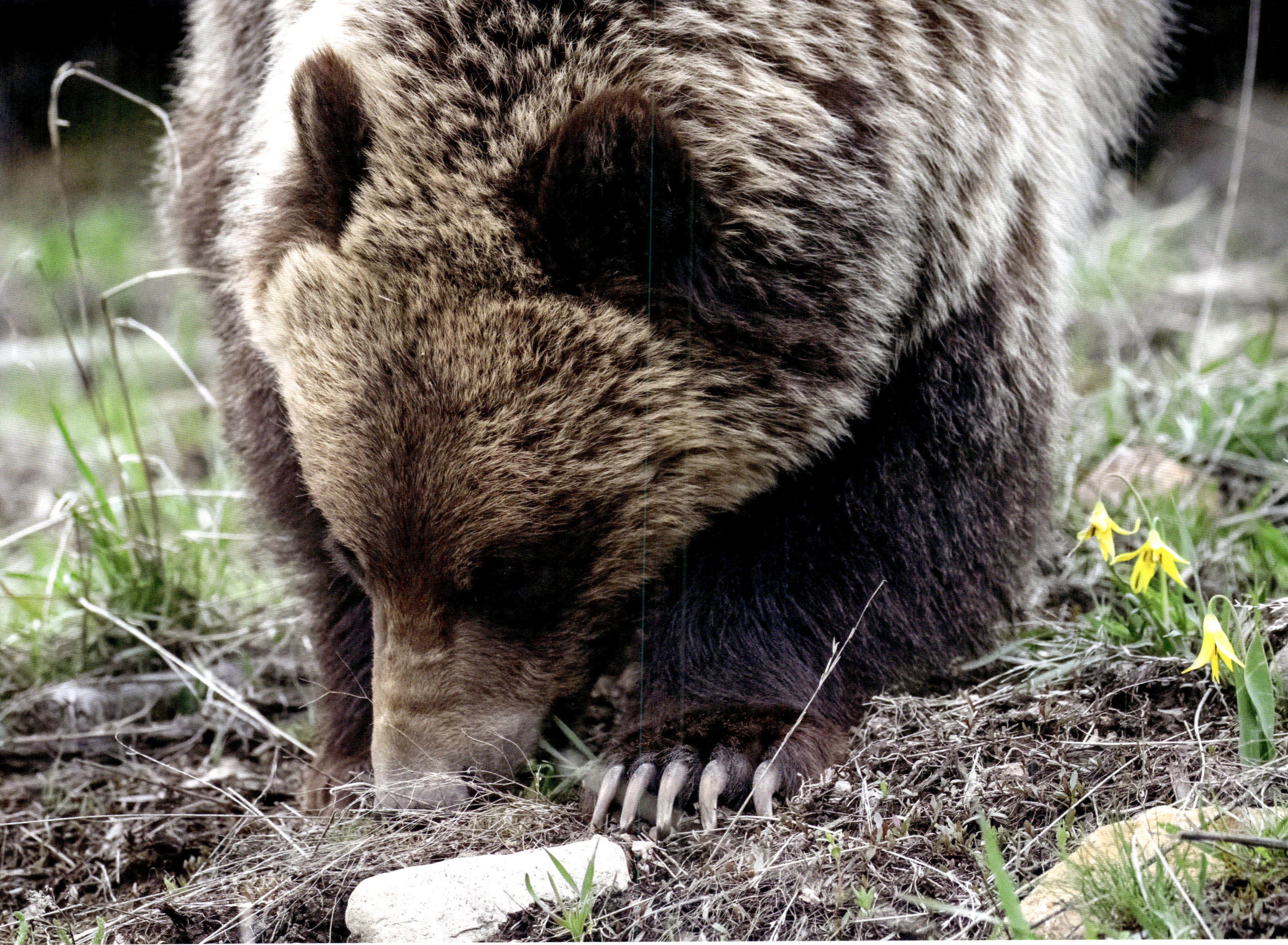

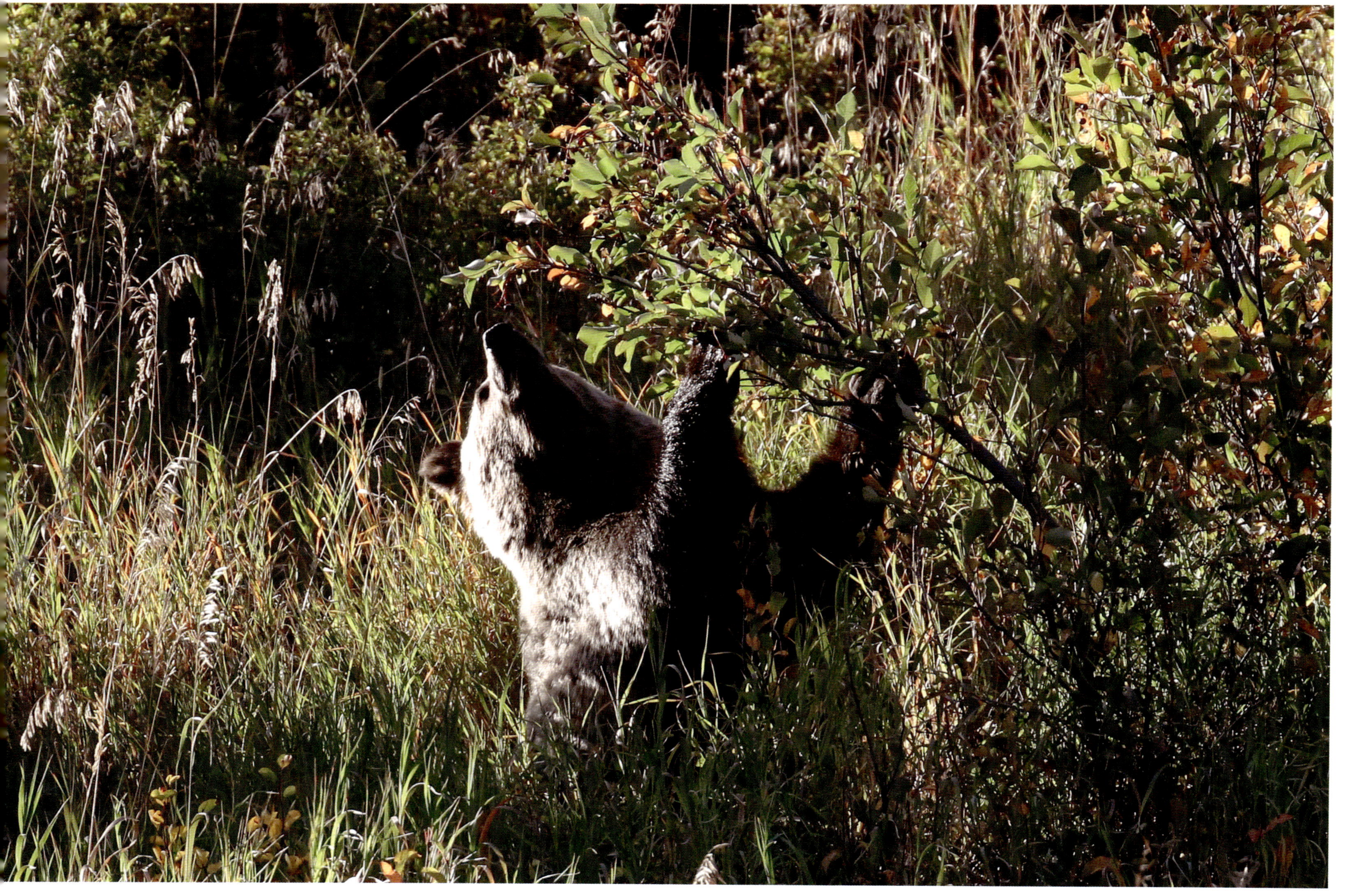

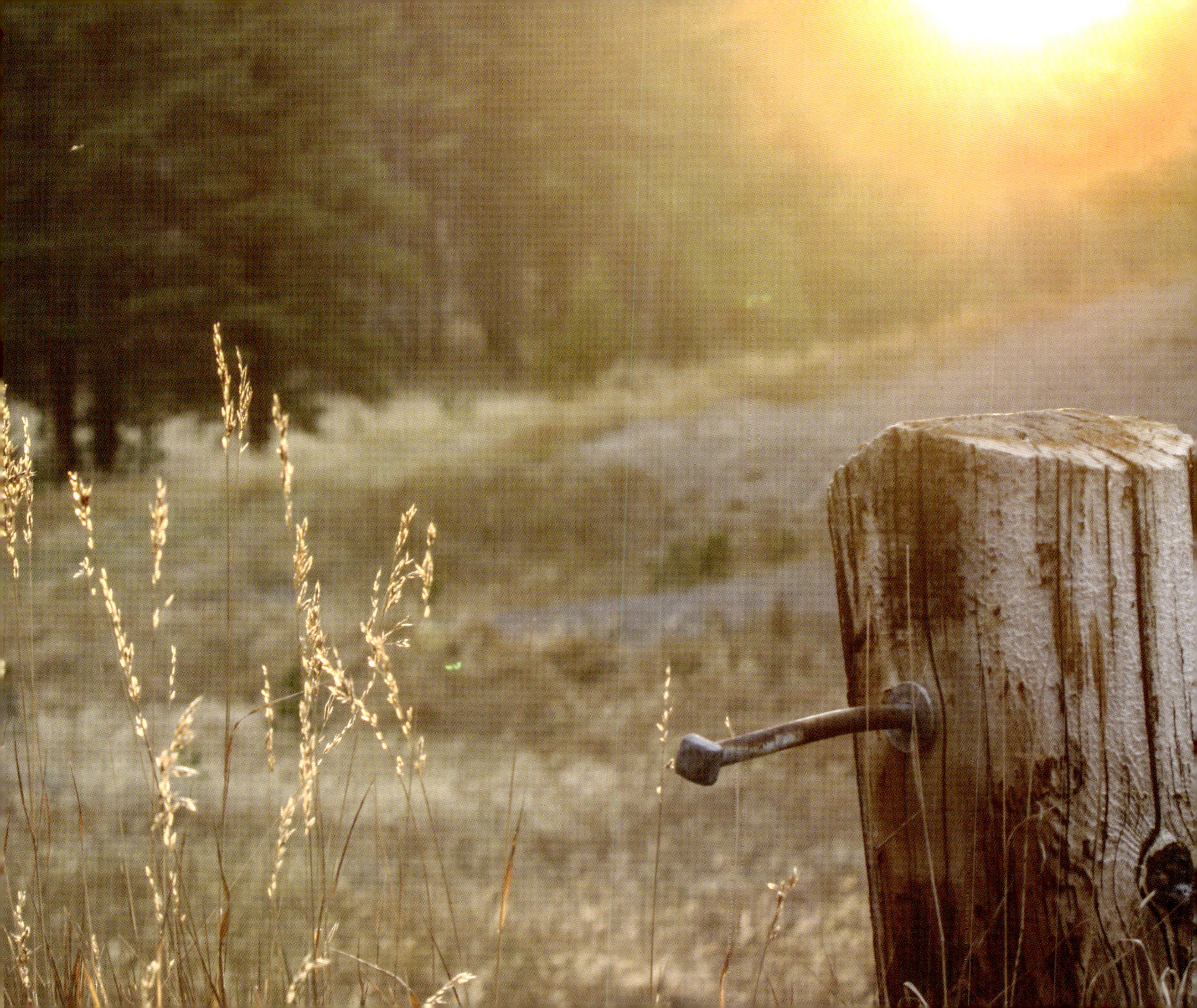

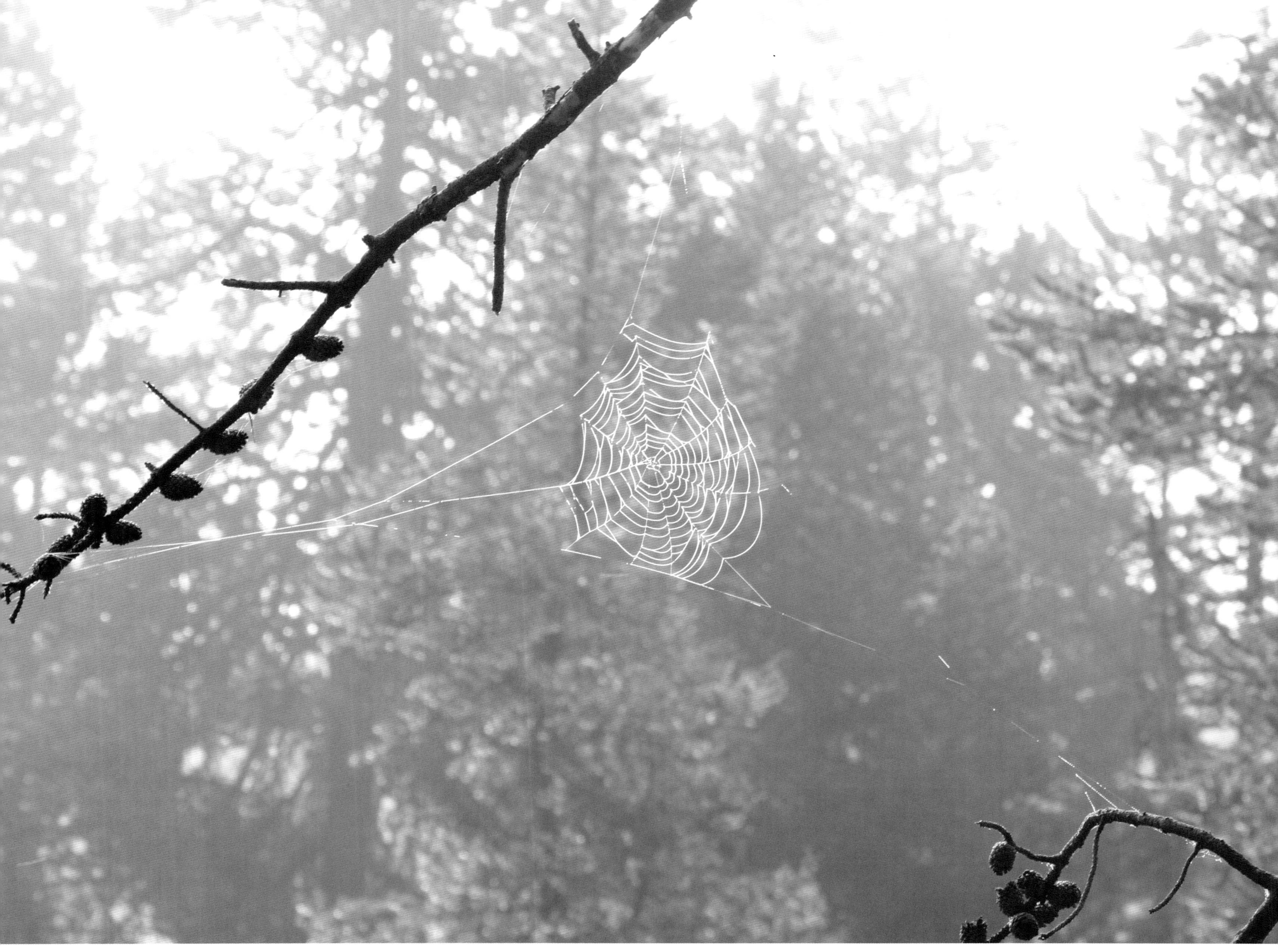

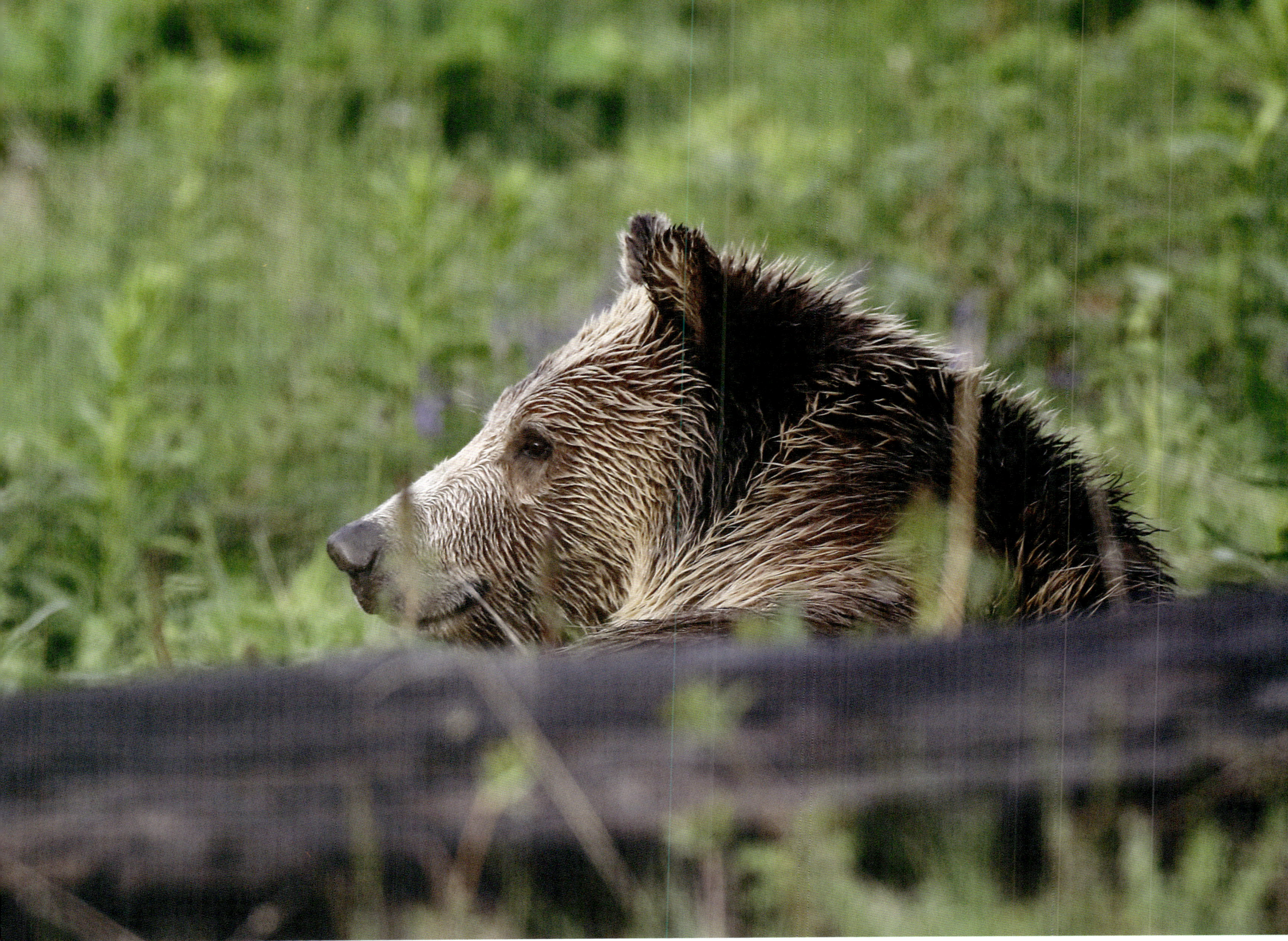

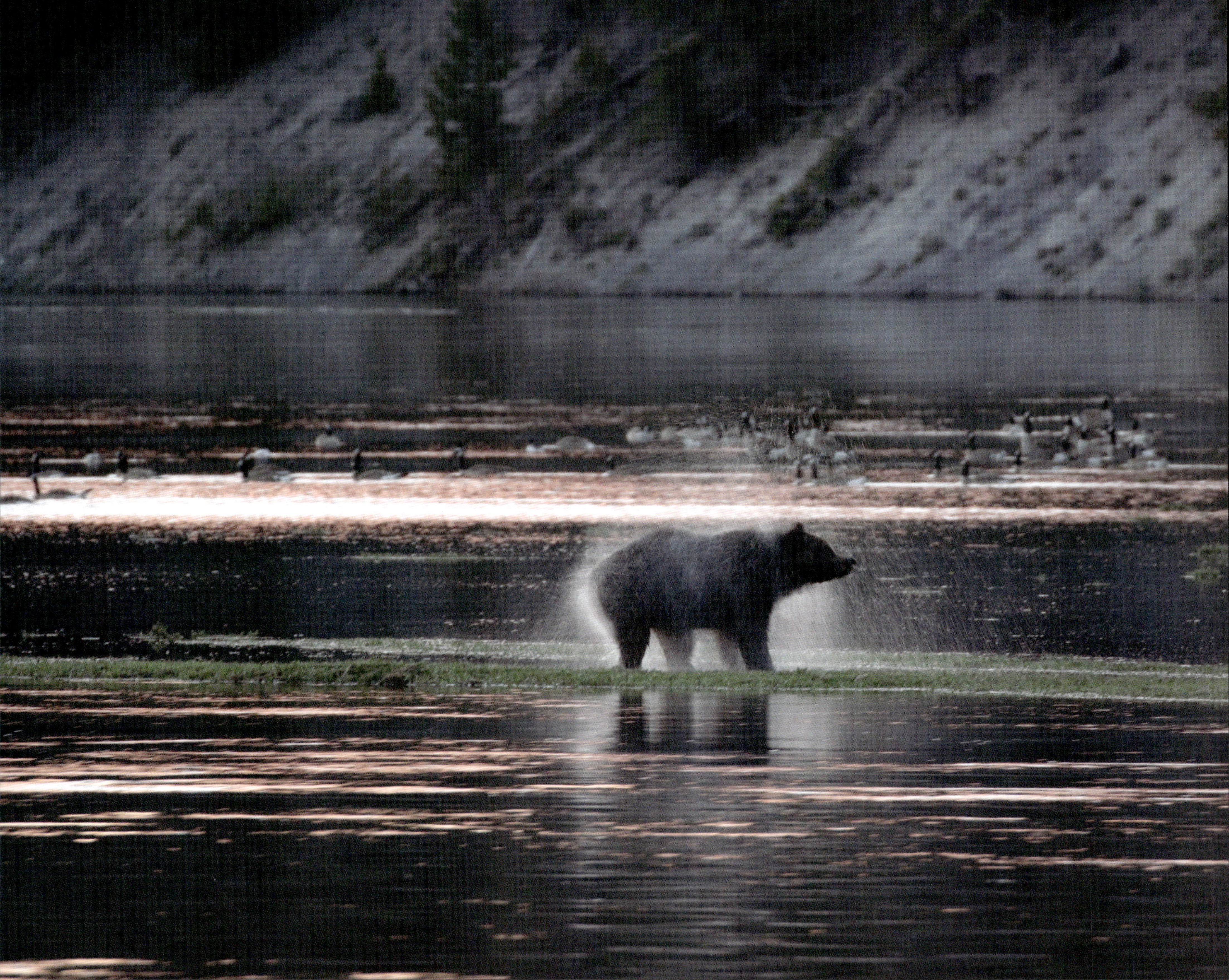

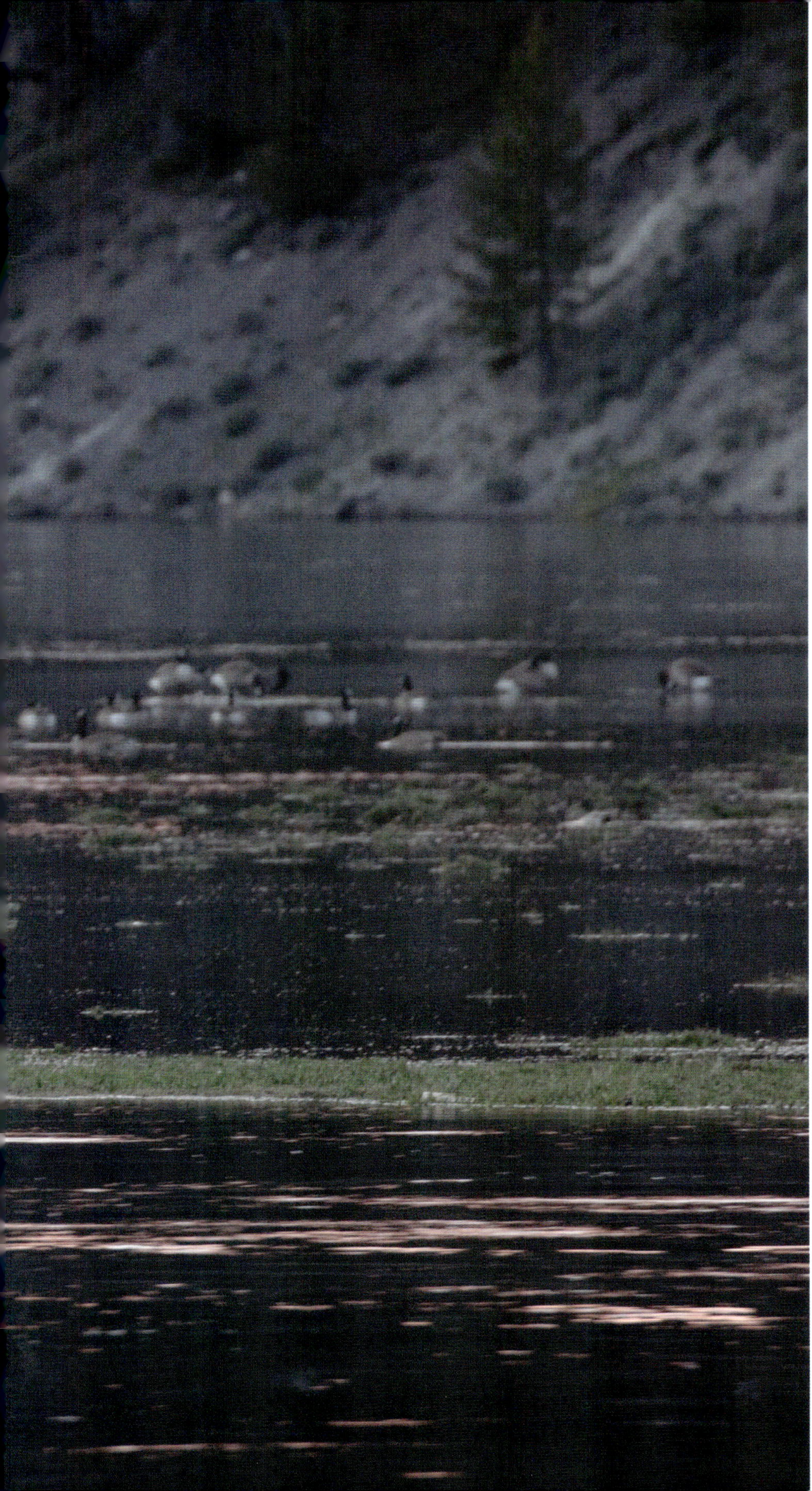

Those who have packed far up into grizzly country know that the presence of even one grizzly on the land elevates the mountains, deepens the canyons, chills the winds, brightens the stars, darkens the forest, and quickens the pulse of all who enter it. They know that when a bear dies, something sacred in every living thing interconnected with that realm . . . also dies.

—John Murray

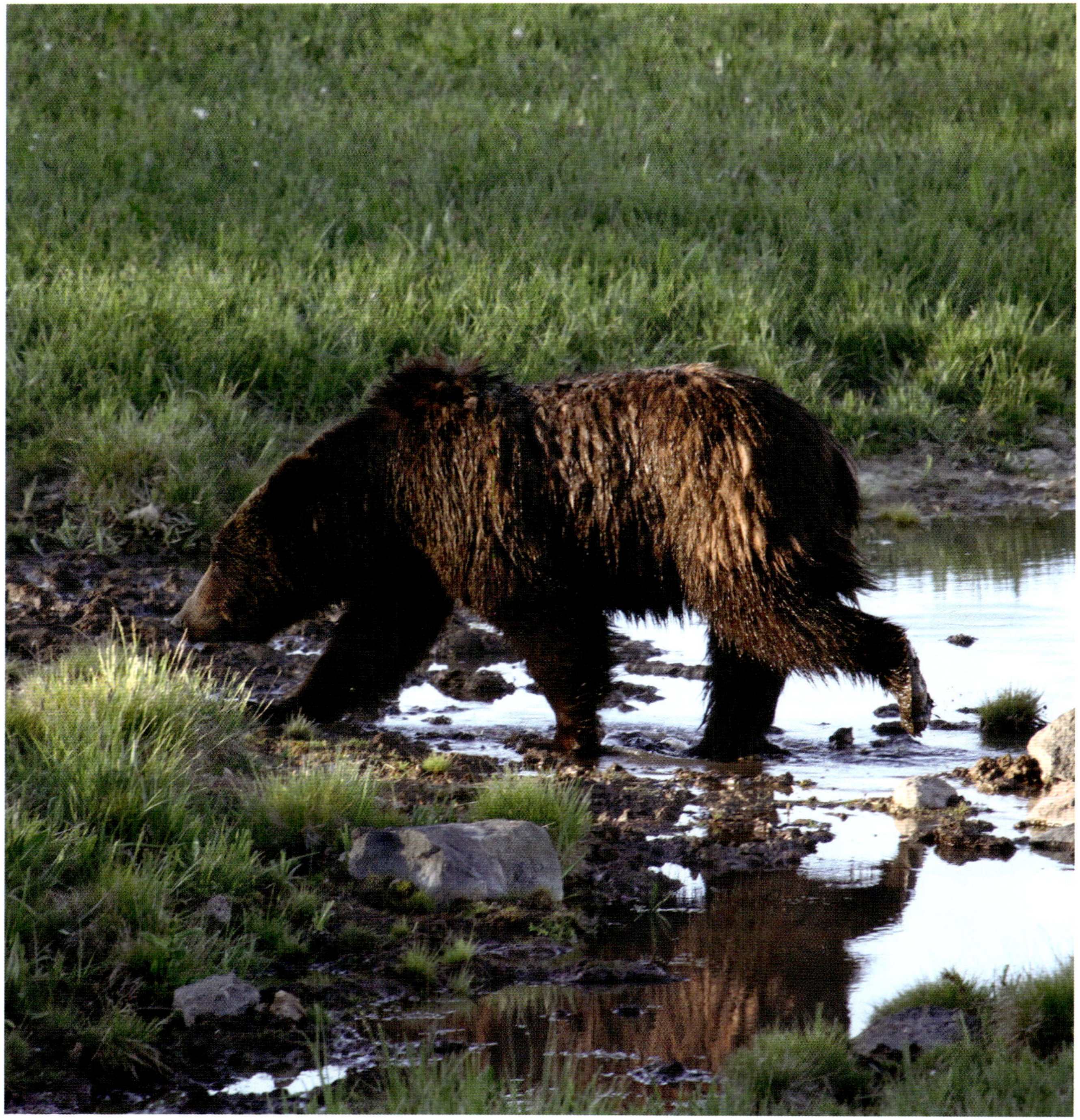

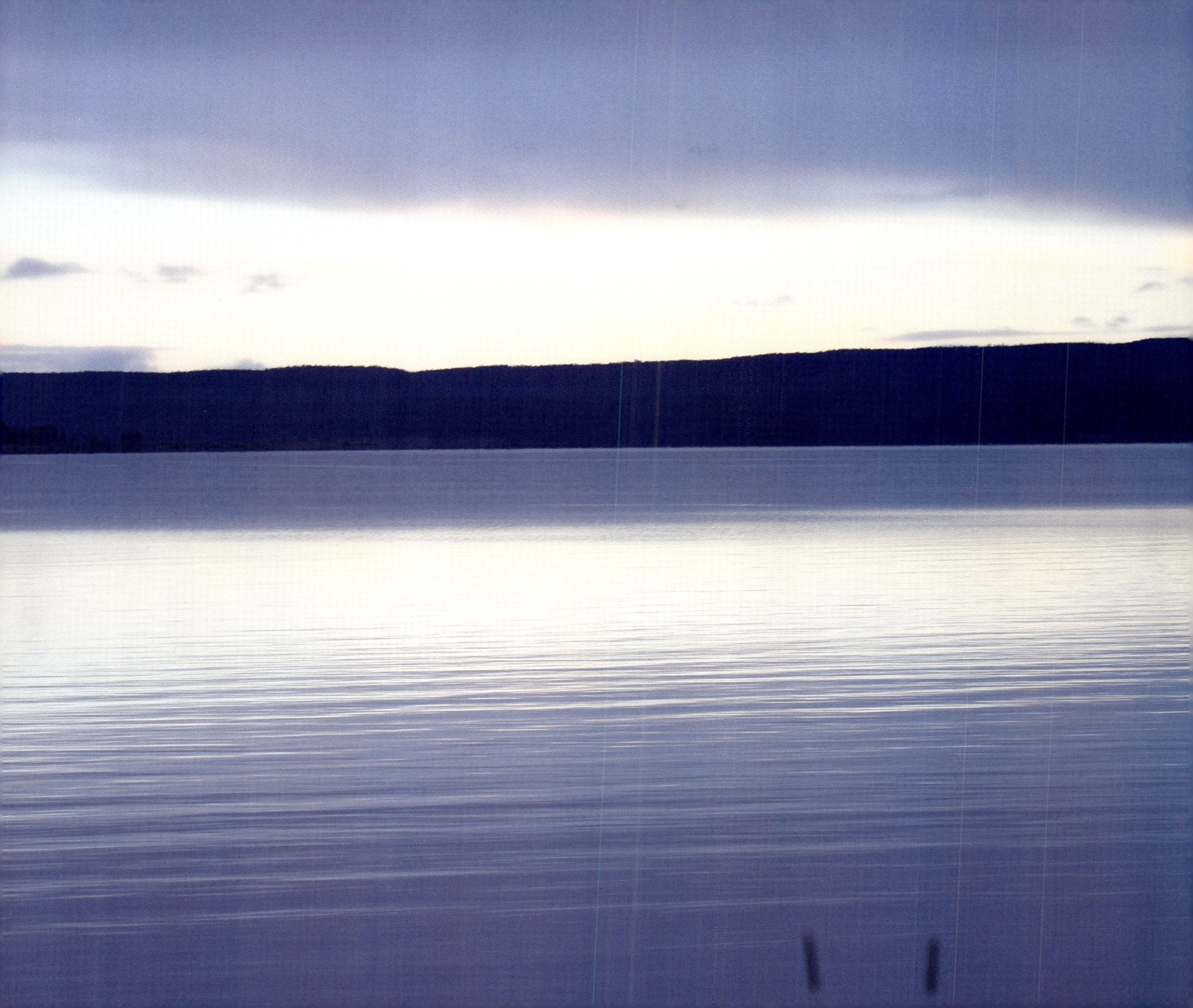

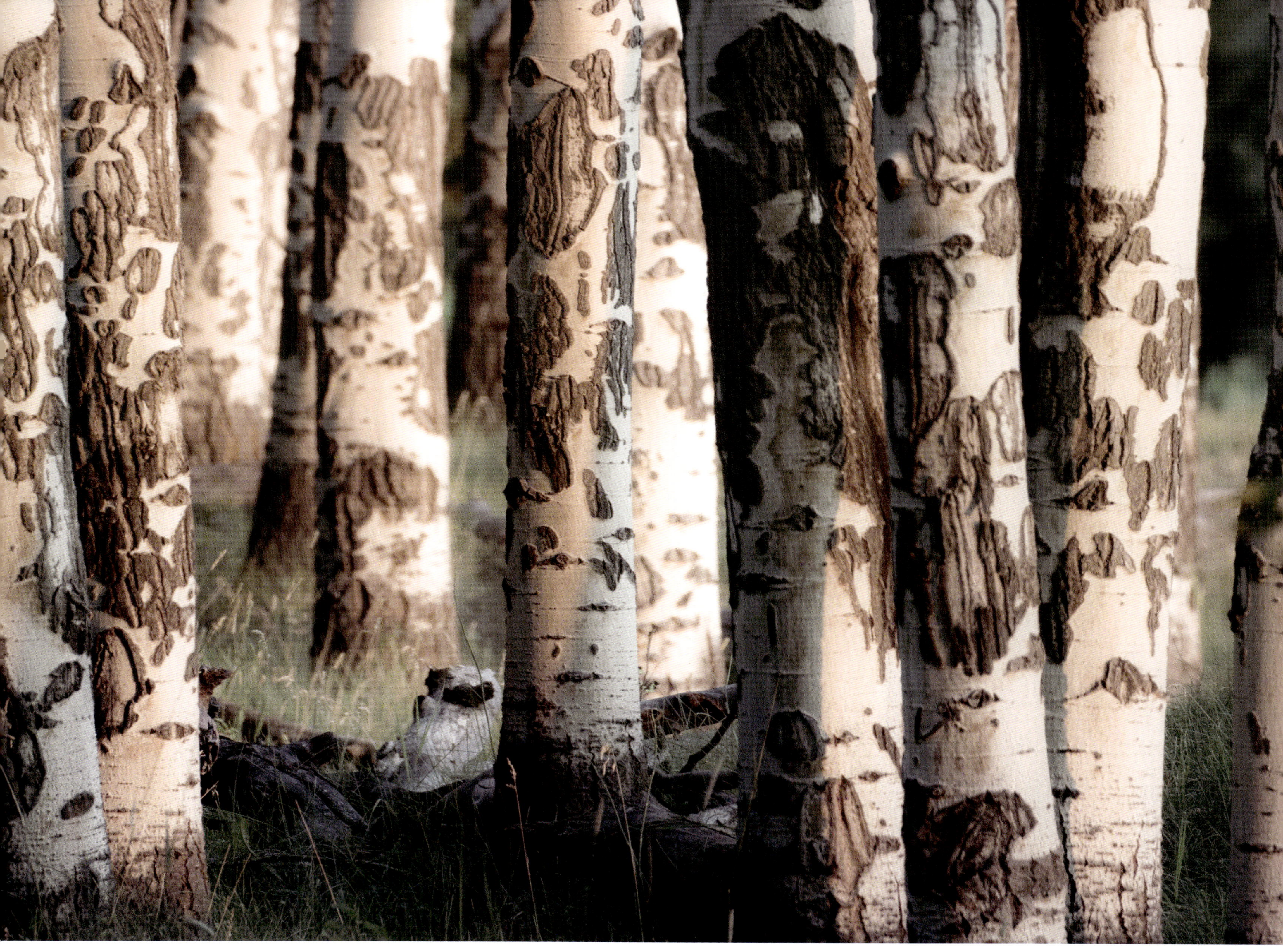

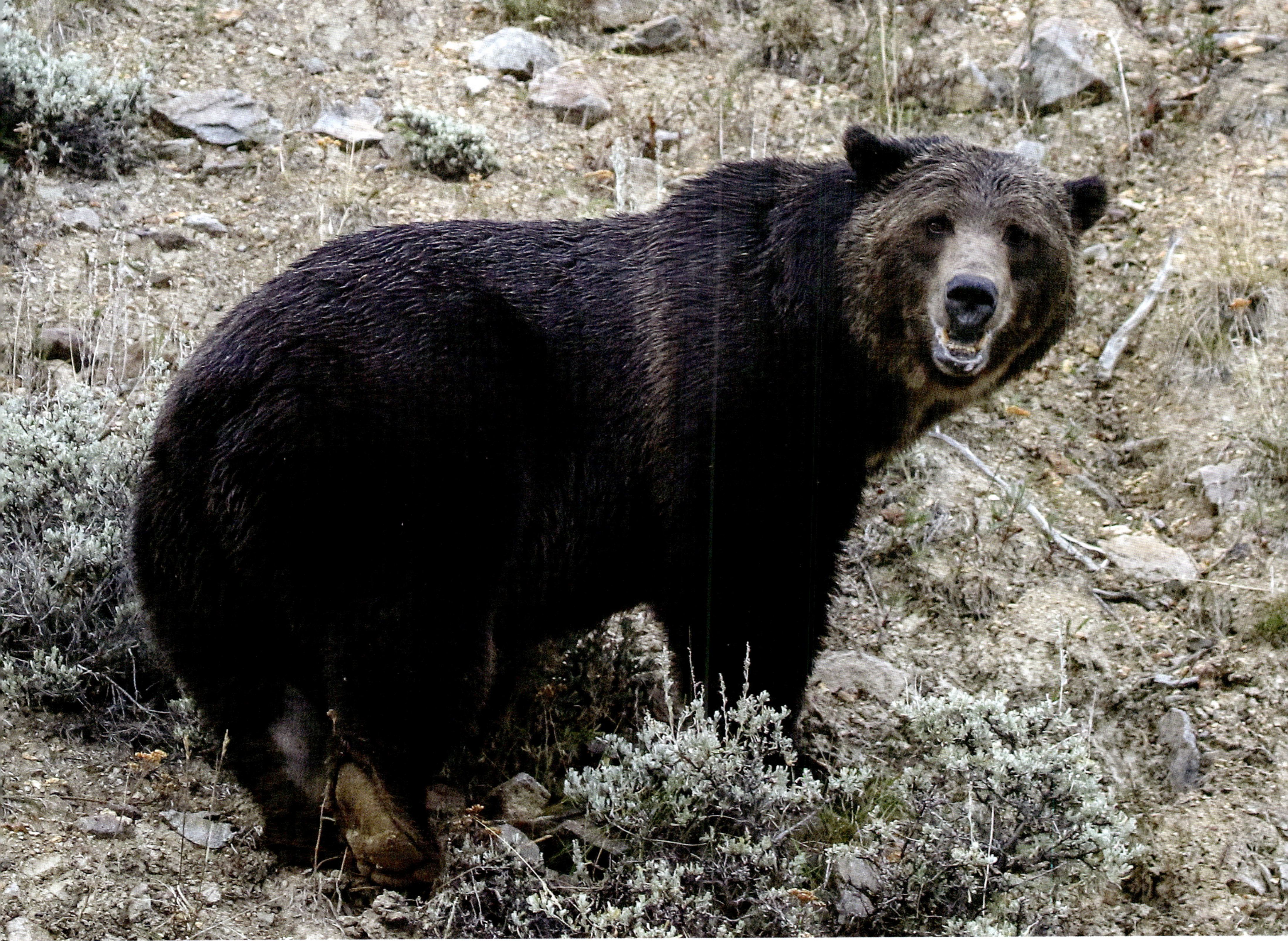

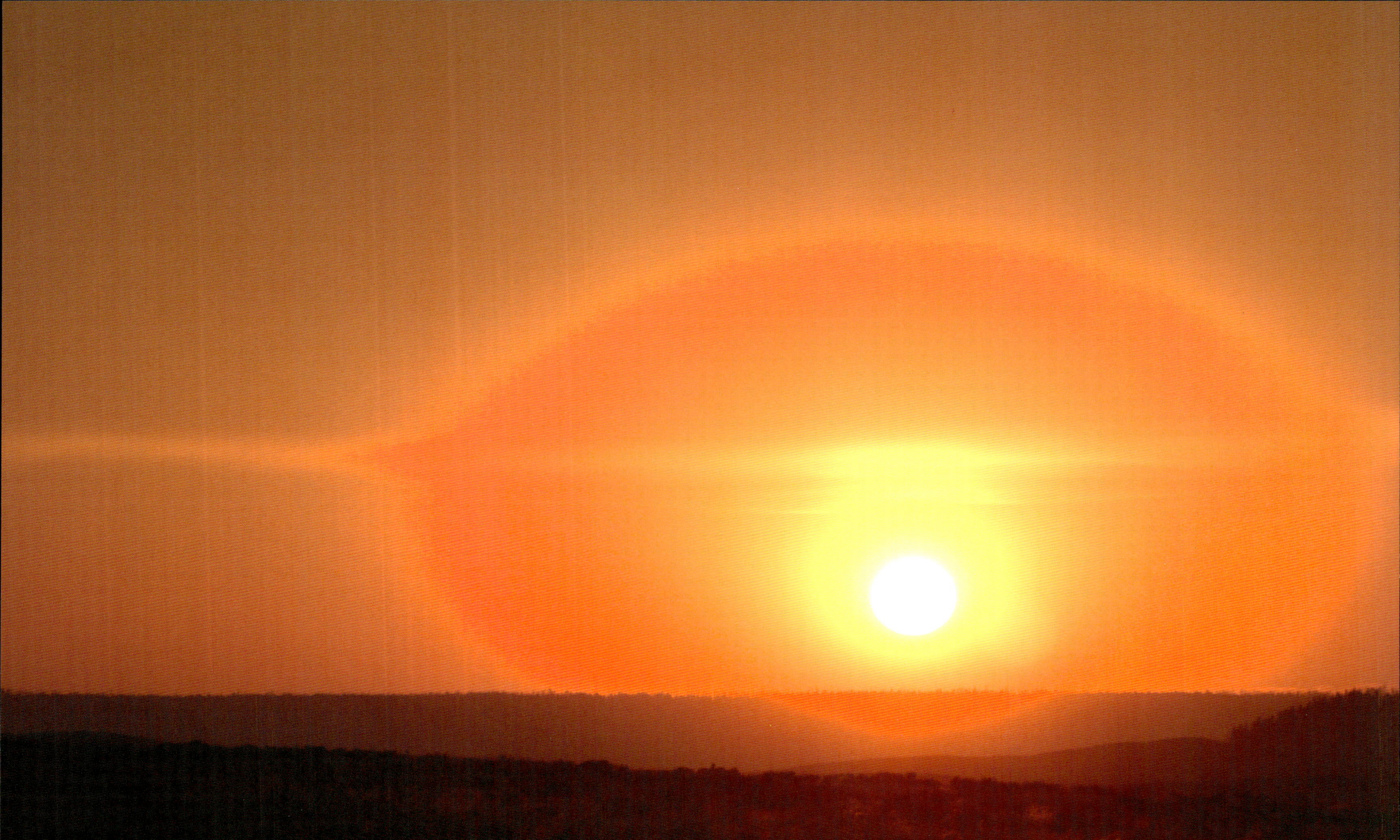

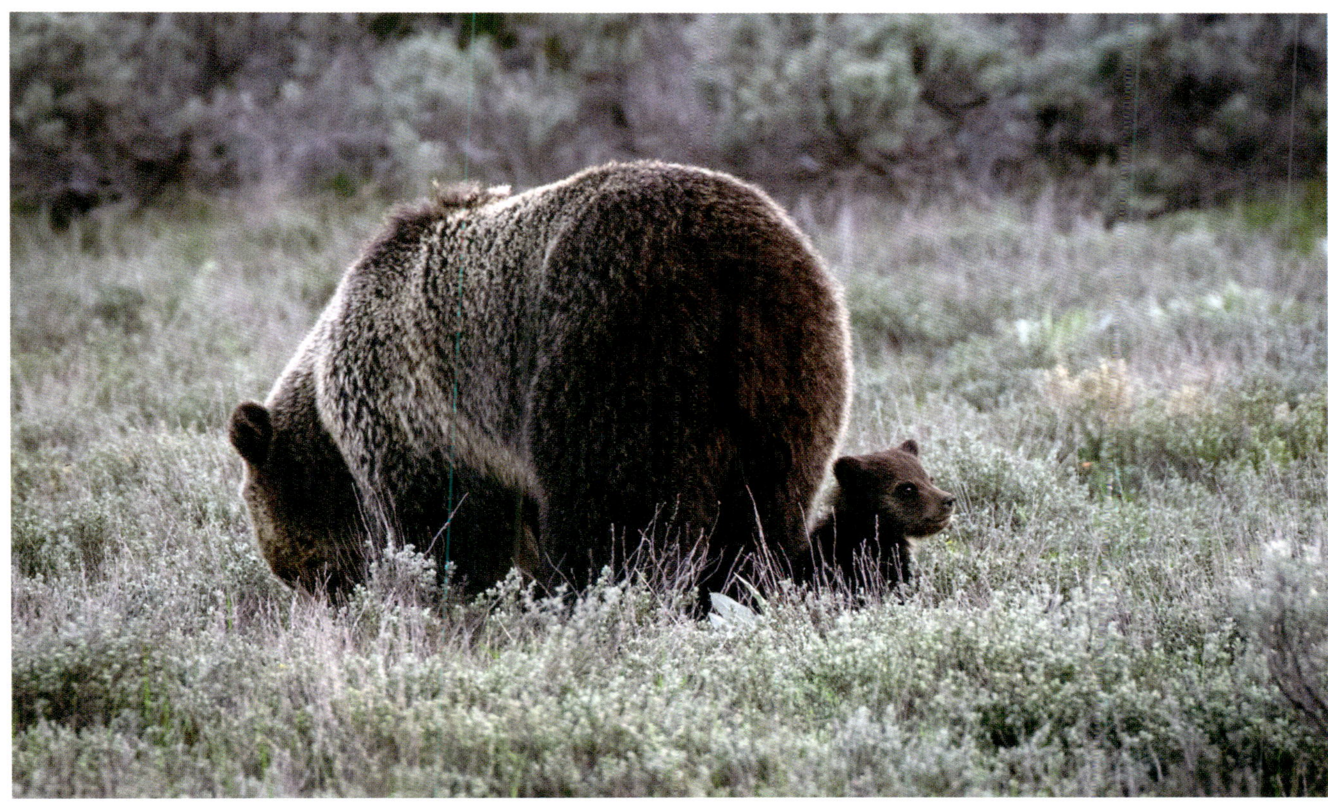

Historically, grizzlies ranged from Alaska to Mexico, with at least 50,000 bears living in the western half of the contiguous United States. With European colonization, the bears were shot, poisoned, and trapped to the brink of extinction.

—Lydia Millet

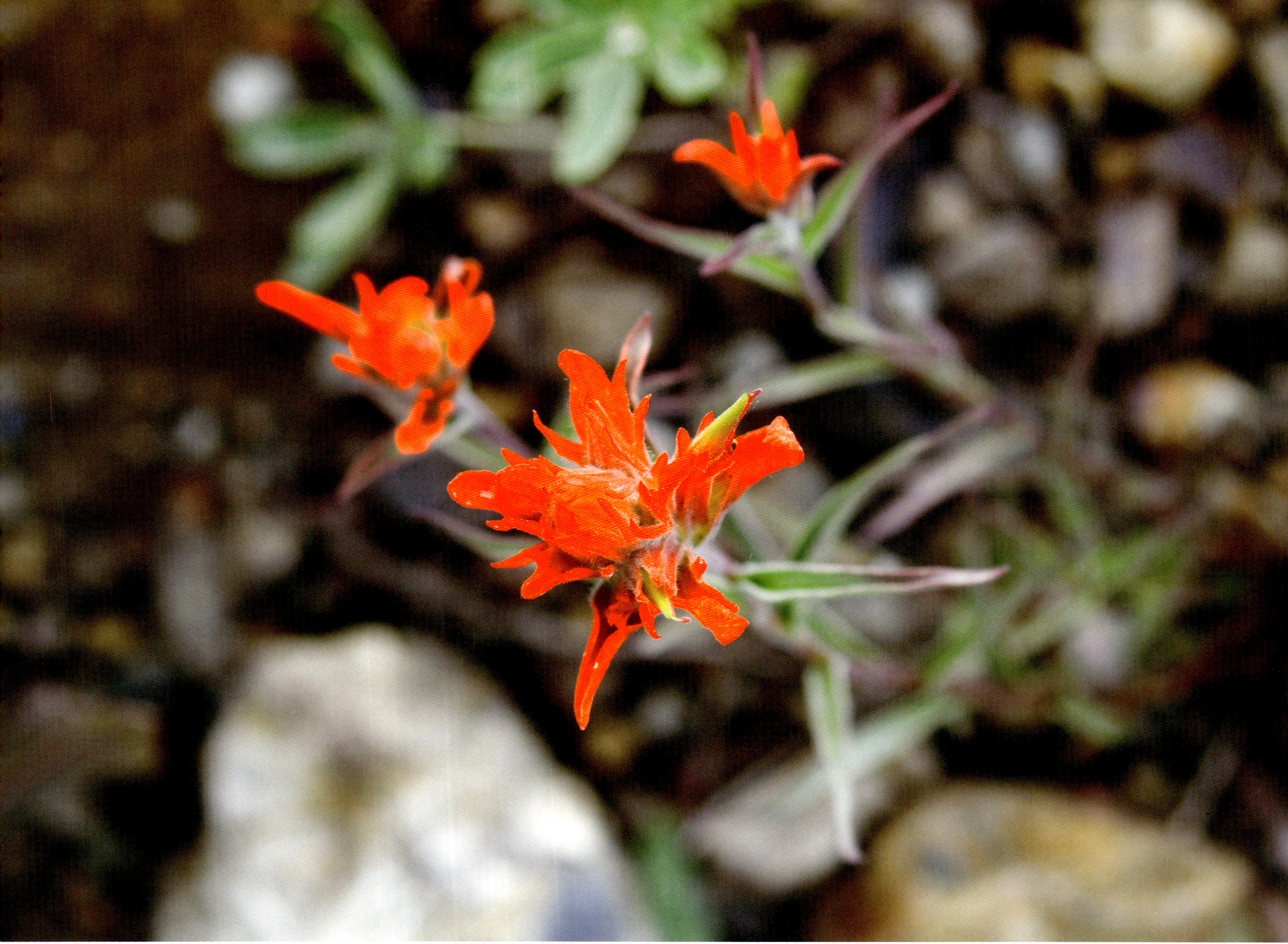

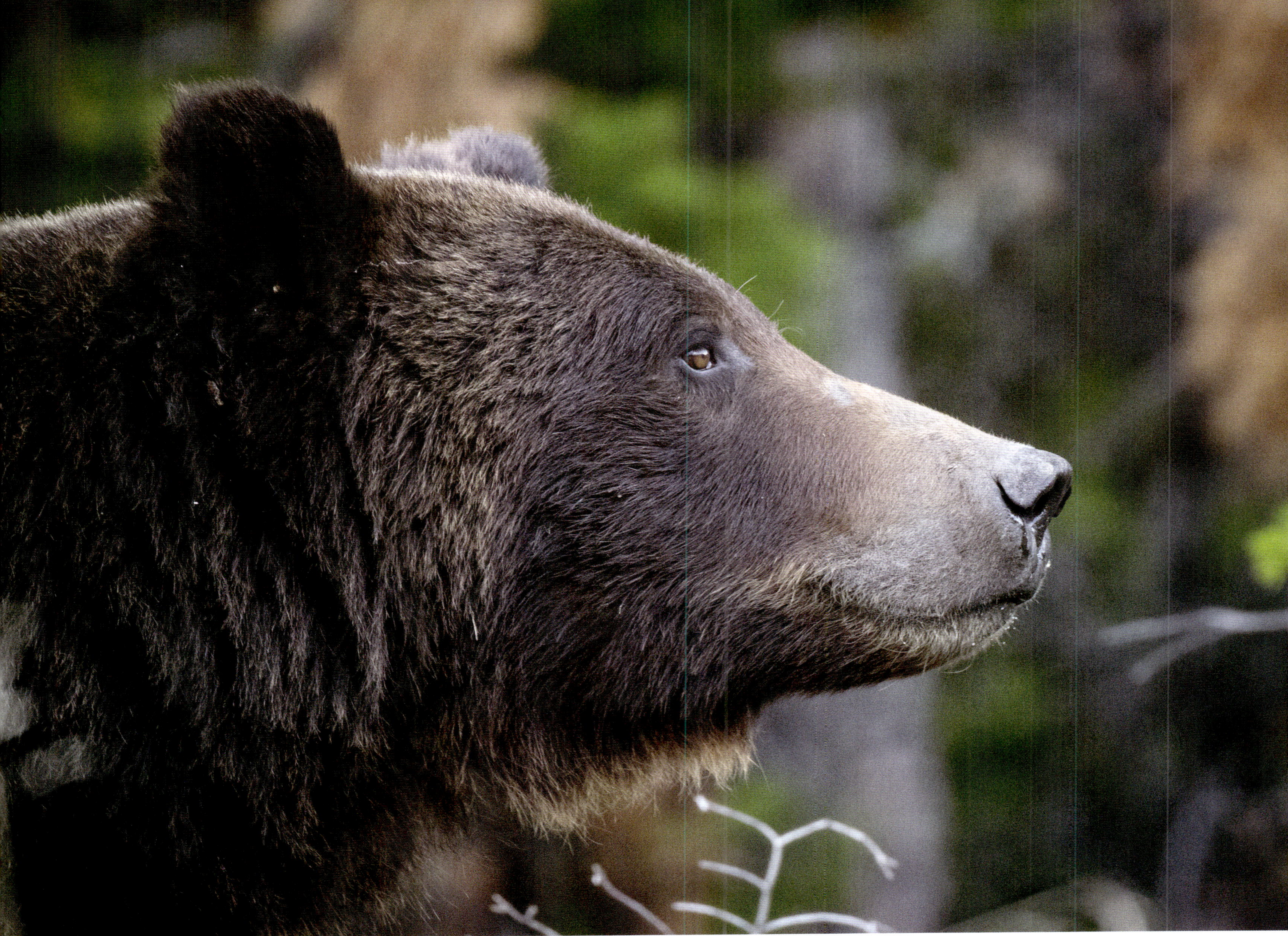

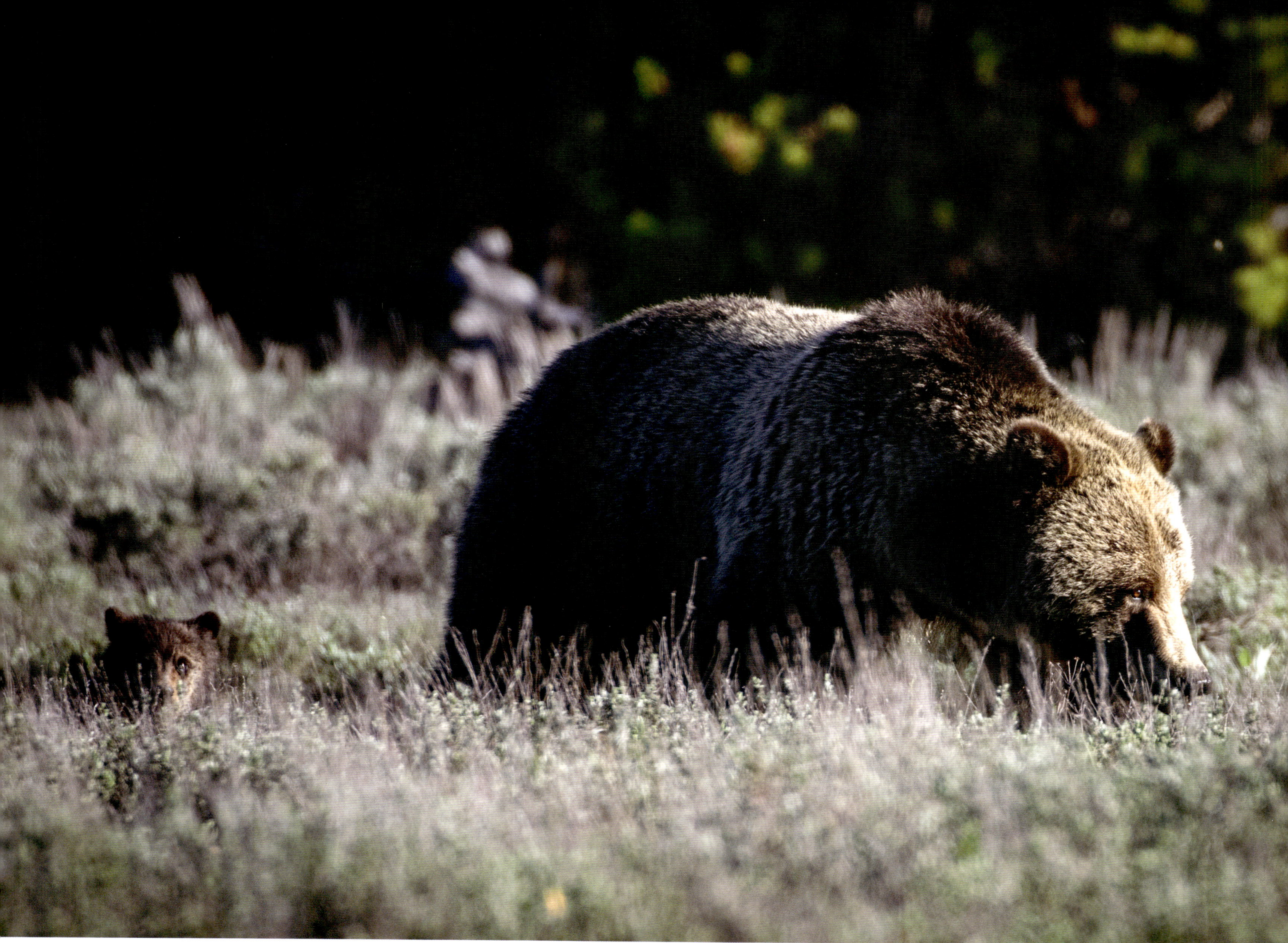

From 1983 to 2001, human causes accounted for one-third of all cub and yearling mortality. Beginning around 2000, the decrease in cub survival was most apparent in areas of high density. So in short, cub survival is likely a density dependent factor.

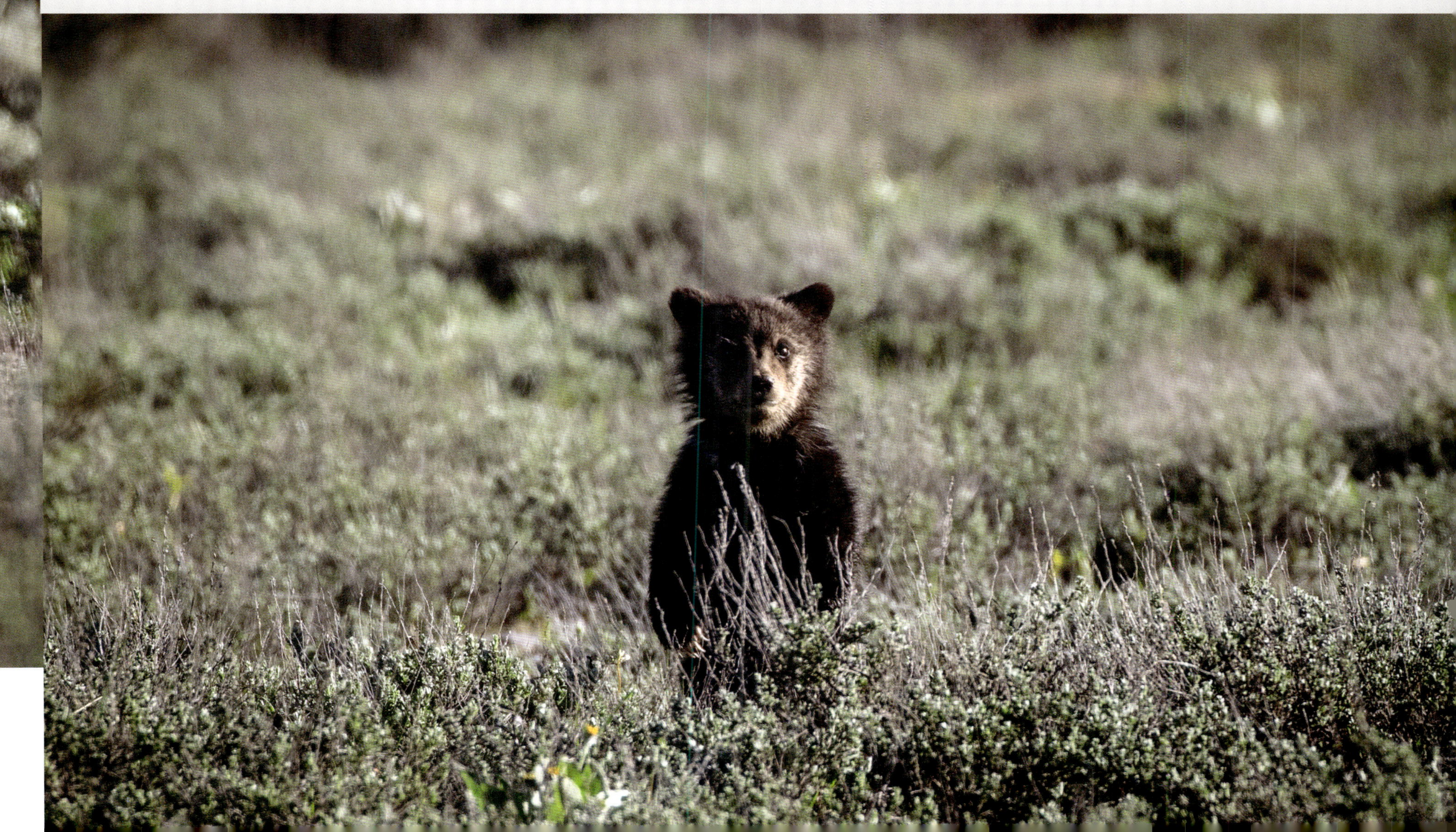

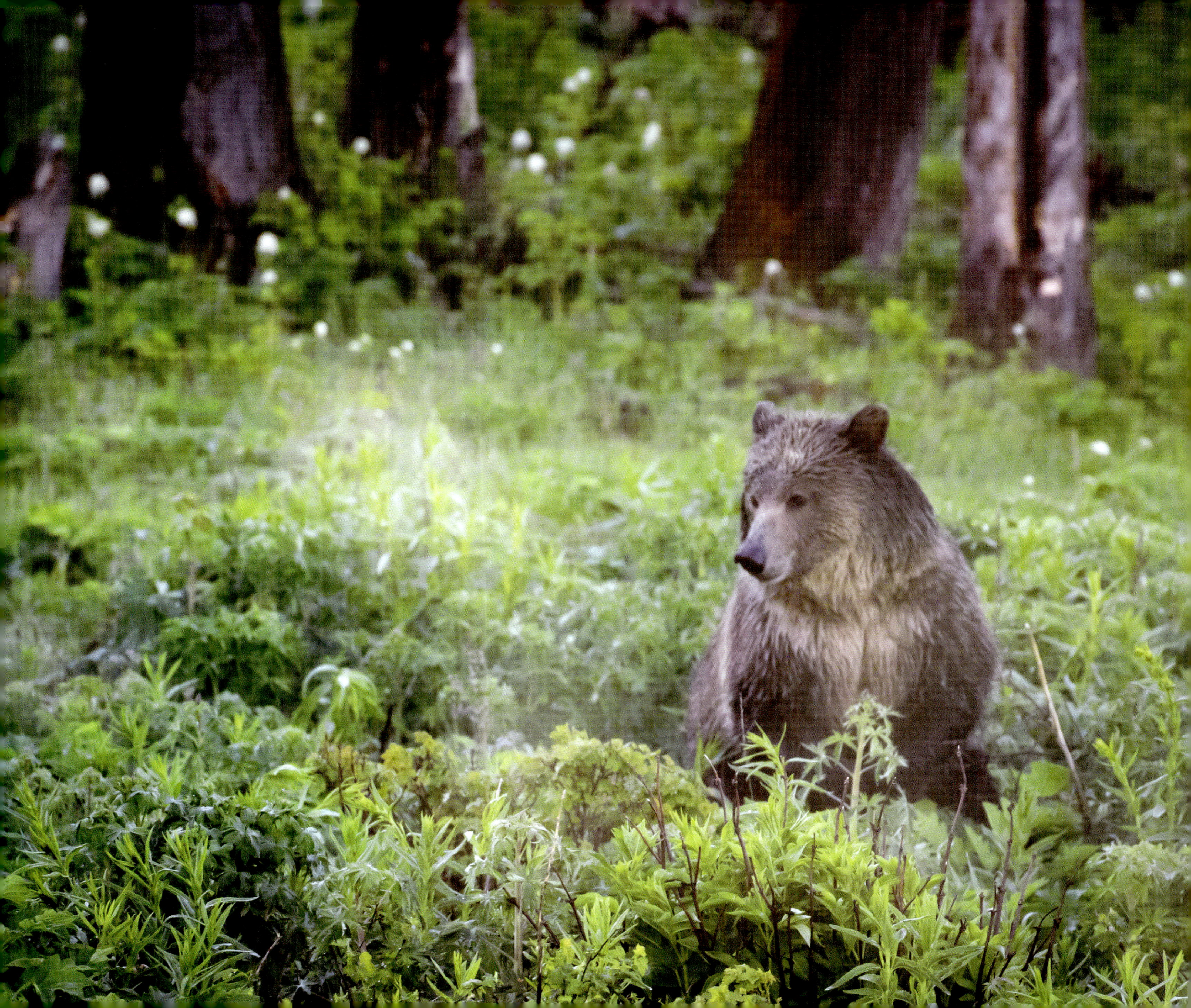

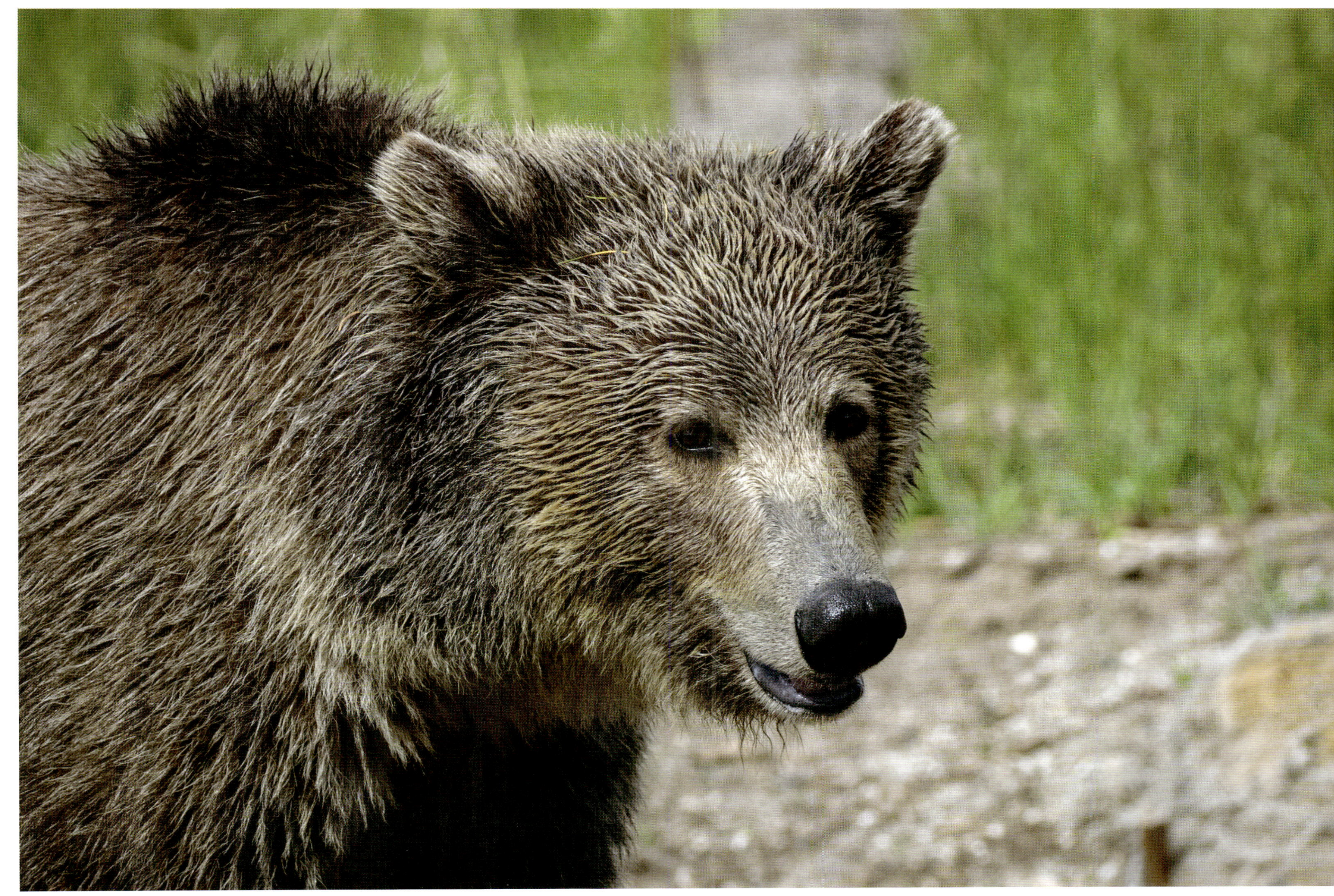

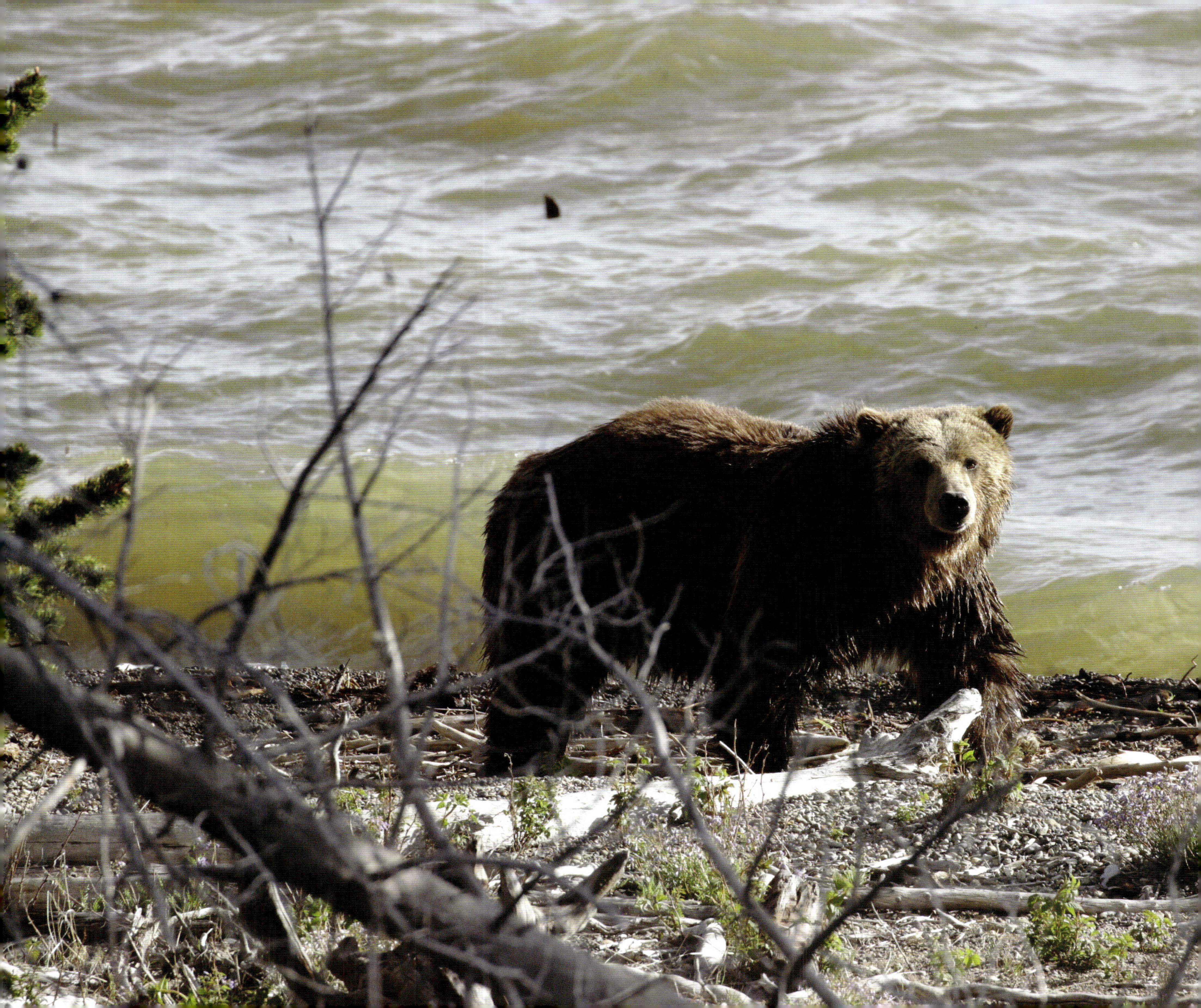

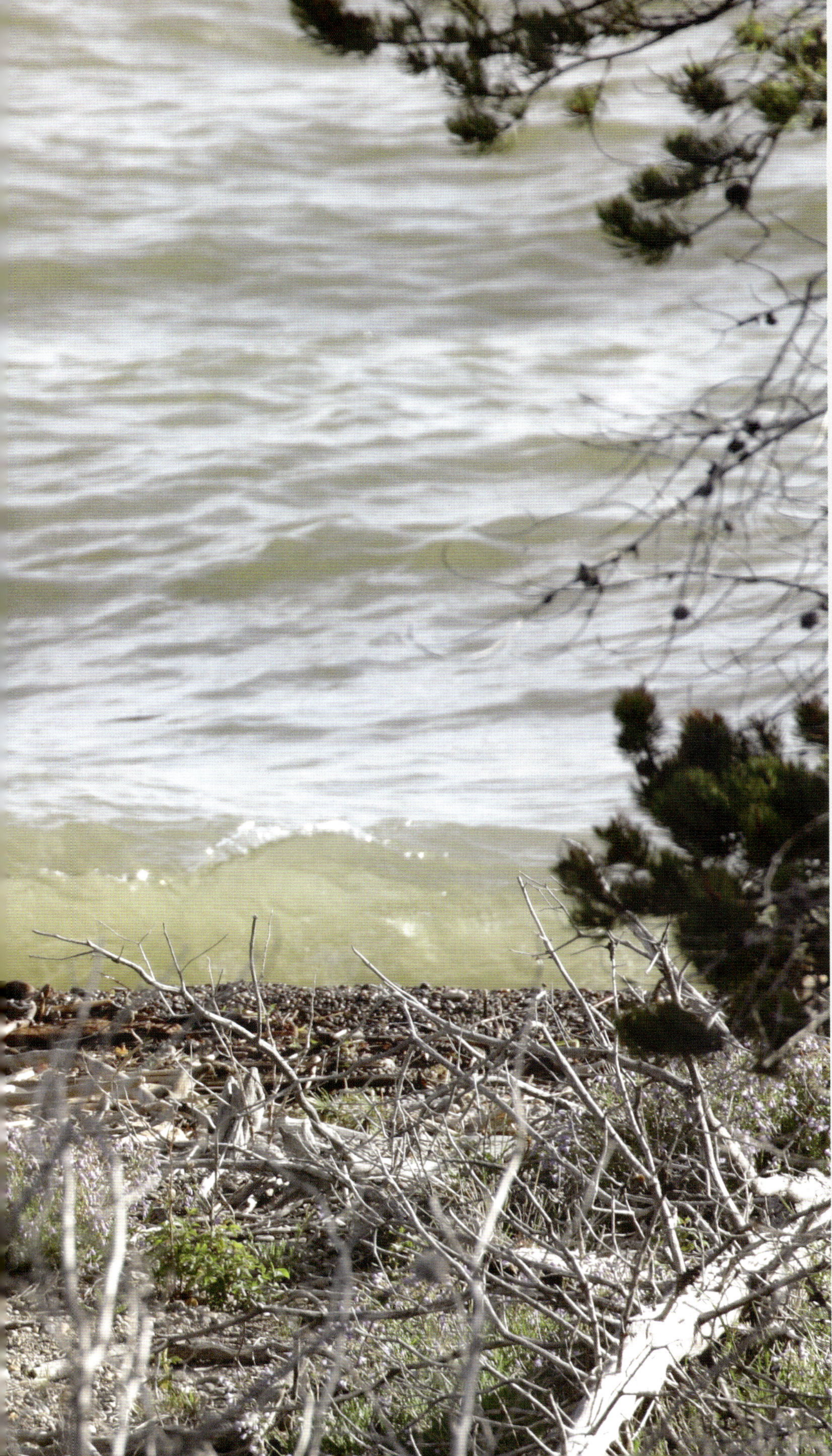

During the 1960s and early 1970s, the dumps were closed in Yellowstone, which resulted in the removal of garbage for bears to feed on as a reliable food source. This in turn resulted in smaller litters and decreased survival of cubs and adult females. Between 1983 and 2001, there was an increase in survival for independent bears two years and older. This is most likely attributed and in response to management efforts, which also lead to an increased population.

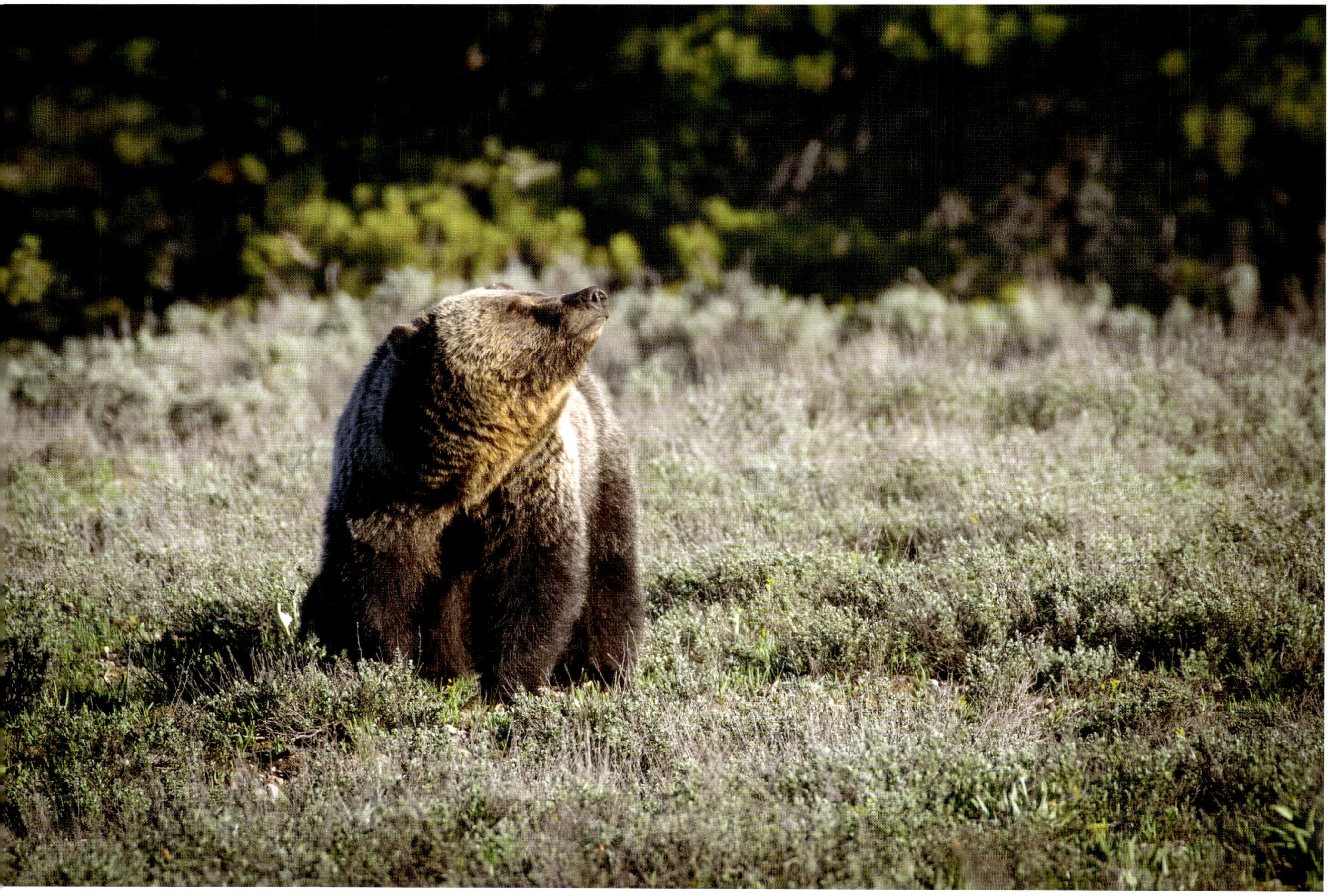

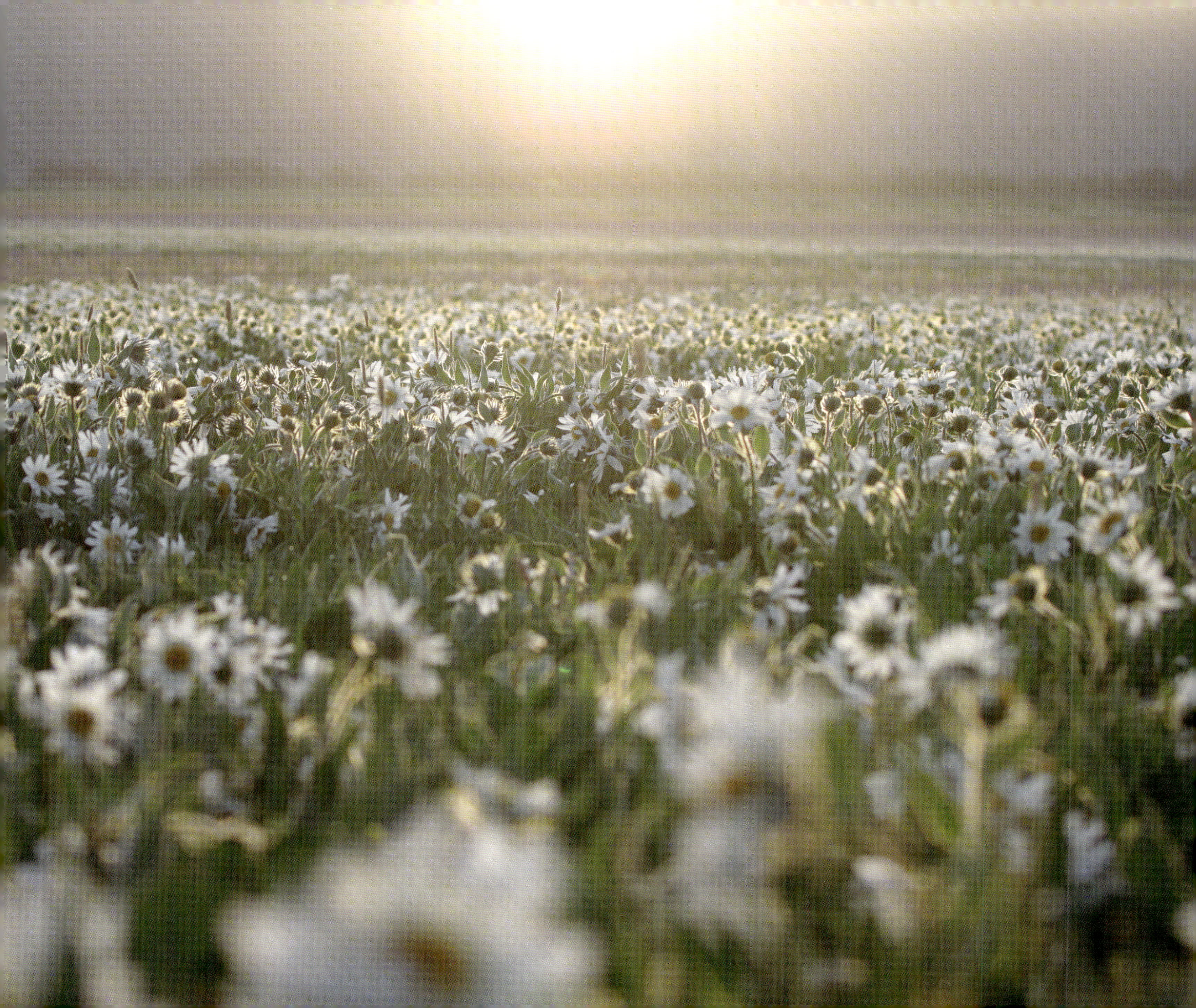

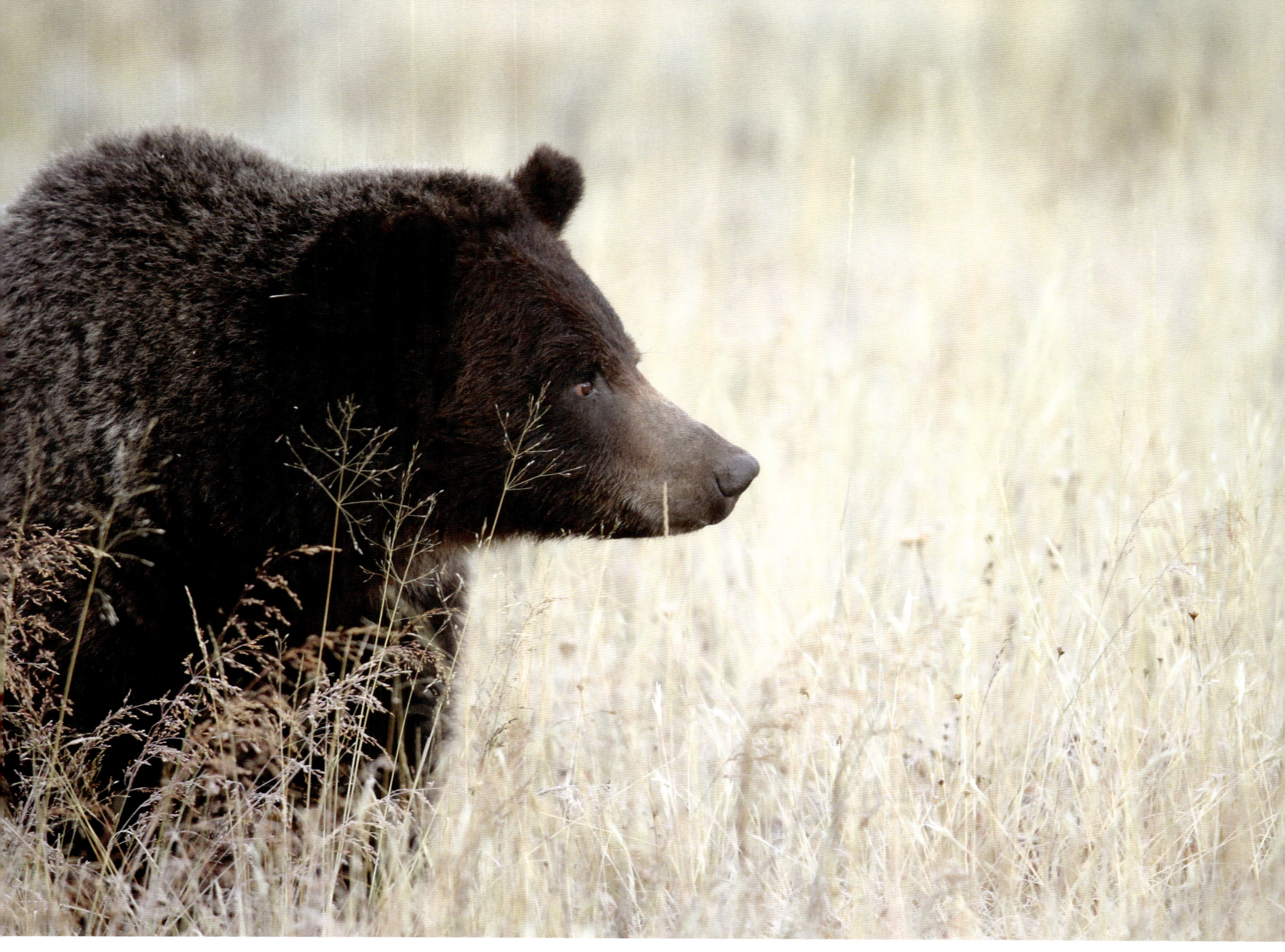

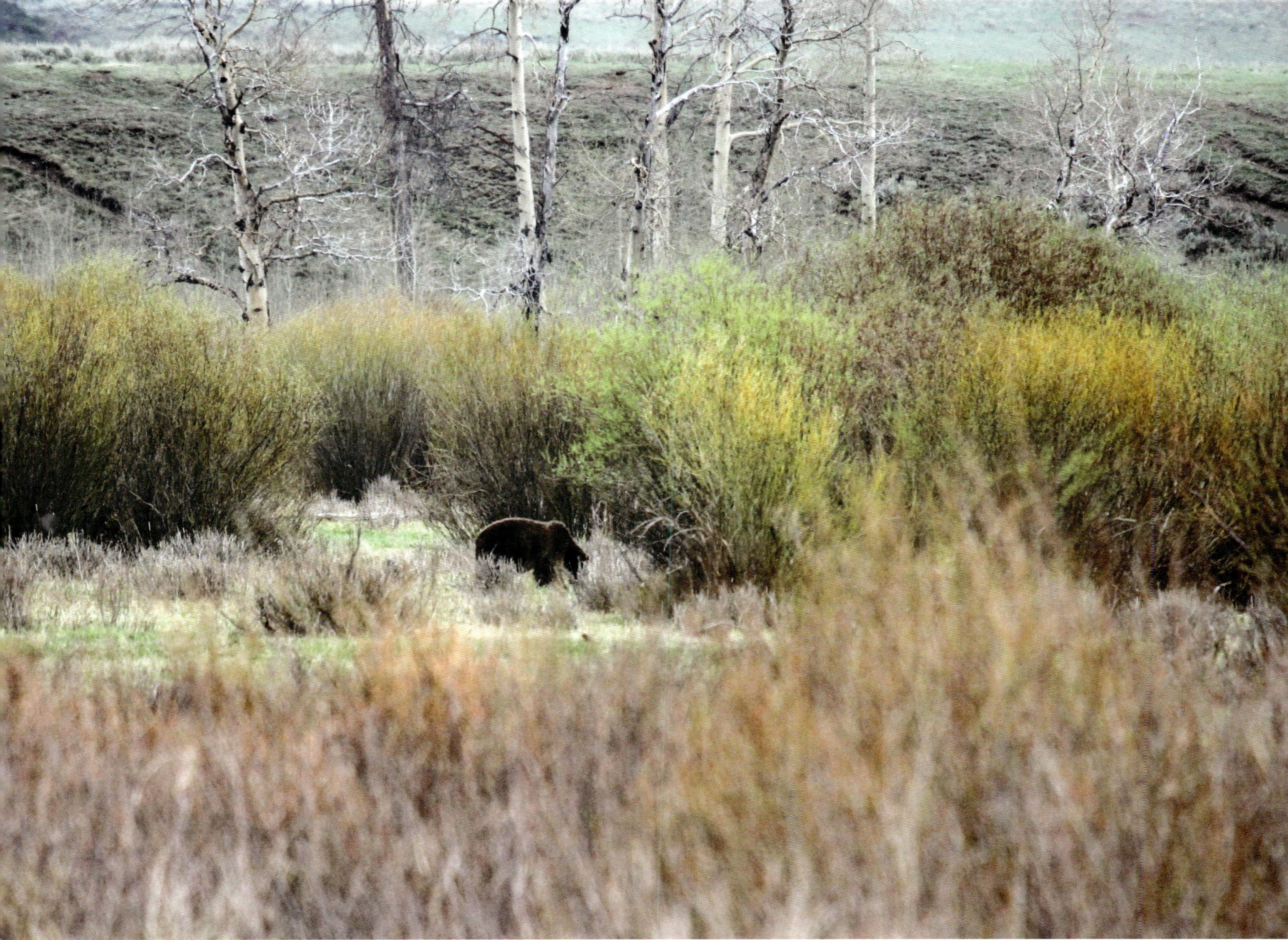

The track of a grizzly is a powerful thing. For not only do they show where a bear has been, but the trajectory to where they're going; a snippet into the lives of a creature where we have only seen the tip of the iceberg. There is much more for us to learn from these wonderful creatures as time progresses; we should never cease to learn and we should never stop exploring.

—Tyler Brasington, Bear Biologist

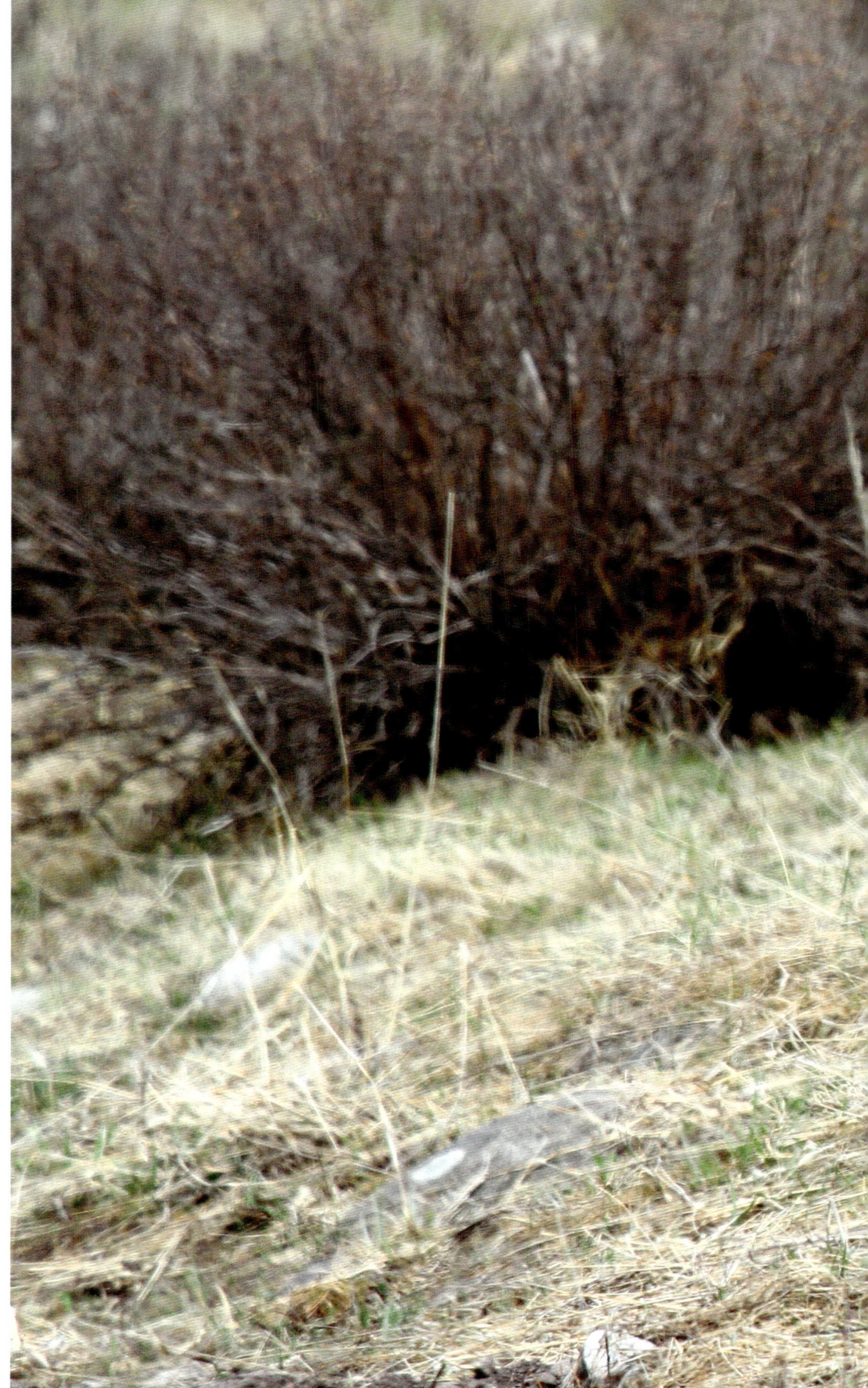

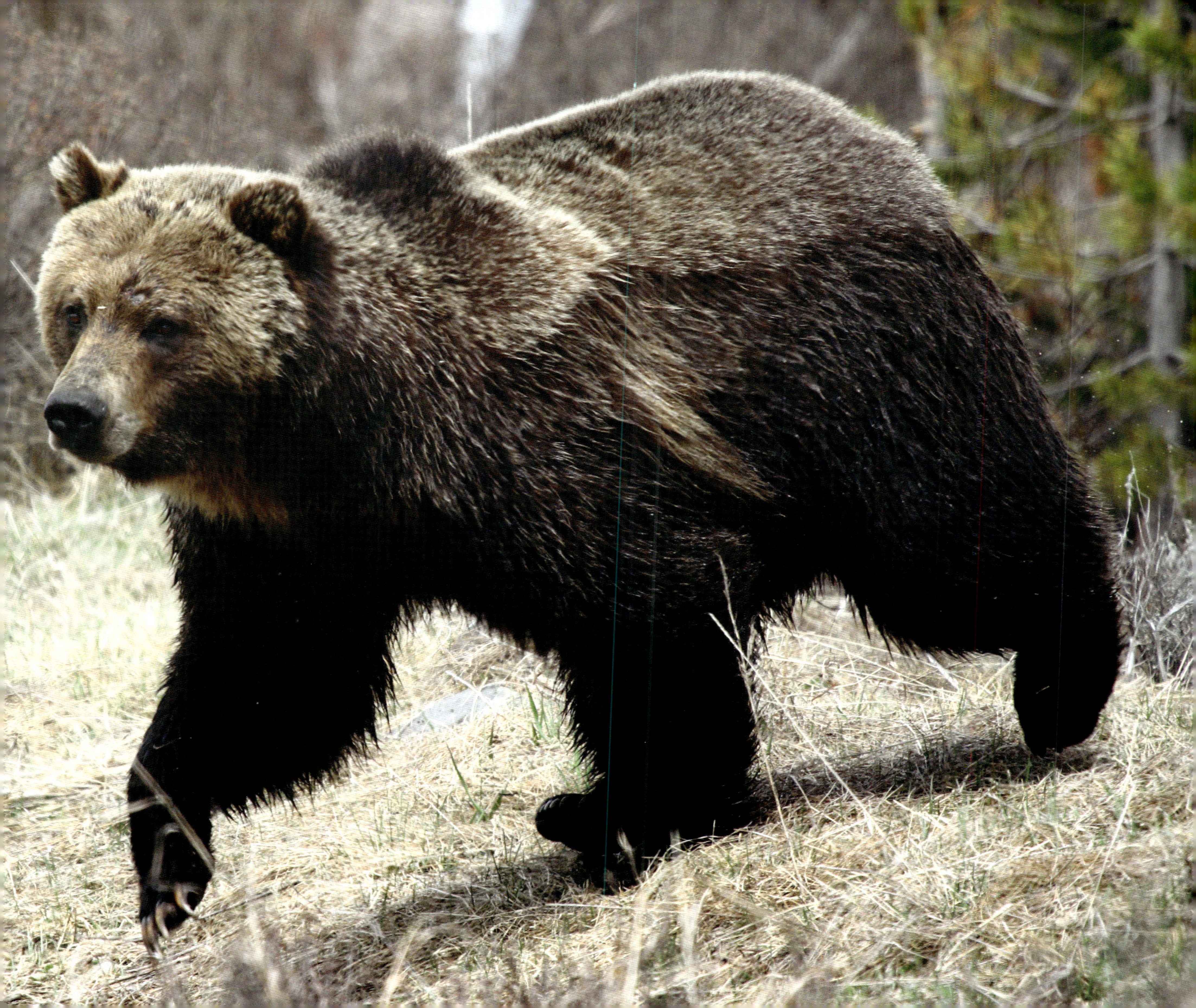

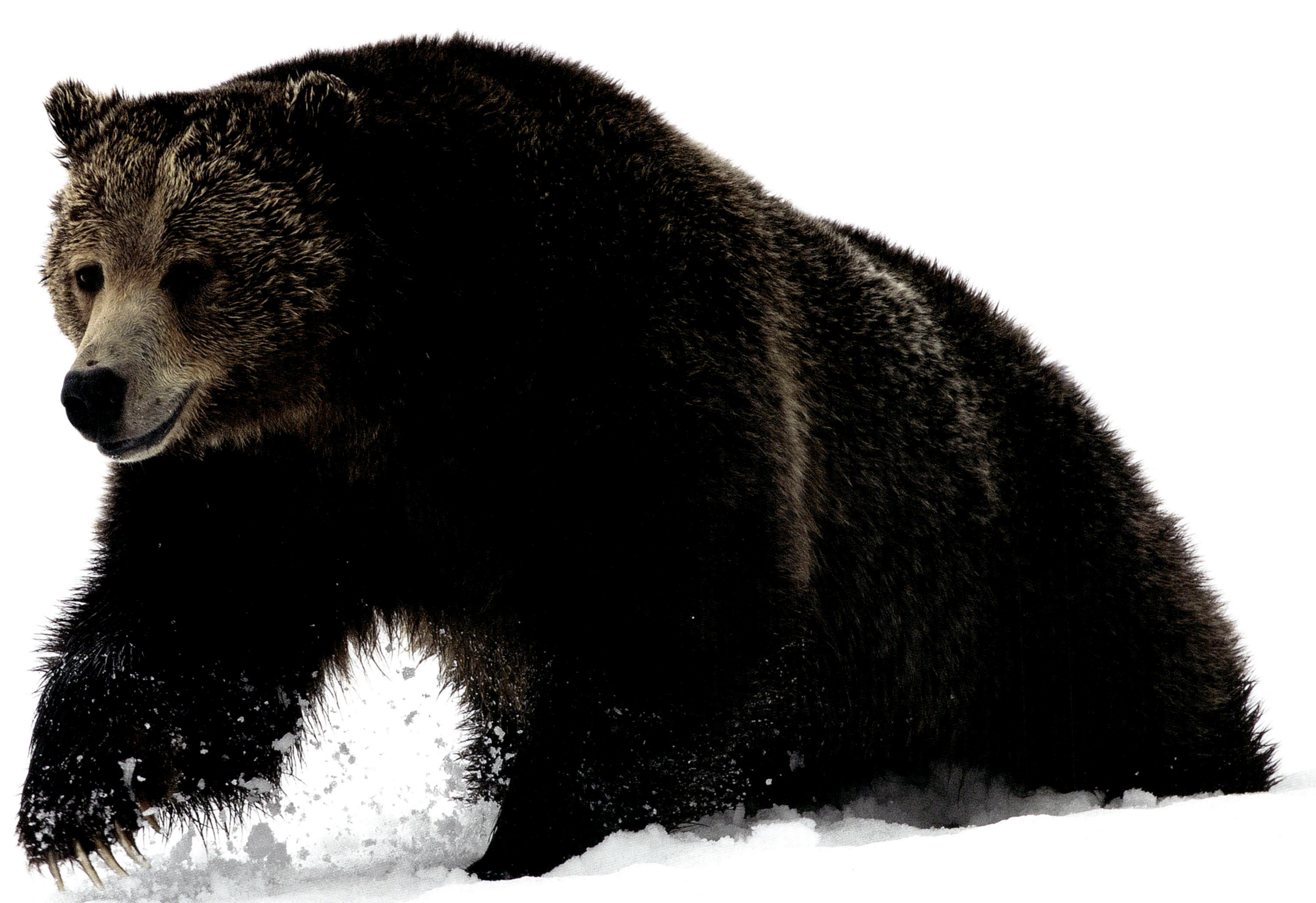

HIBERNATION

GRIZZLY BEARS HEAD FOR HIBERNATION IN THE FALL MONTHS in response to food shortages, cold weather, and snow.

For many years, bears were not considered to be true hibernators. This is because, in comparison to mammals that are considered to be true or deep hibernators, a bear's body temperature remains within twelve degrees of their normal body temperature. Animals considered to be true hibernators have to warm up before they can actually move. For bears, the warmer body temperature allows them to react to danger more quickly.

Many scientists now consider bears to be super hibernators. While in hibernation, their respiration decreases to one breath every forty-five seconds and their heart rate drops to between eight and nineteen beats per minute. Their thick pelts keep them warm, and they usually do not need to move around very much. Sometimes a bear will leave their den during the winter, but they generally do not eat, drink, defecate, or urinate during hibernation.

Denning for bears in the Greater Yellowstone Ecosystem generally lasts around five months. Occasionally a bear will use the same den from the last year, but most of the time they will have to dig a new one because spring run-off causes the old ones to collapse. Digging a den usually takes from three to seven days, and a den consists of an entrance, a short tunnel, and a sleeping chamber. The entrance and the tunnel are quite small, just big enough for a bear to squeeze through. This is to minimize heat loss.

The floor of the den is covered with bedding material such as evergreen boughs or dead leaves and plant parts, depending on what is available in the immediate area. The bedding material has many air pockets which

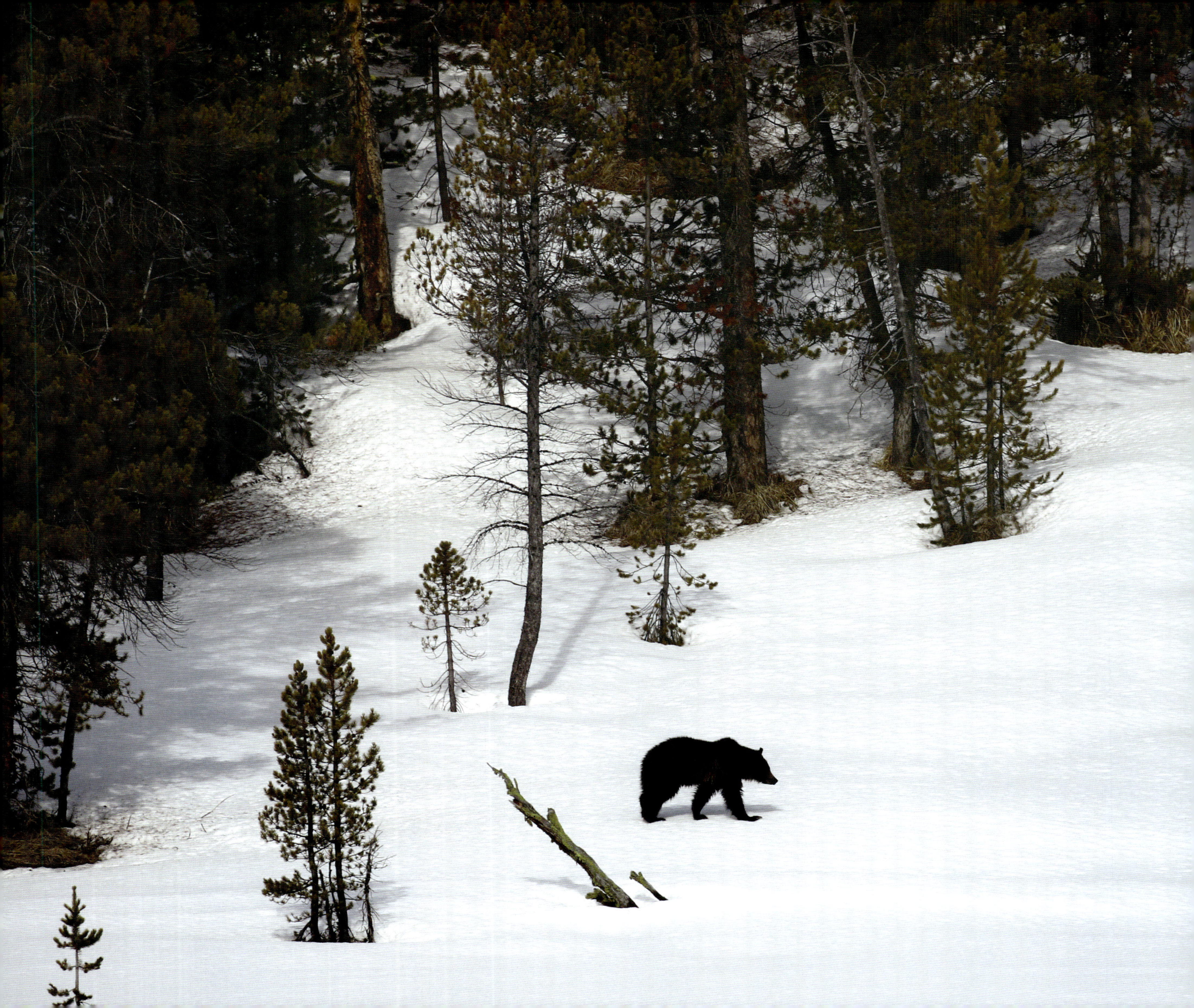

trap heat around the bear to help it to keep warm. Pregnant females will usually den first, followed by females with cubs, subadults, and then adult males.

Even though bears can mate from May to July, implantation usually does not take place until November to December. Cubs are born in the den in late January or early February. At this time, they are about eight inches long and are blind and have no hair. These cubs do not hibernate—they sleep next to mom and nurse and grow. By ten weeks of age, they range from ten to twenty pounds.

In spring, the grizzlies emerge from their dens when the weather warms up and food is available. In the Greater Yellowstone Ecosystem, this can start in February, and by the end of May, most bears have left their den. Big males usually leave first, followed by solitary females, then females with yearlings or two-year-olds, and then the mama bears with COY (cubs-of-the-year) are last to leave the den. Females with COY generally stay in the den area for a few weeks, while all others usually leave within one week.

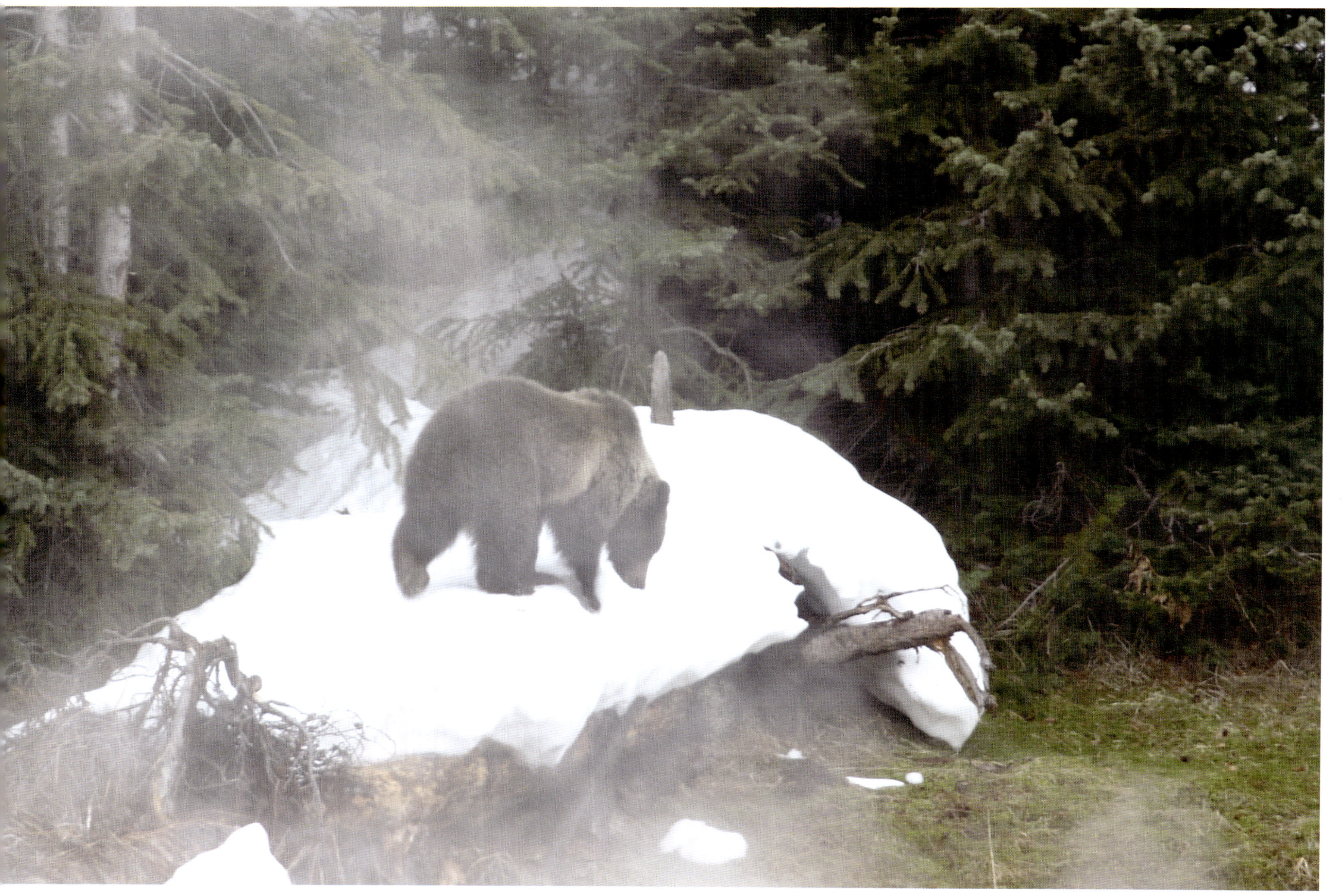

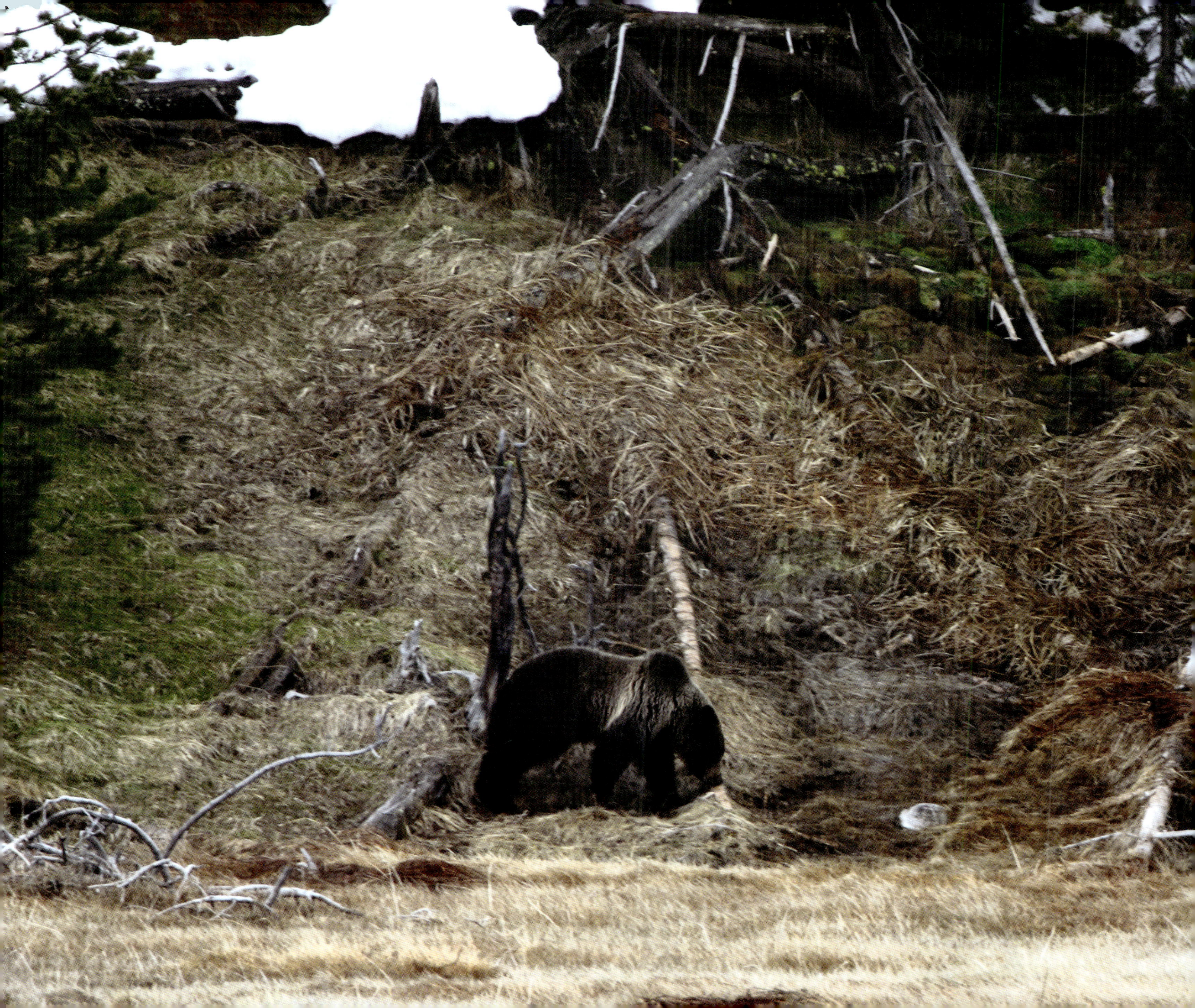

The grizzly bear is considered an American icon, and has become the mascot for many schools around the country.

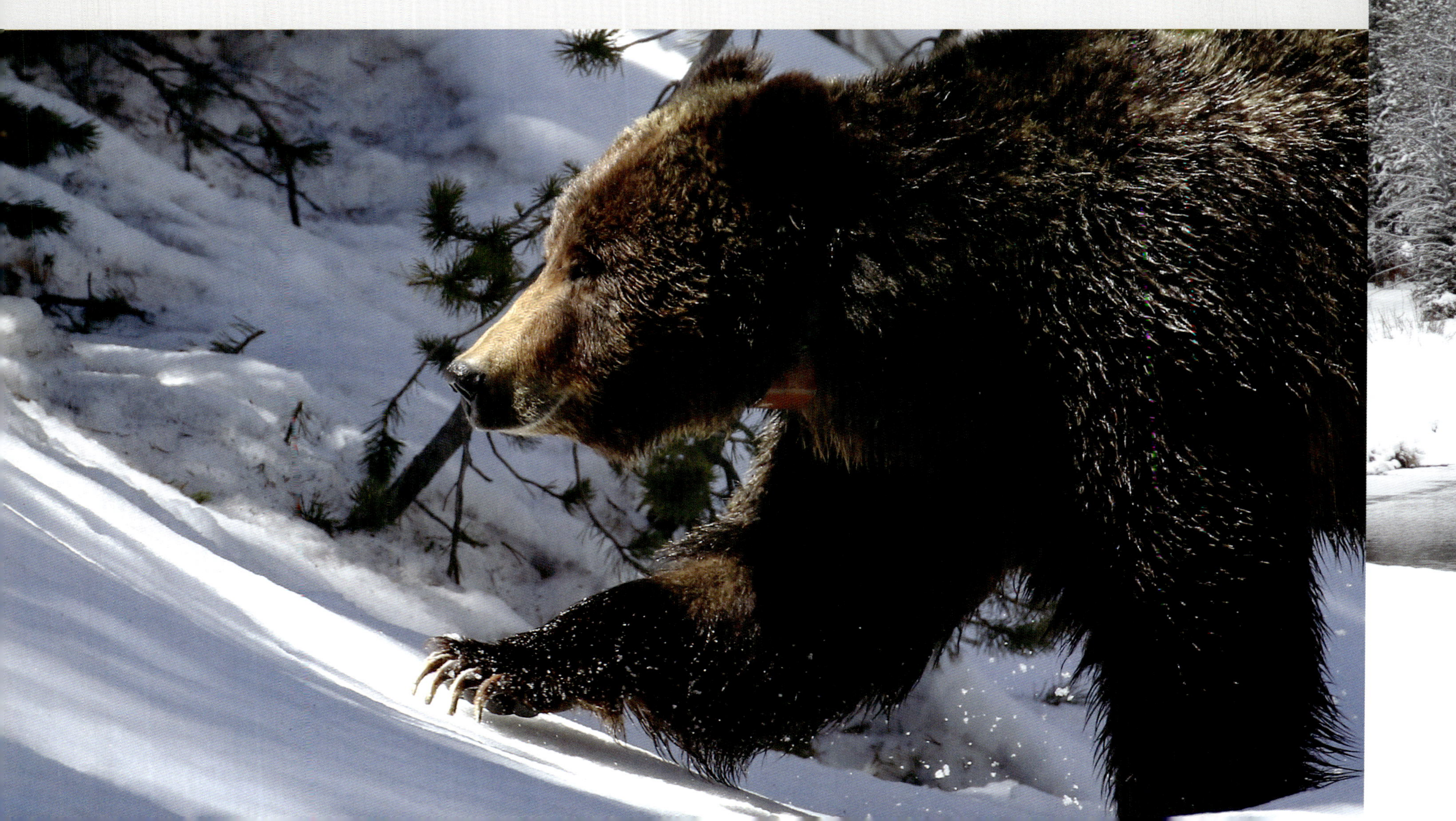

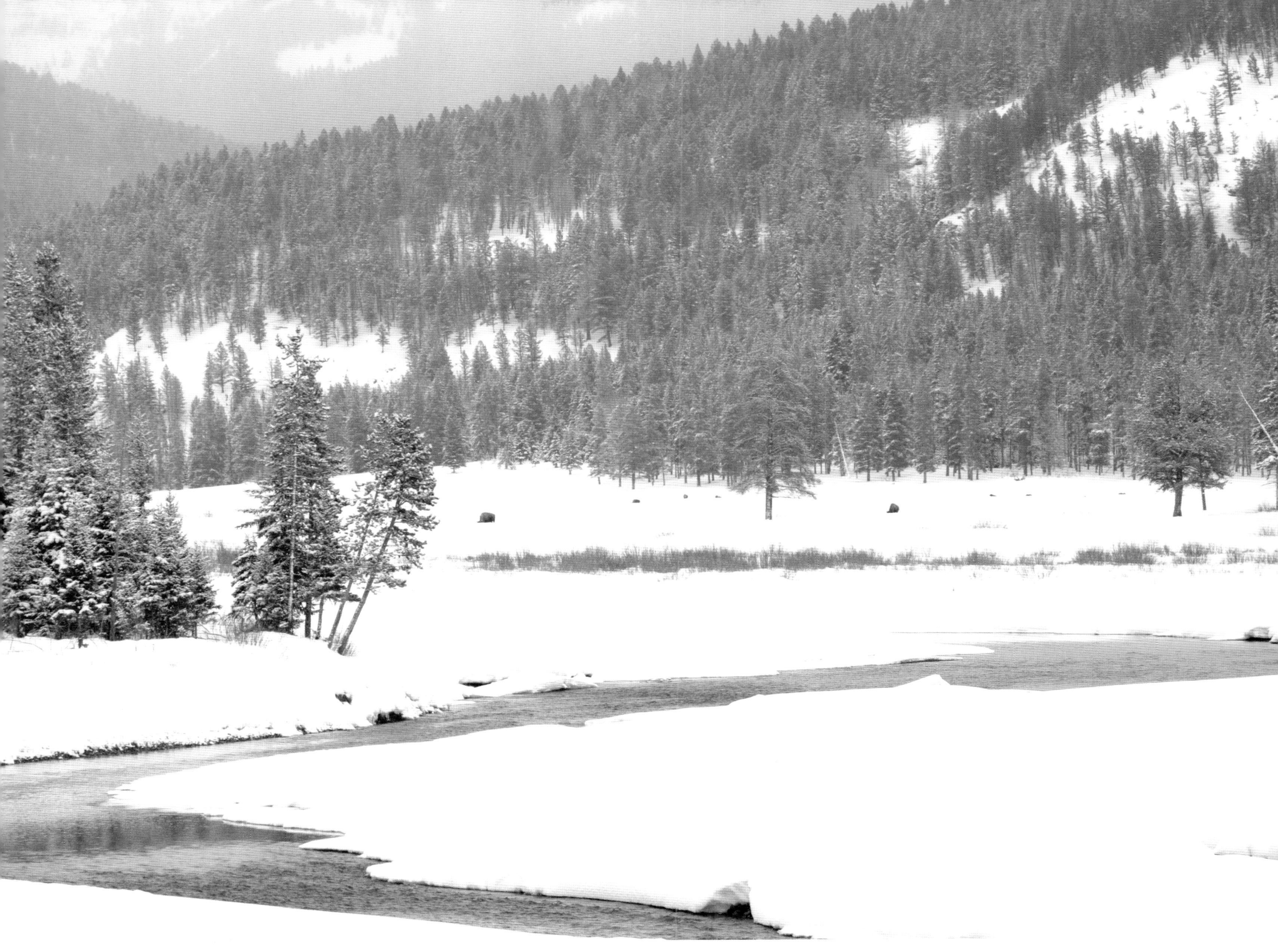

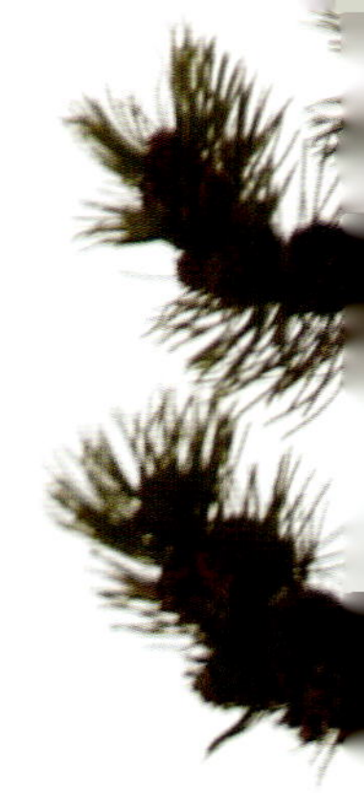

People probably have revered bears for as long as we have interacted with them. And while bear worship is not a part of most modern religions, it is natural for us —even those of us who have never seen a bear—to be awed by such magnificent and formidable creatures.

—*Charles Fergus*

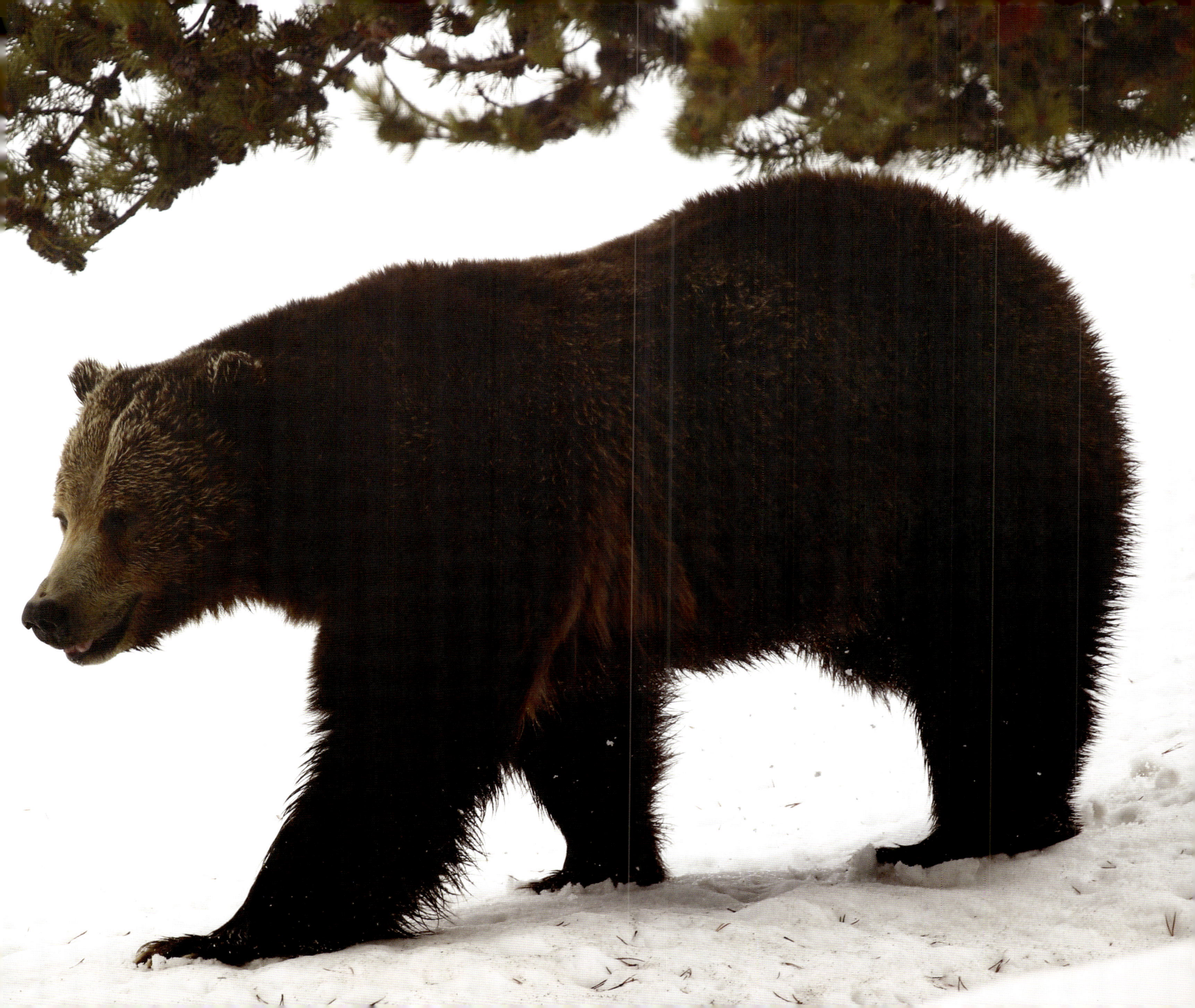

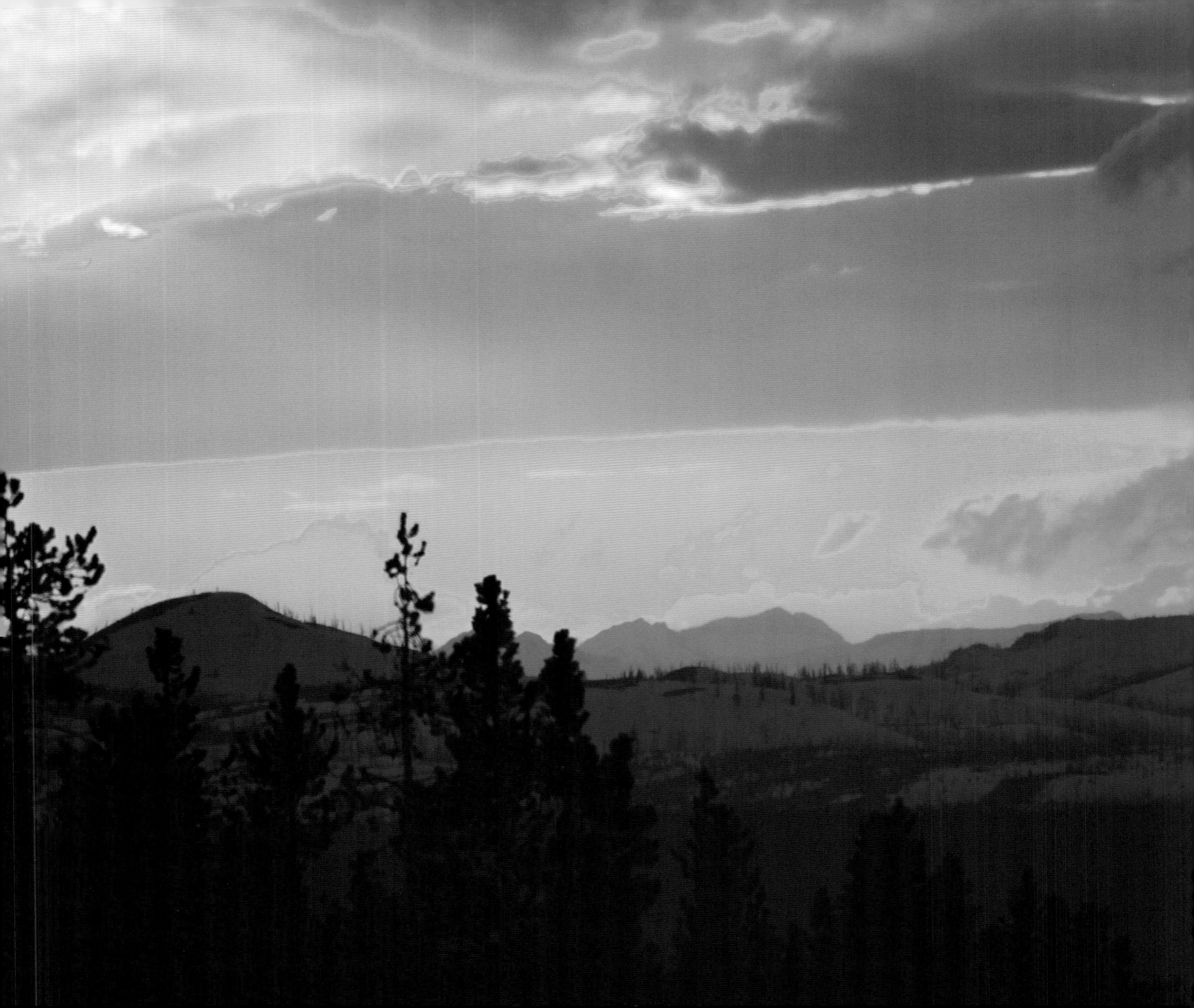

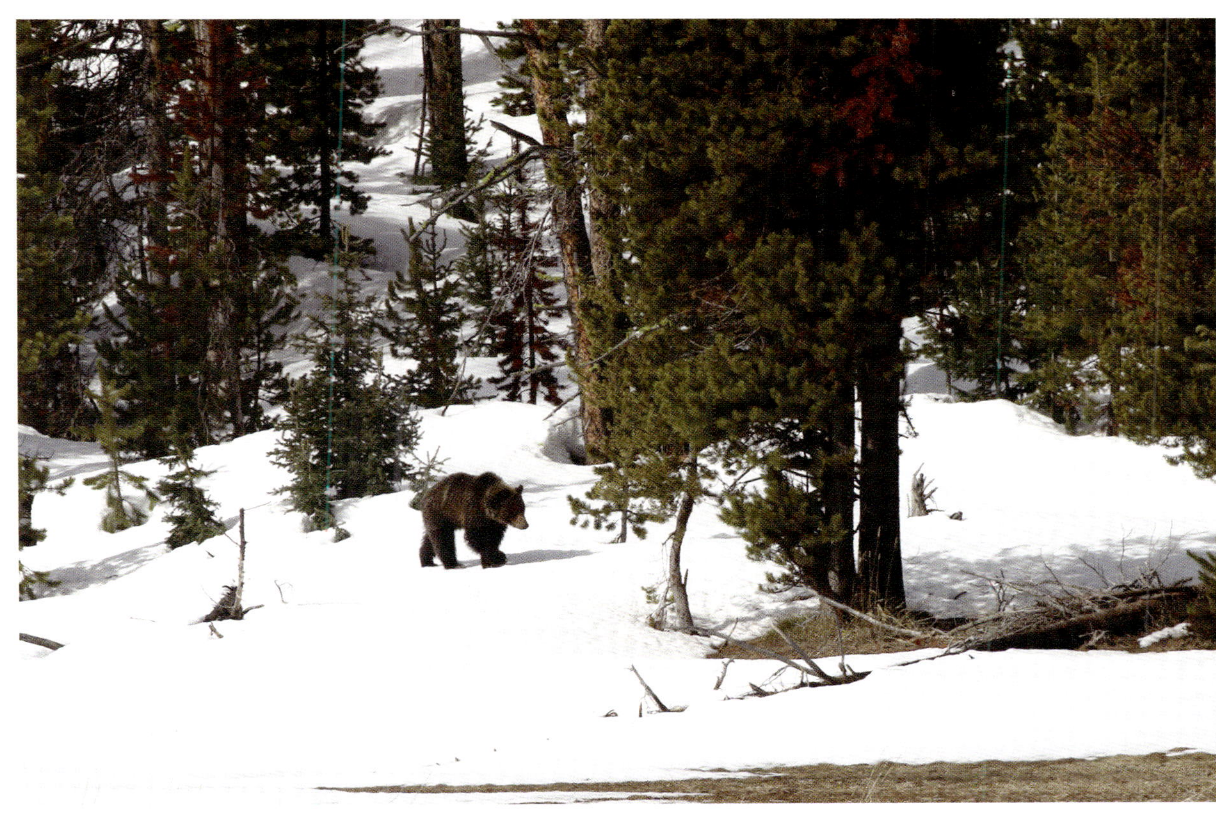

For bears, winter is one night. —*Unknown*

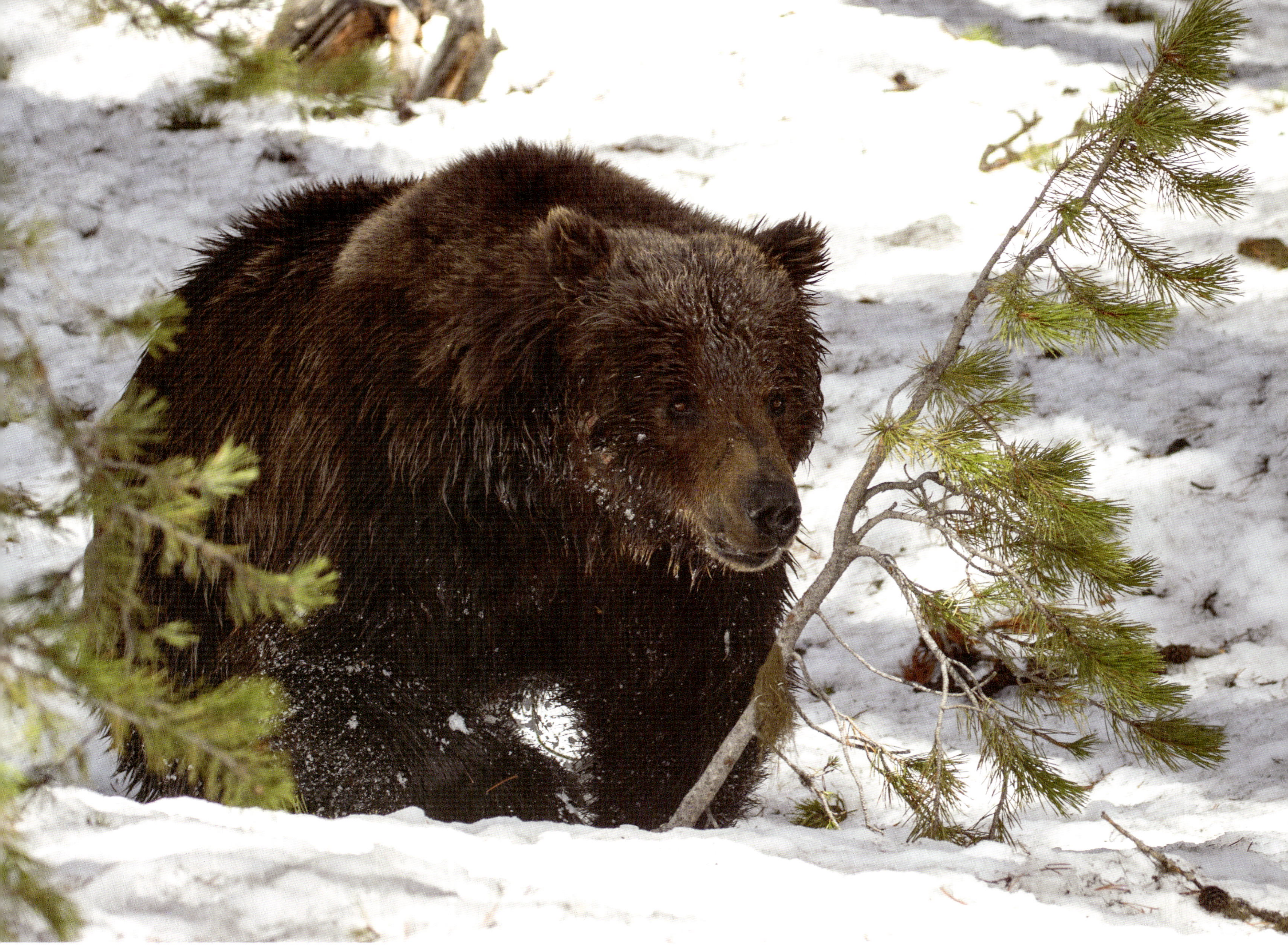

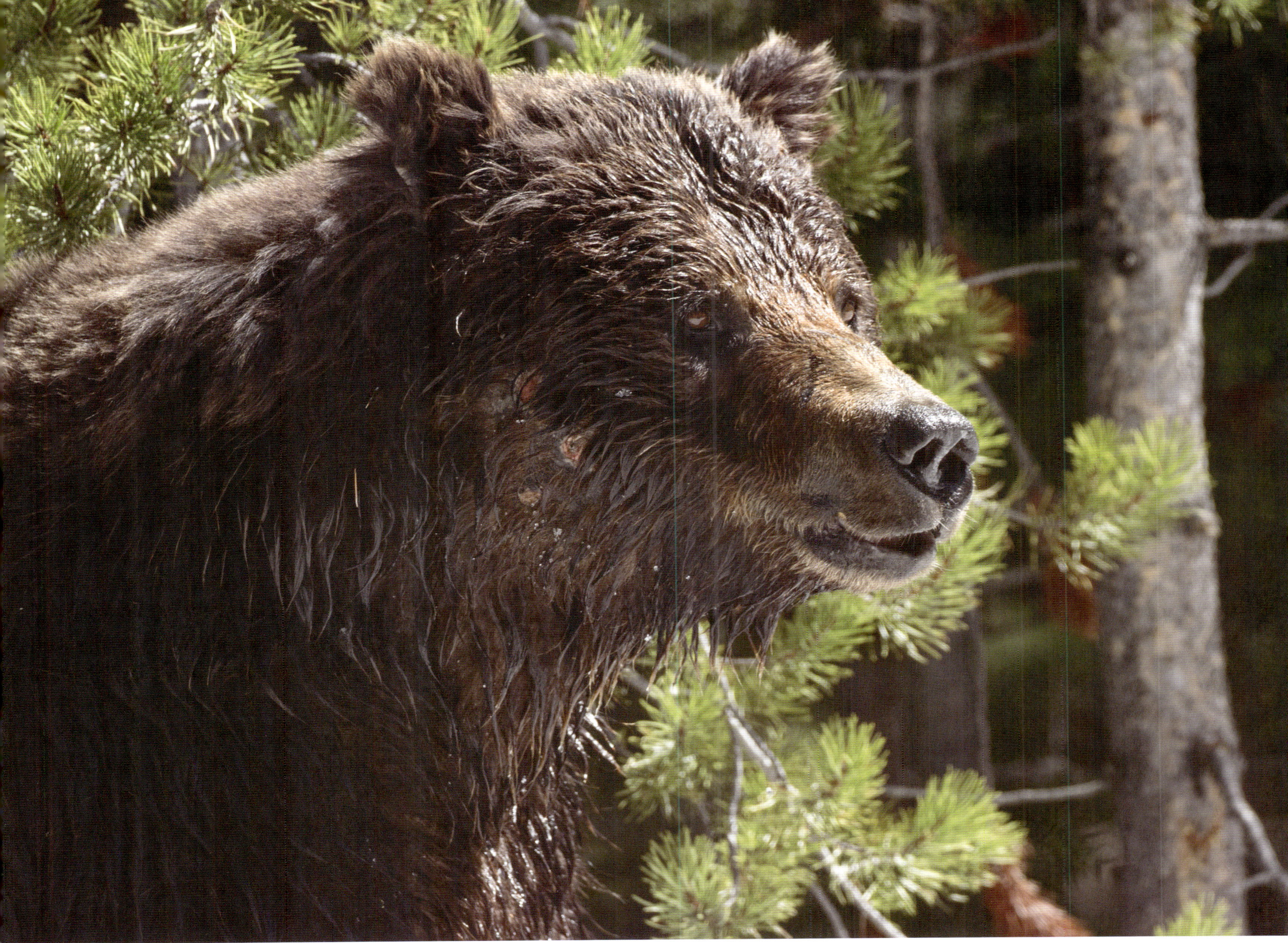

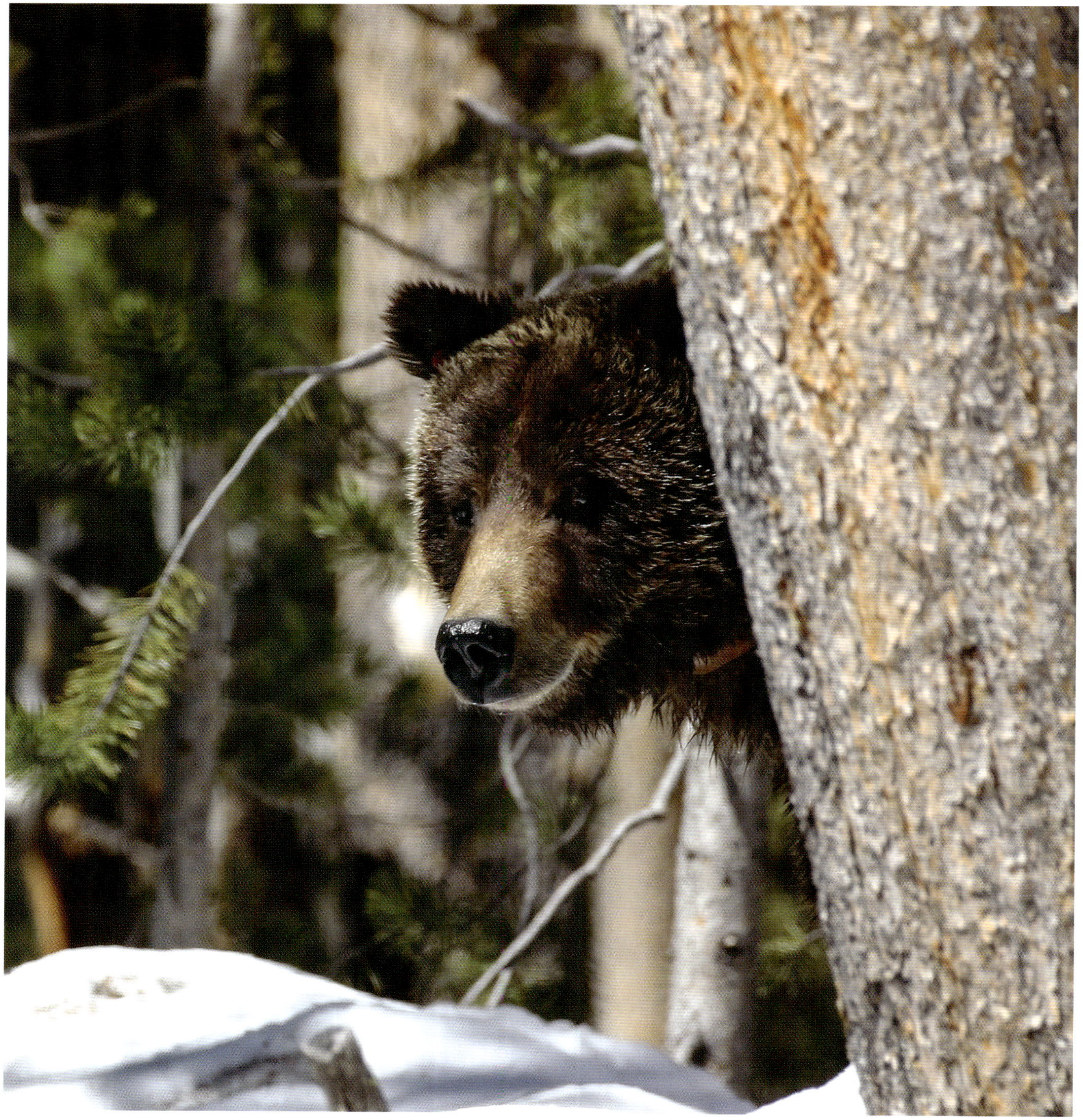

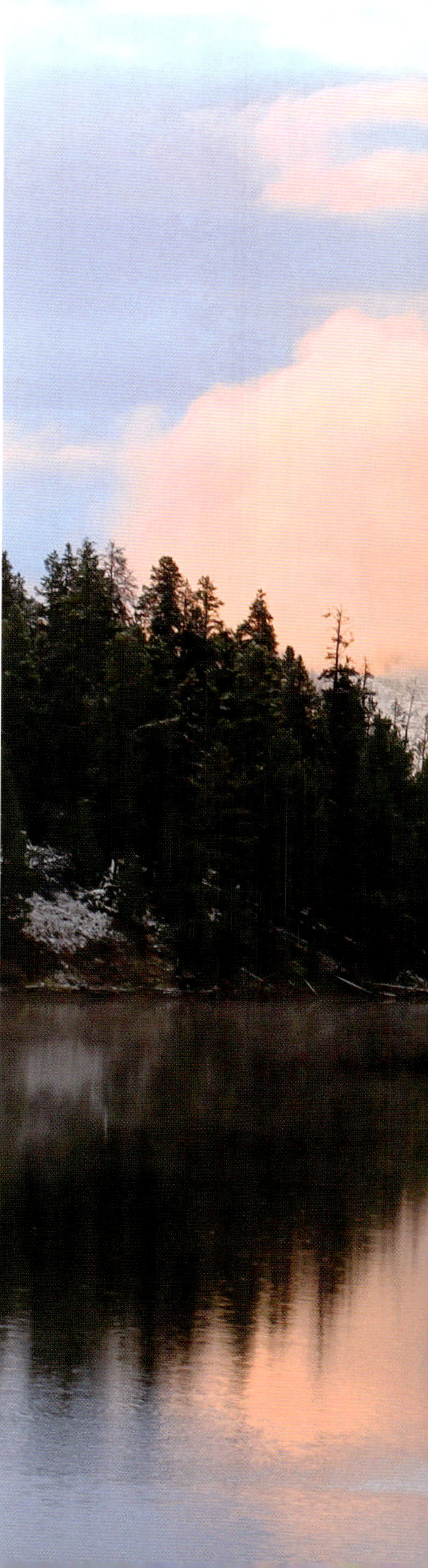

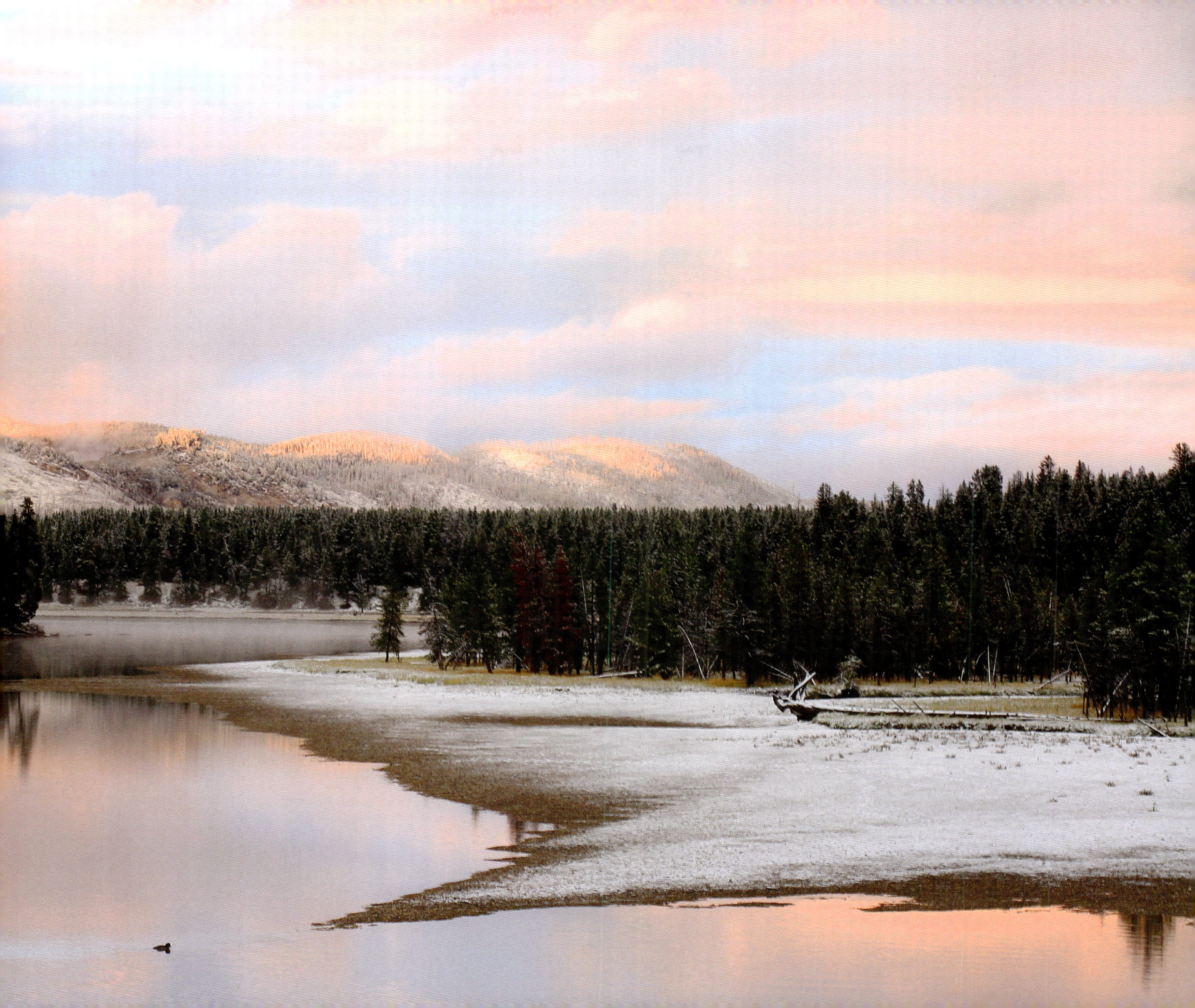

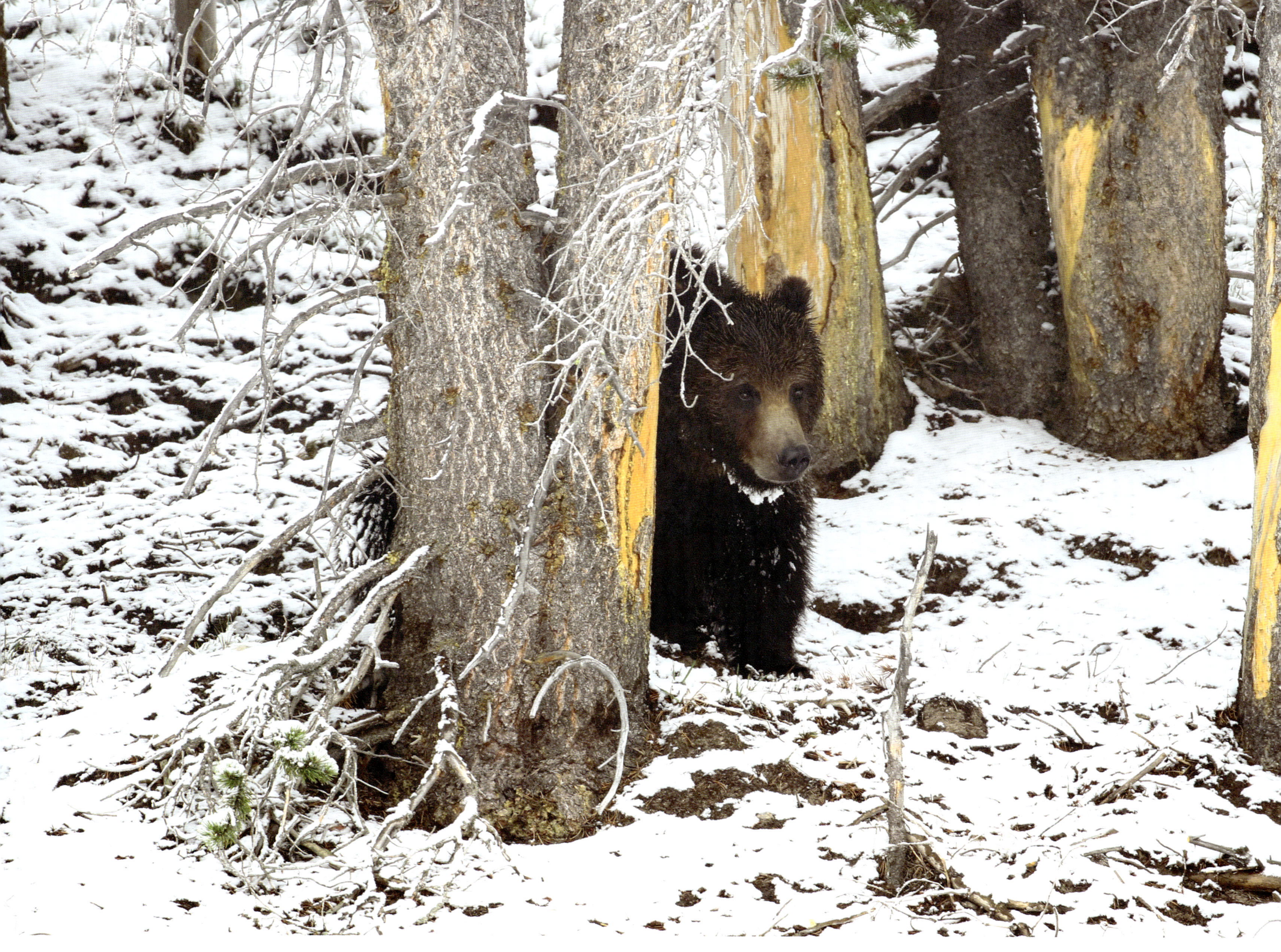

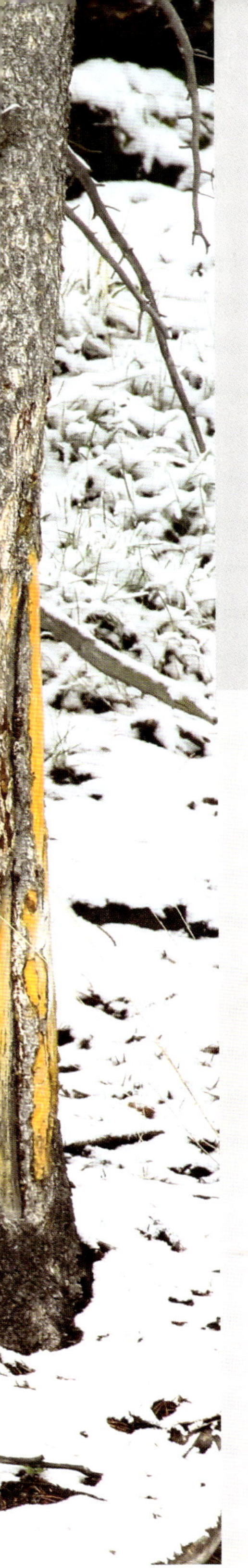

Everyone is responsible for continued success where grizzly bears are concerned. It's important to continue to prevent bears from obtaining human food sources as well as reduce human-and-bear conflicts while hiking, hunting, and simply coexisting. It's up to us to share the landscape responsibly.

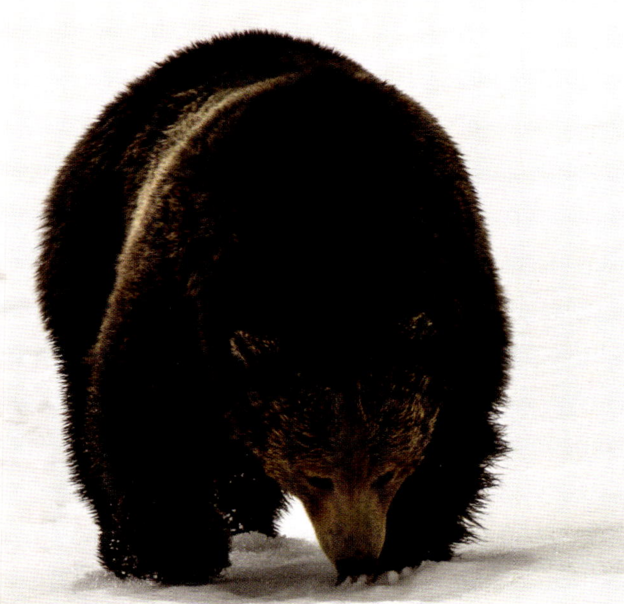

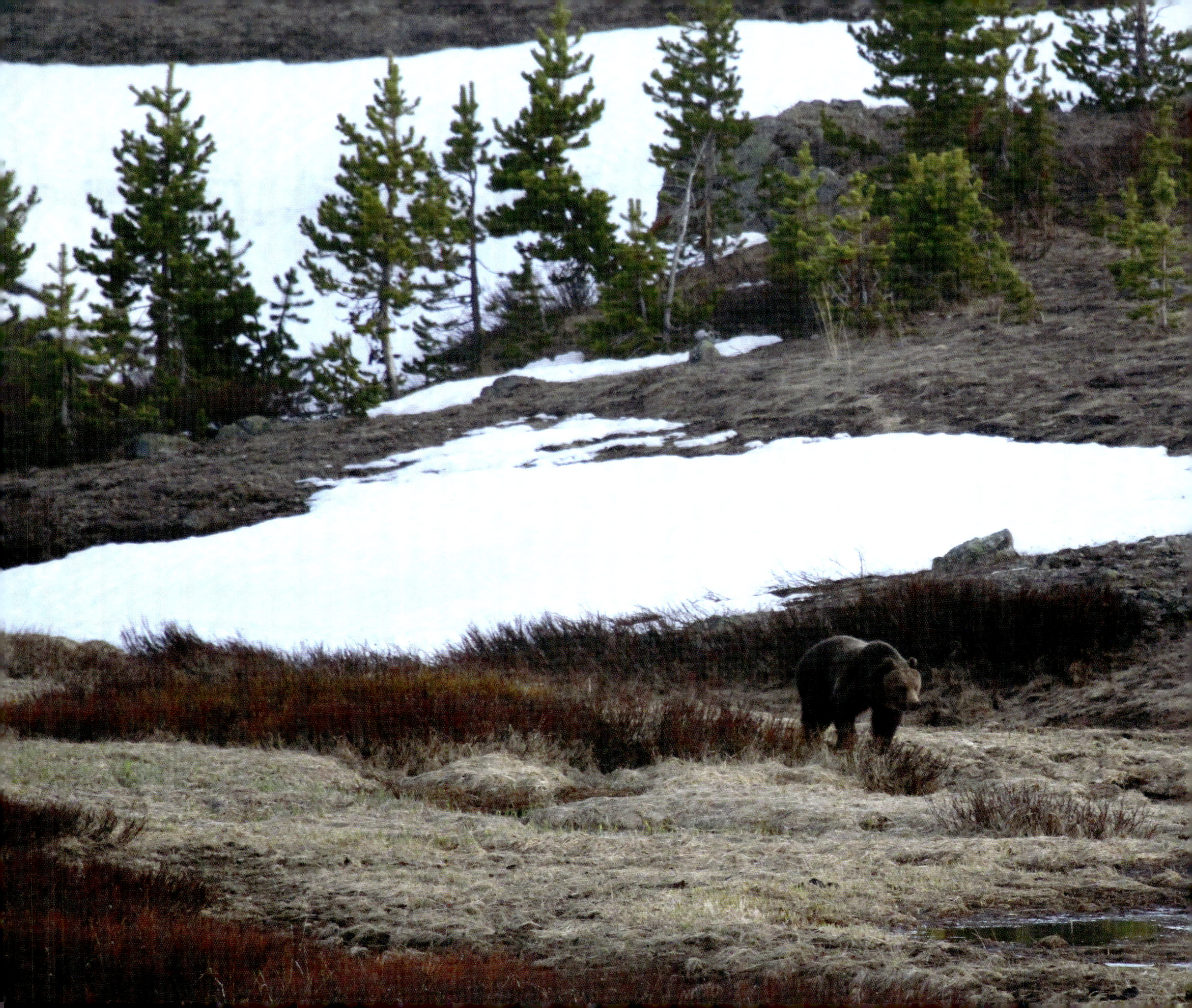

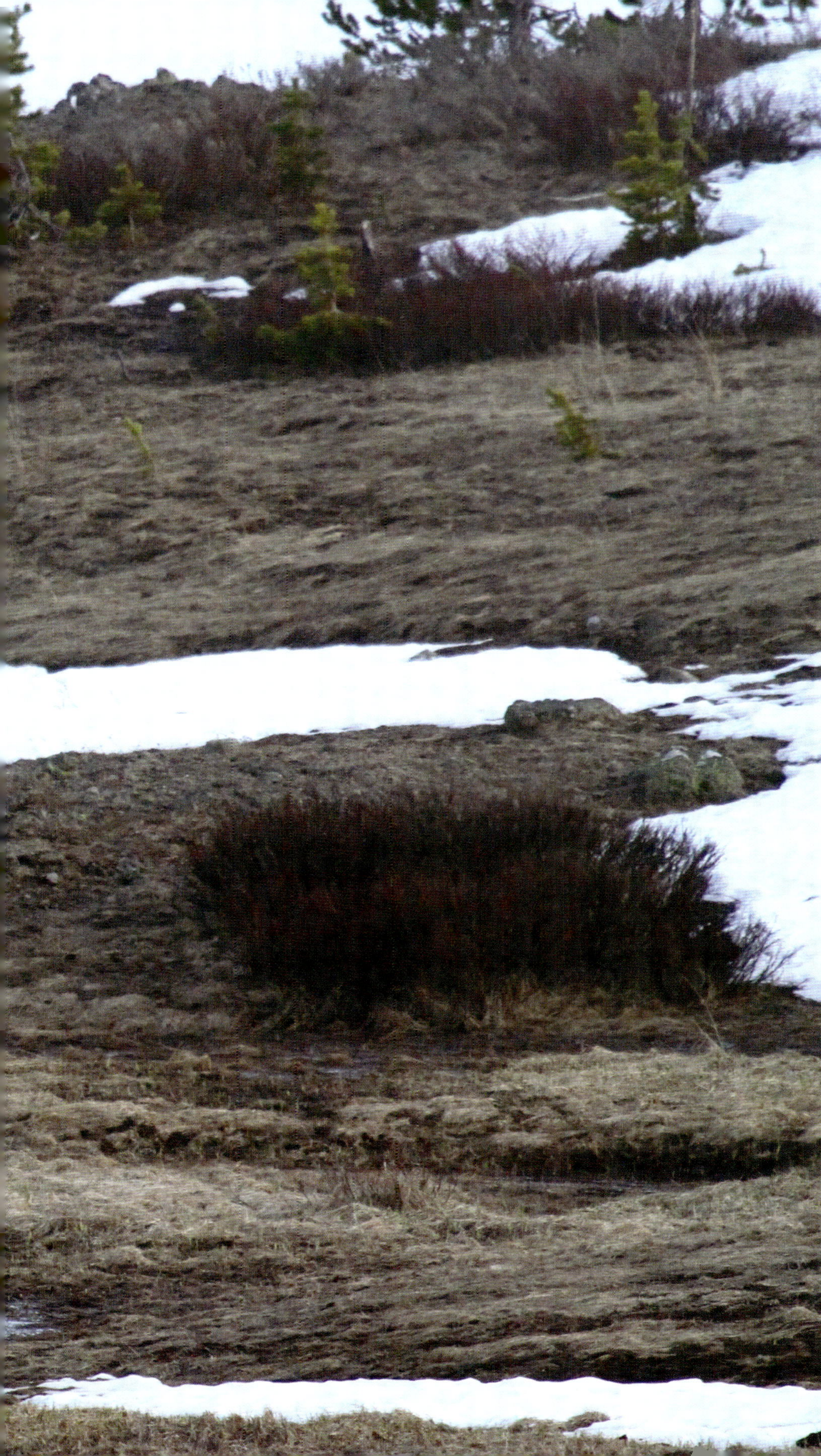

When you are where wild bears live you learn to pay attention to the rhythm of the land and yourself.

—*Linda Jo Hunter*

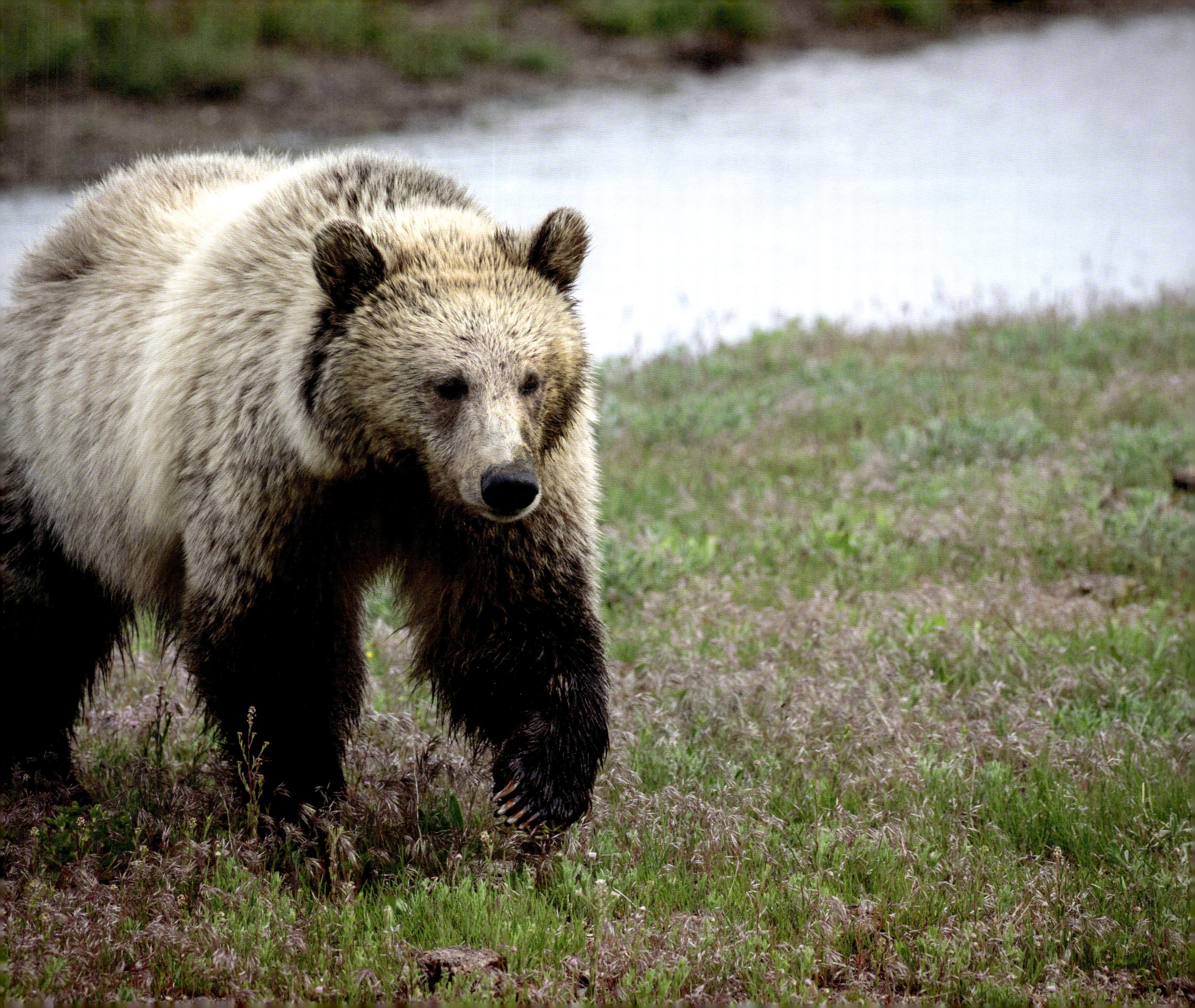

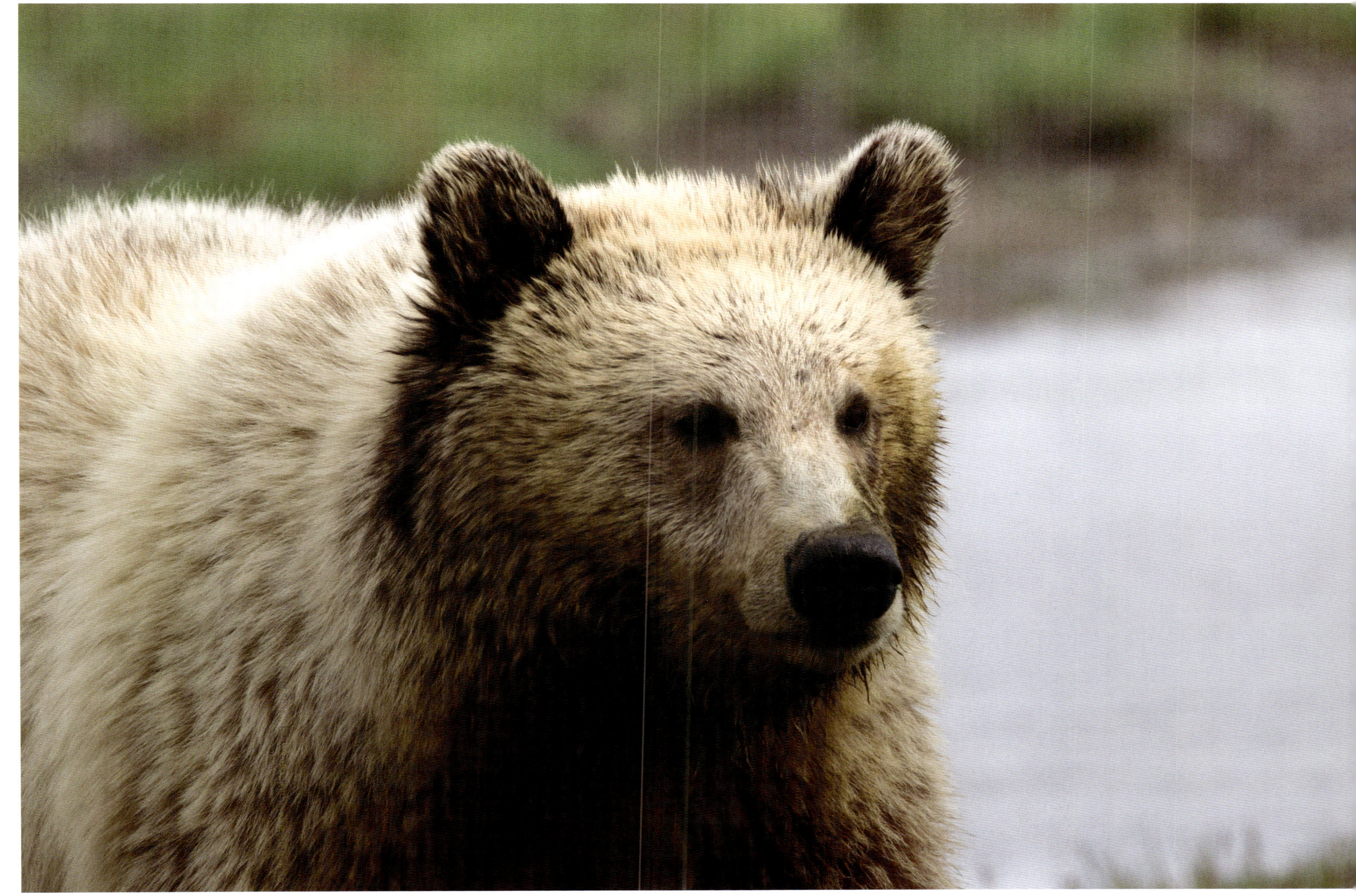

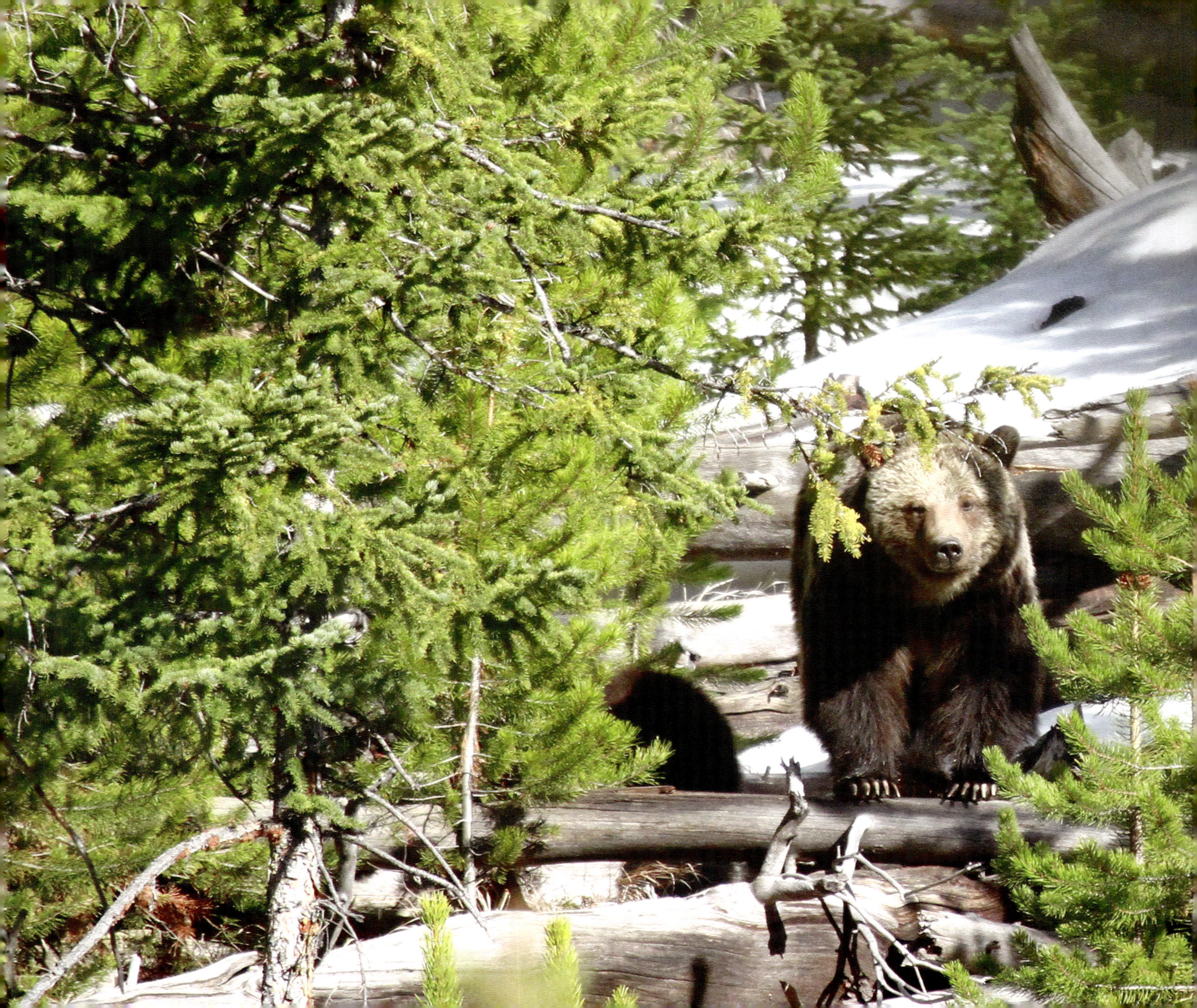

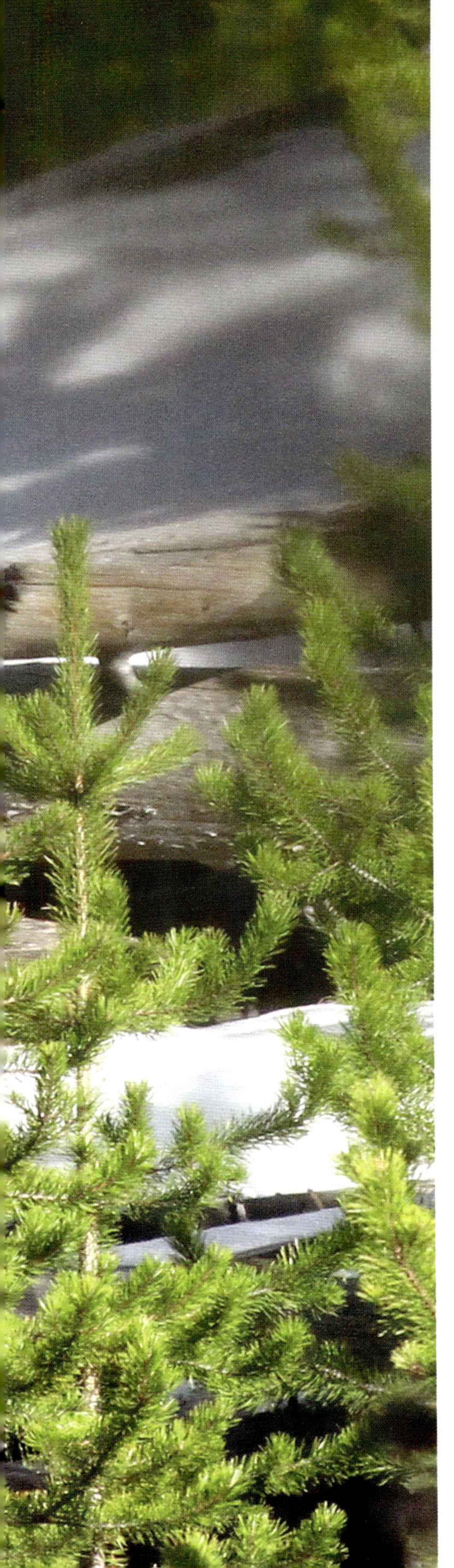

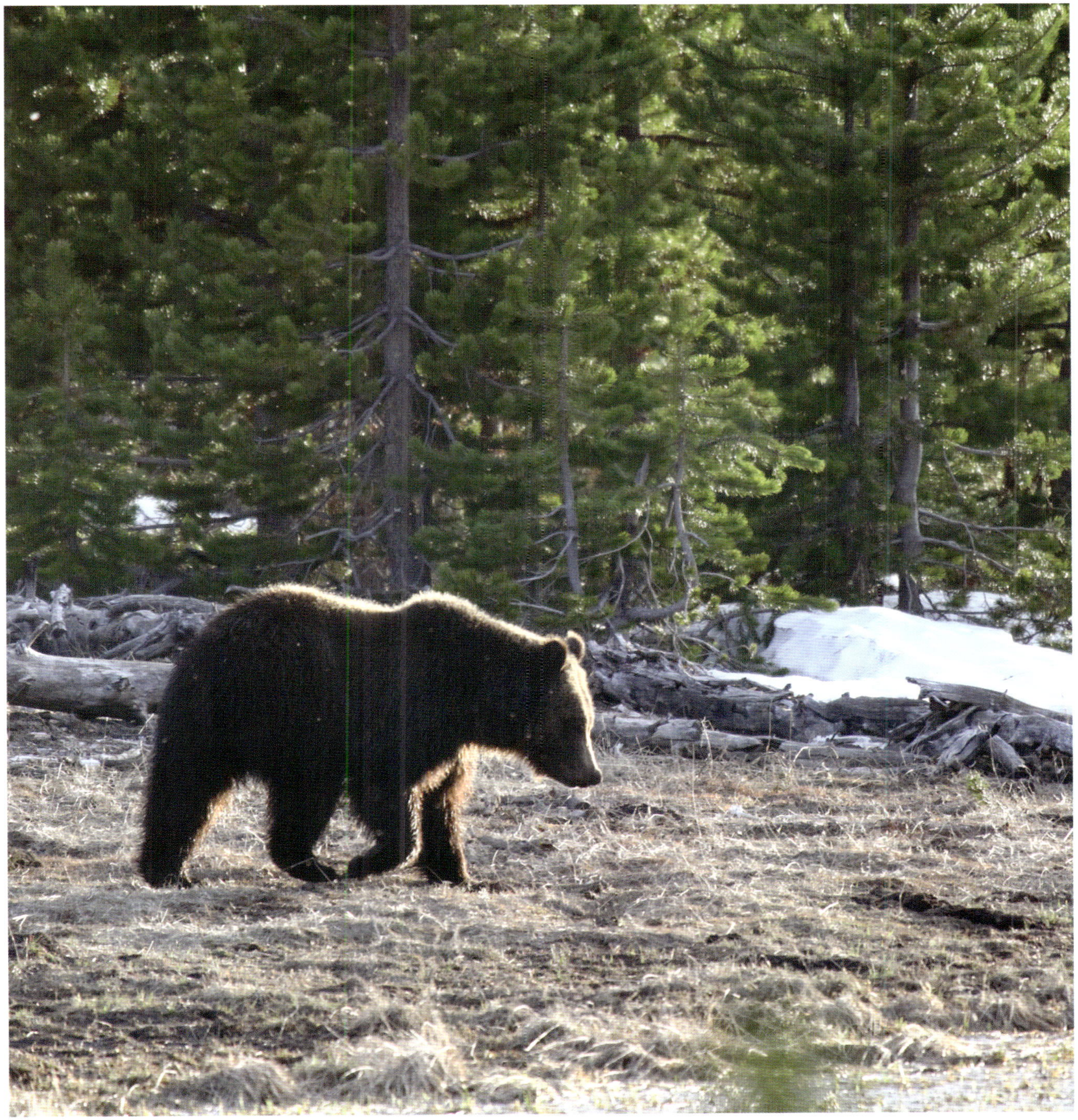

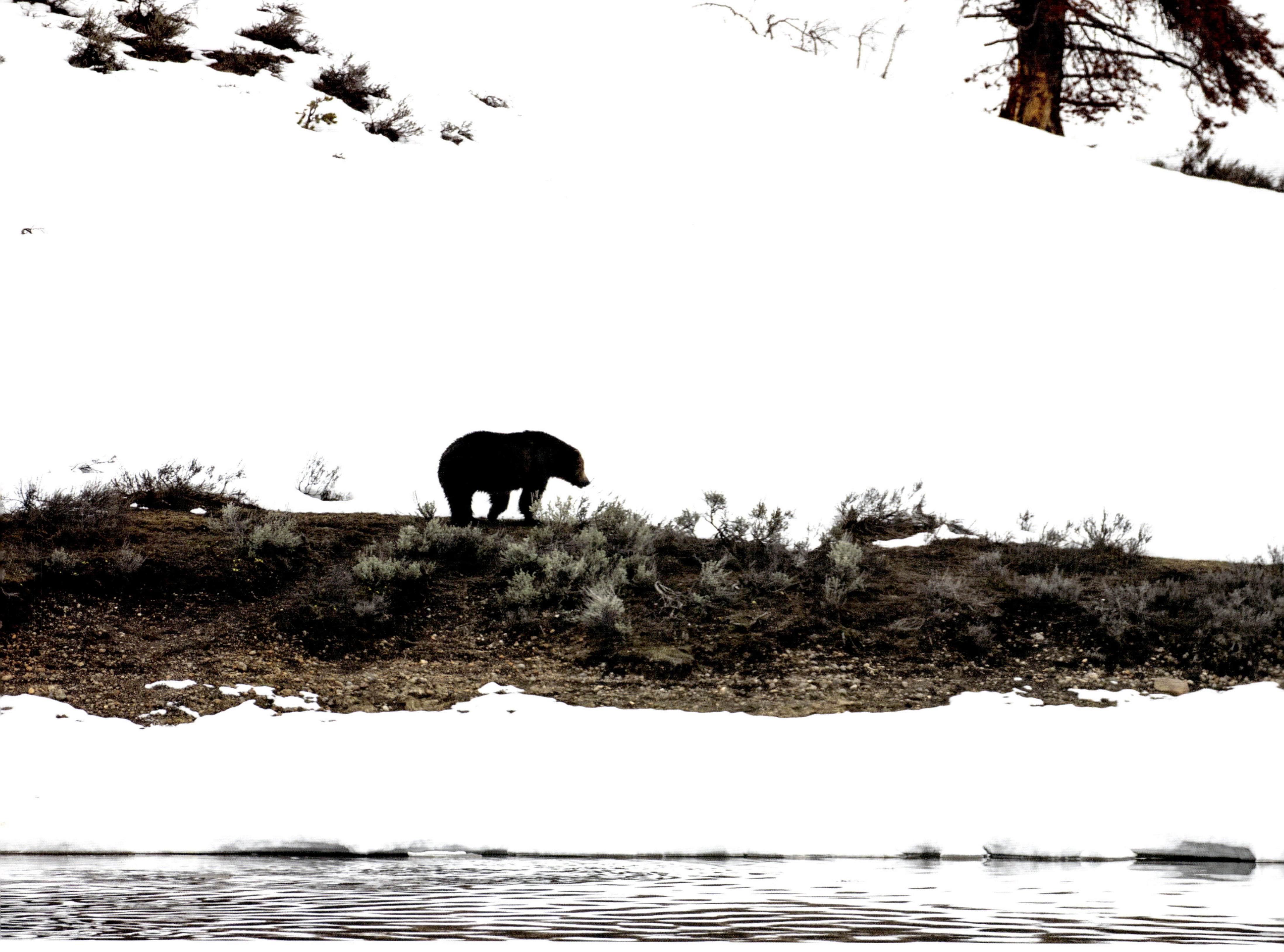

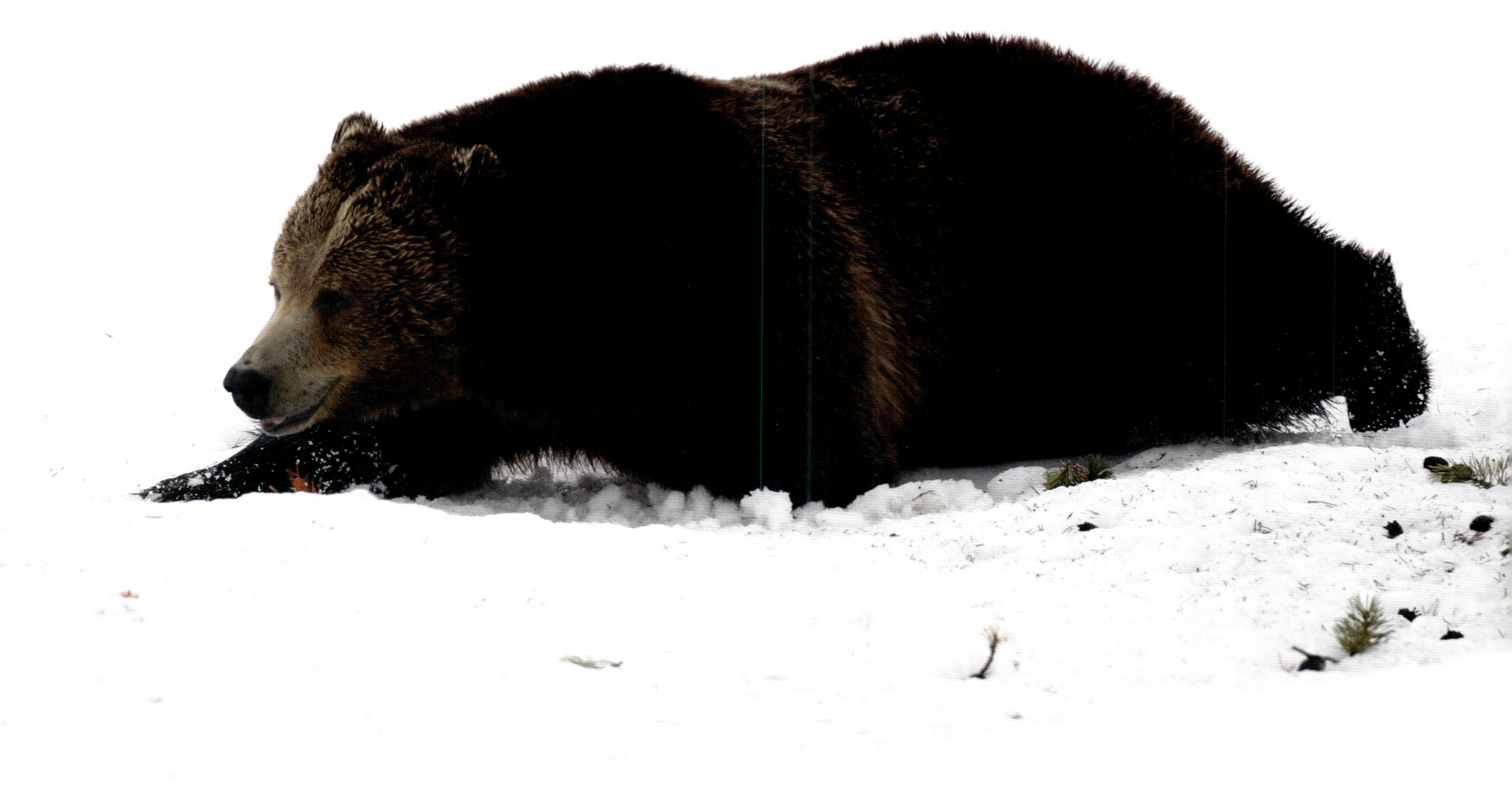

Millions of people visit Yellowstone and the Greater Yellowstone Ecosystem every year with the intention of seeing their first, second, or maybe even their tenth grizzly bear. Regardless of how many they have seen, the hope of finding one, and the excitement when they do, is still the highlight of their trip.

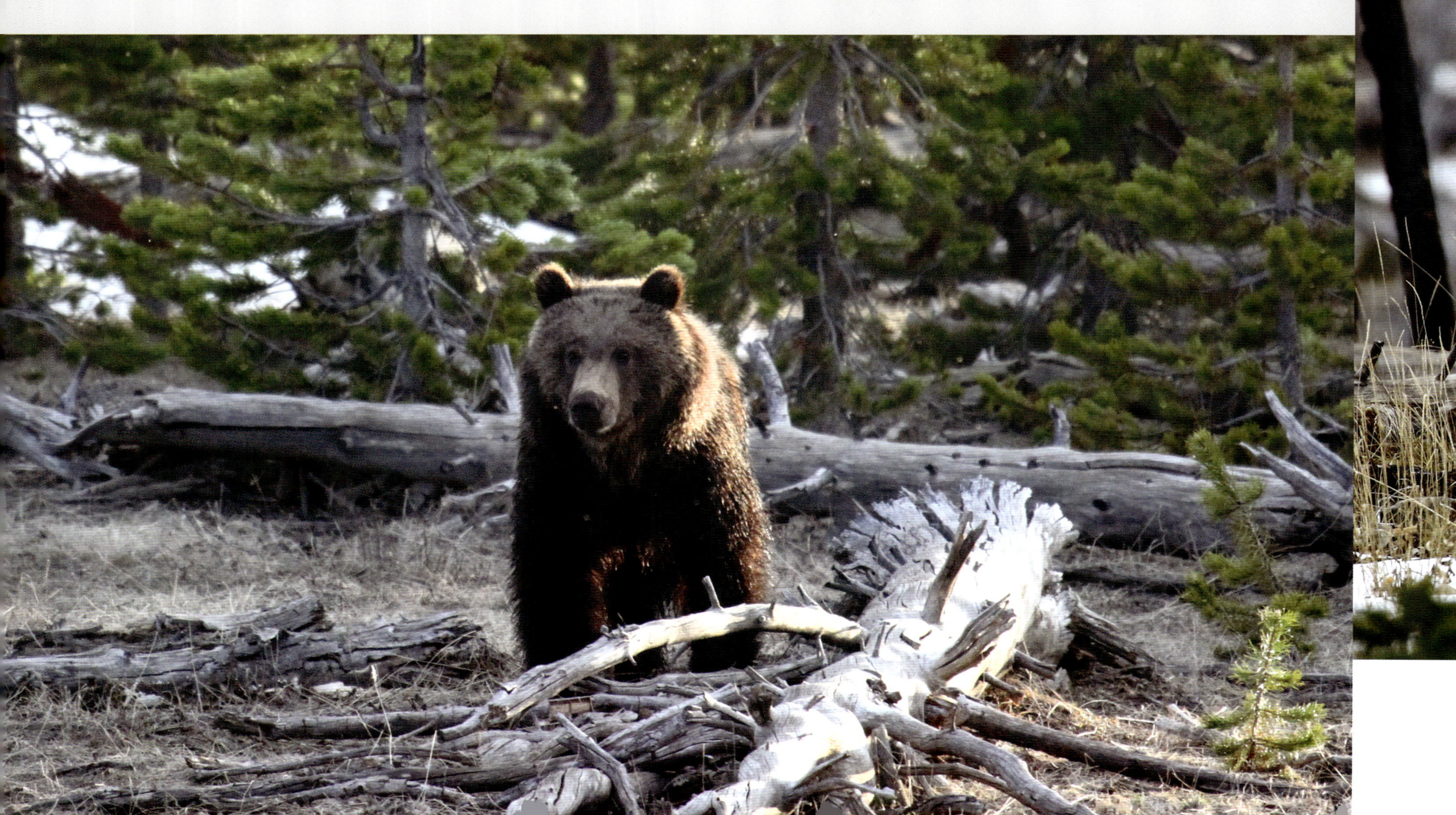

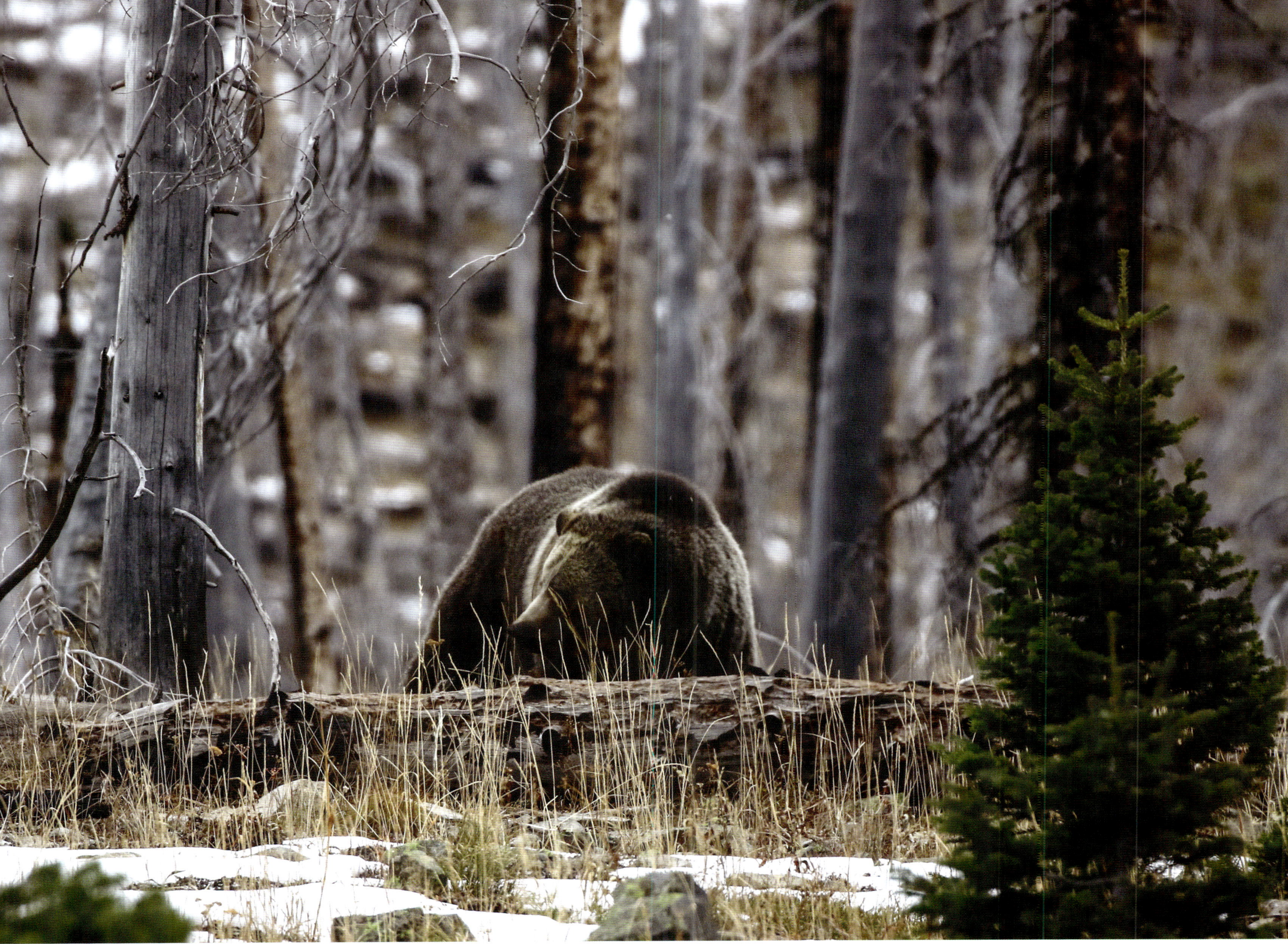

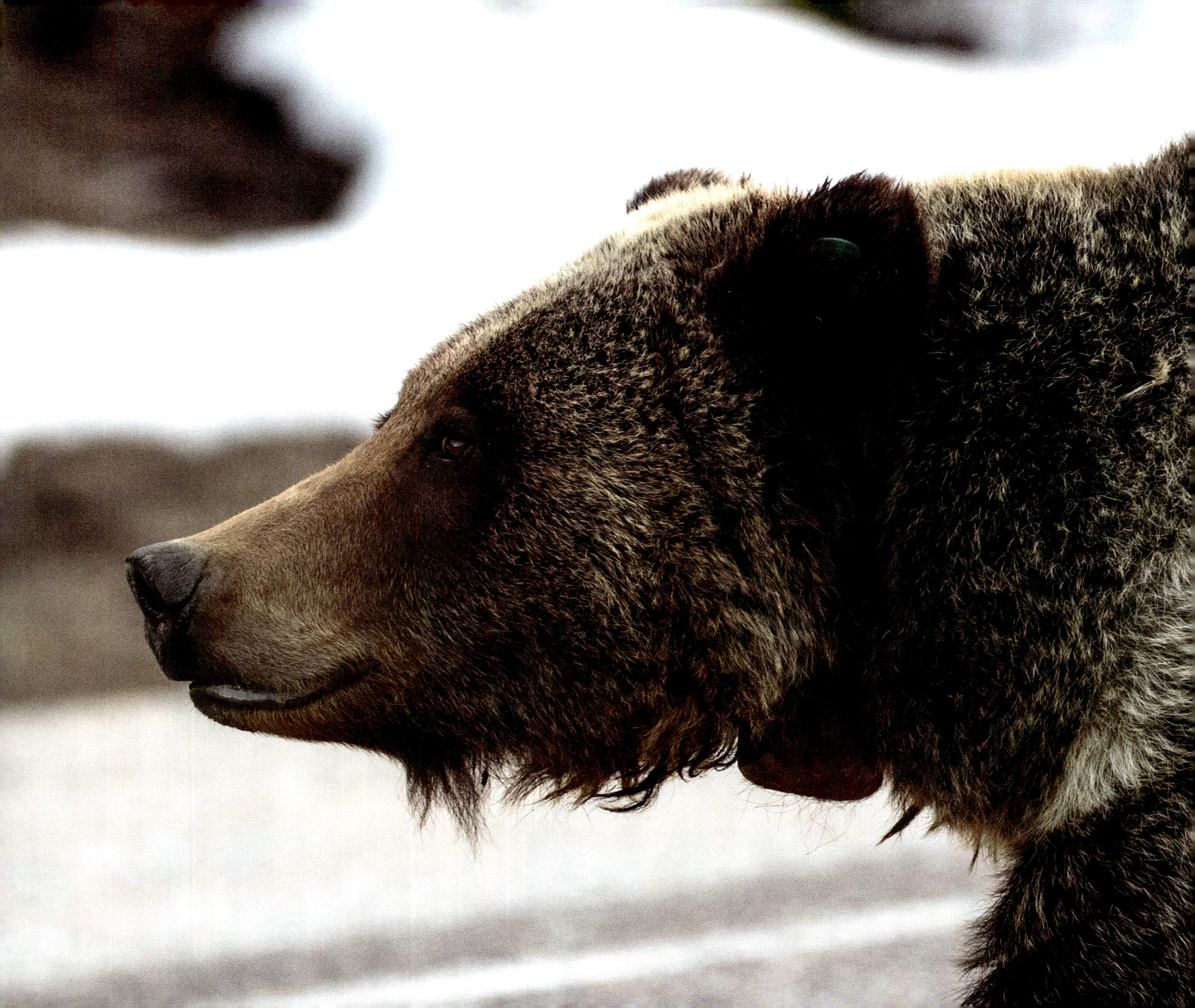

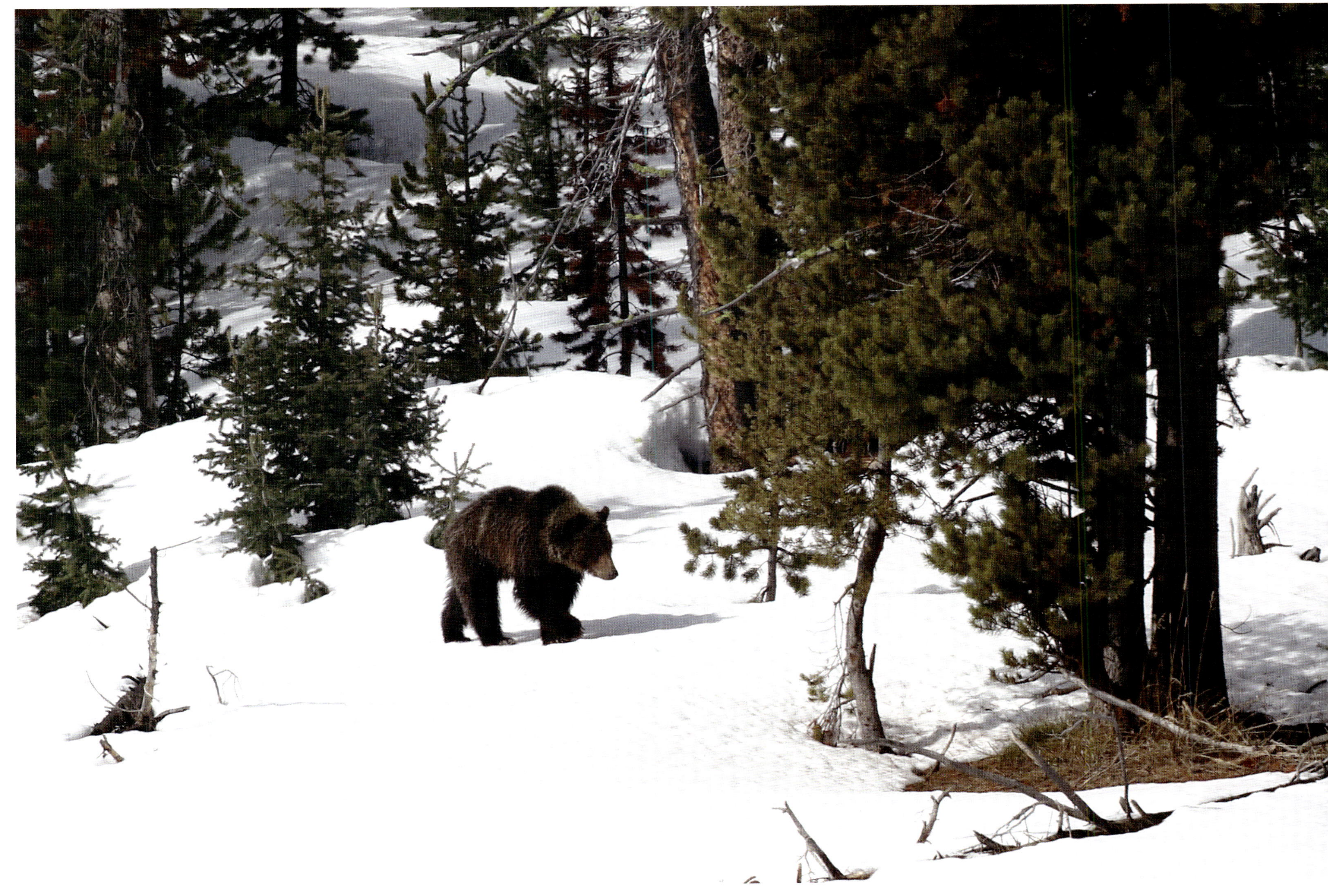

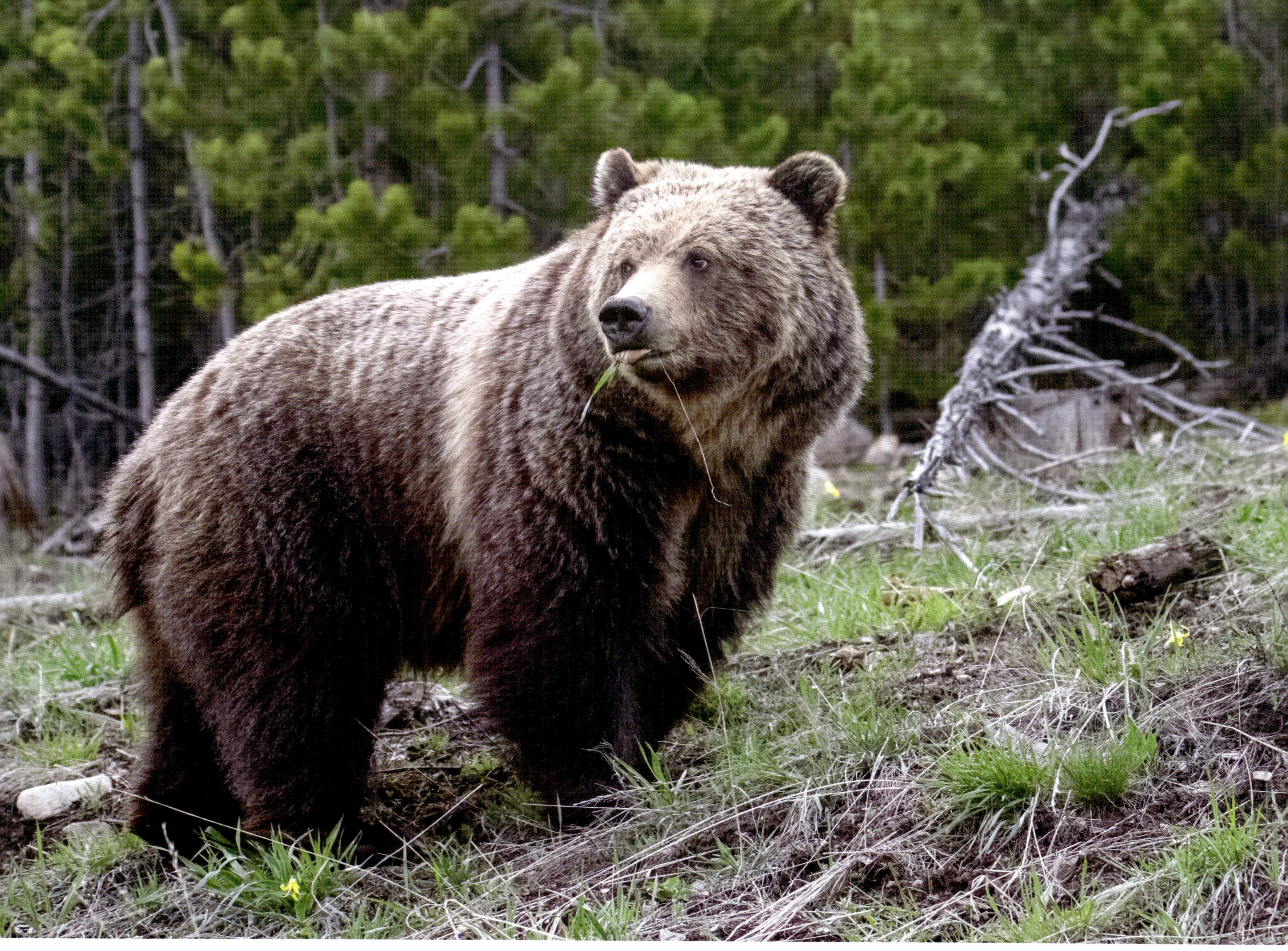

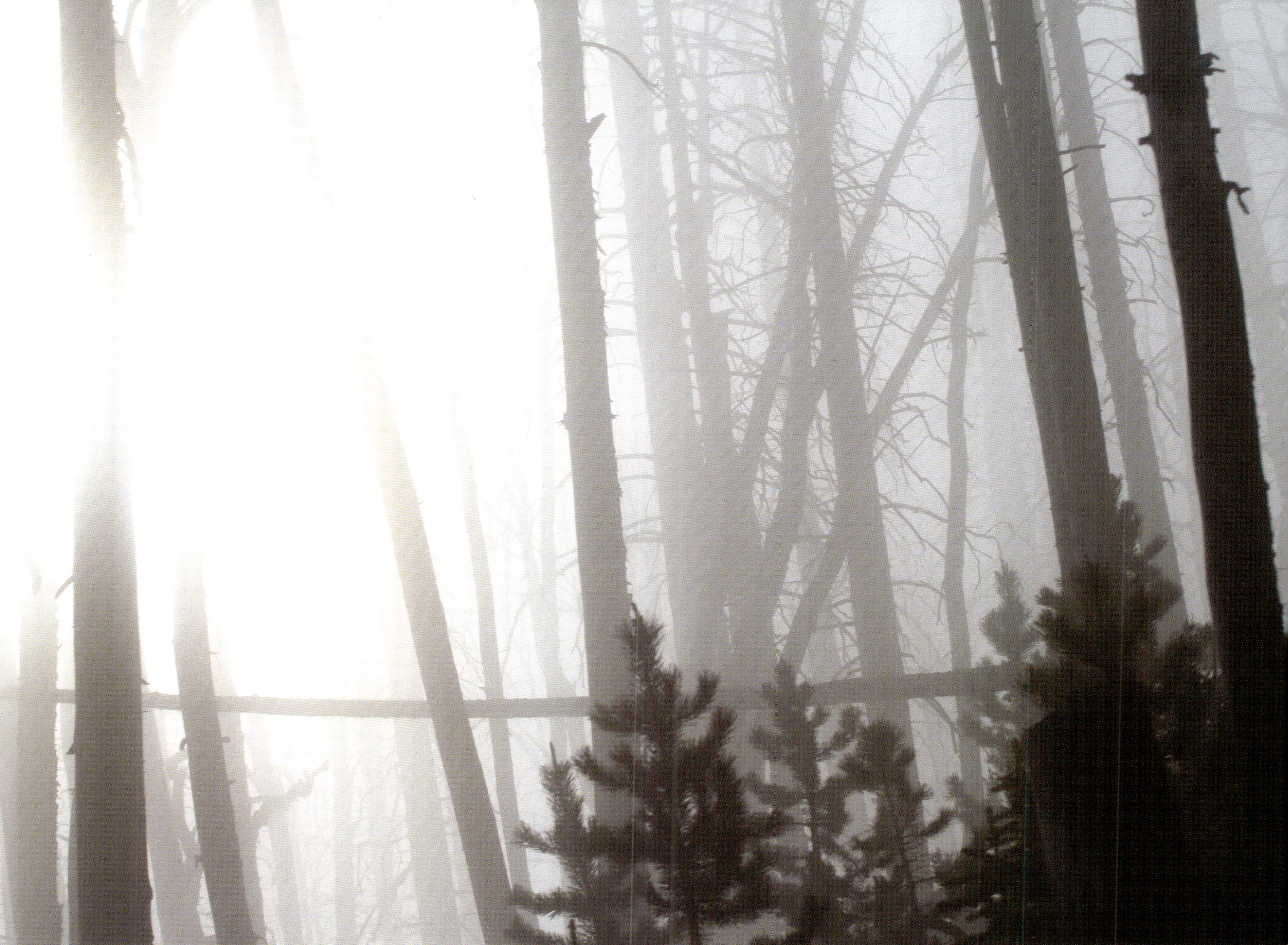

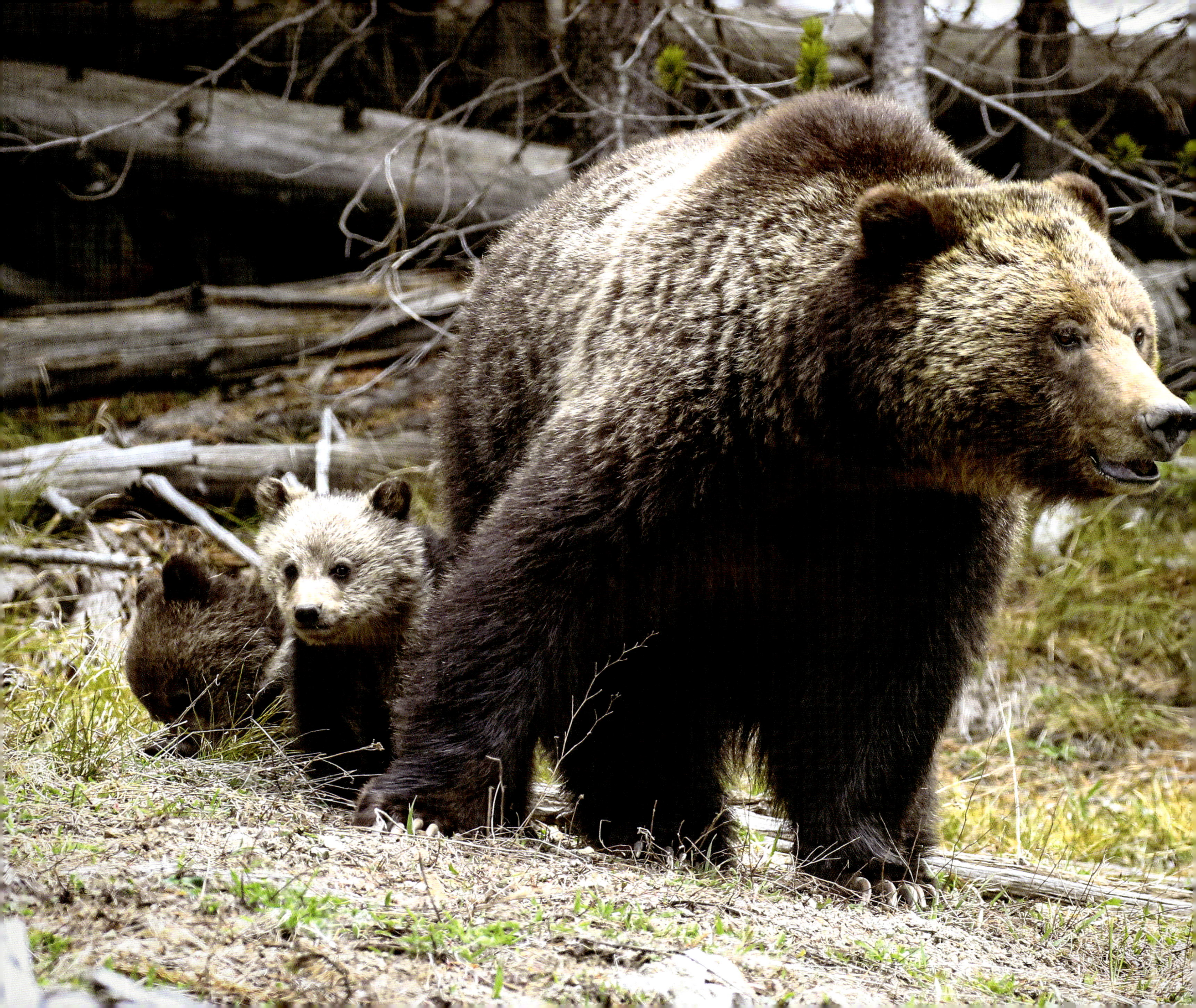

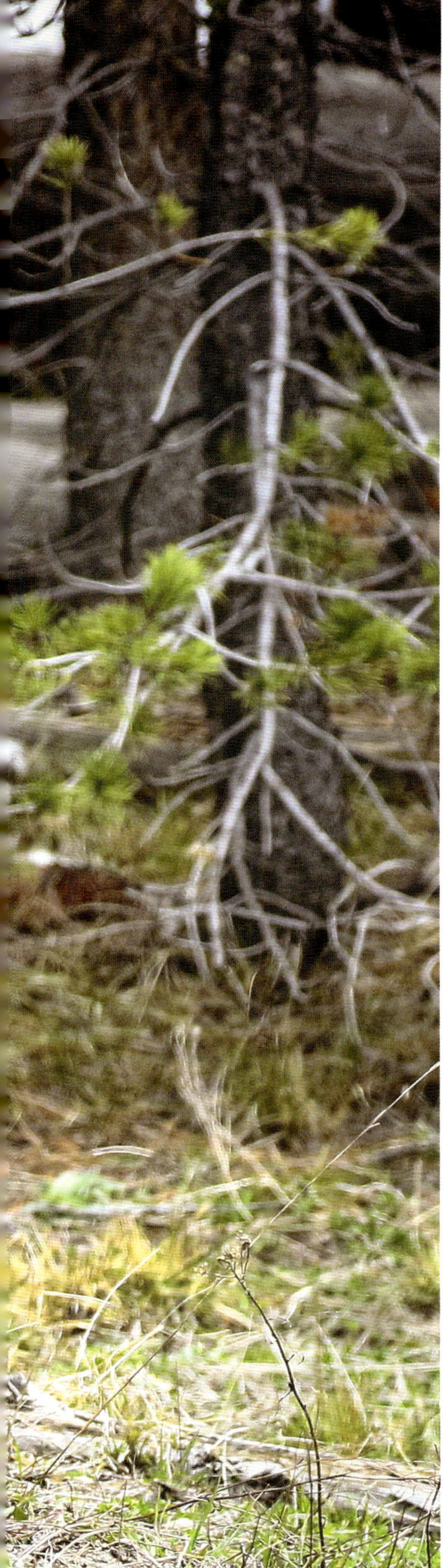

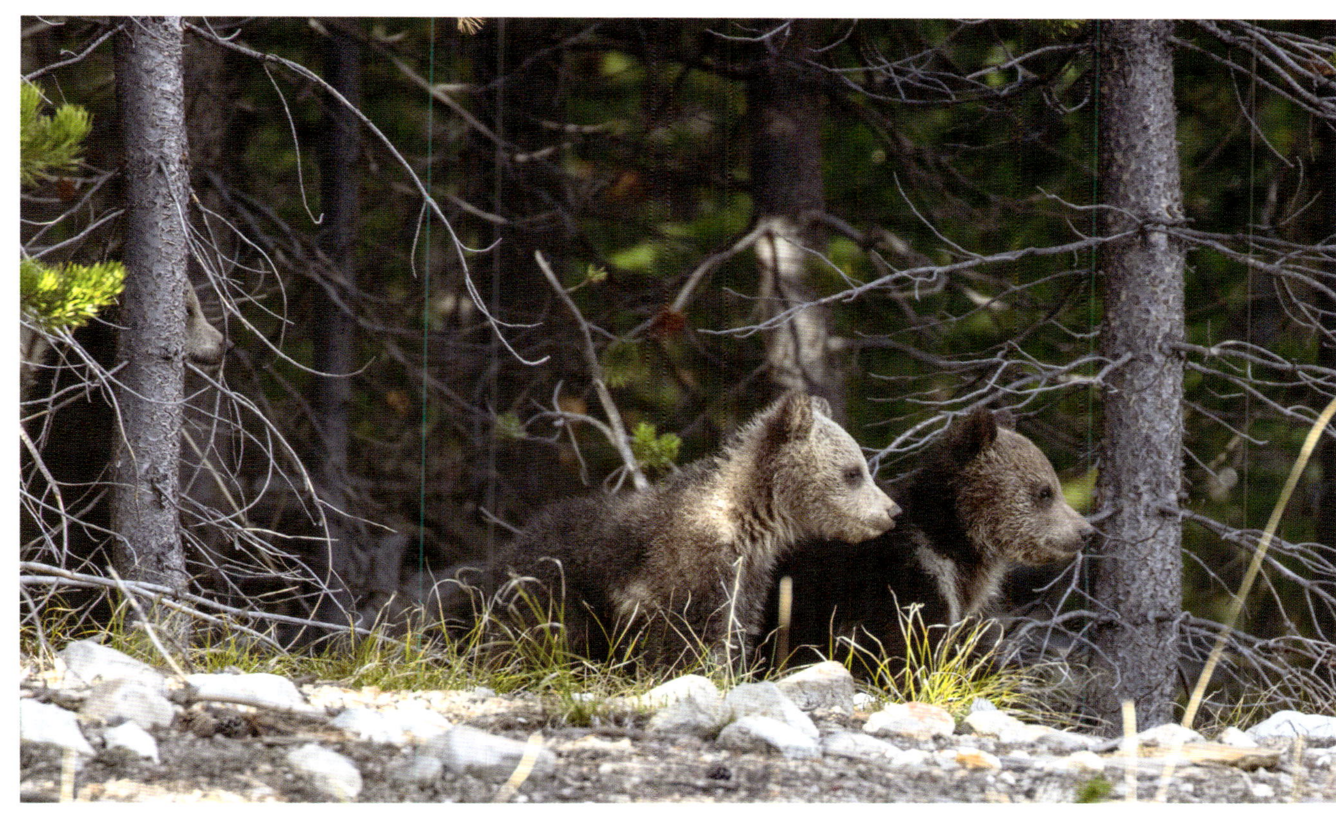

Each bear has a personality. No bear is the same. Just like people, they do have unique and distinguished identities which separate them.

—Tyler Brasington, Bear Biologist

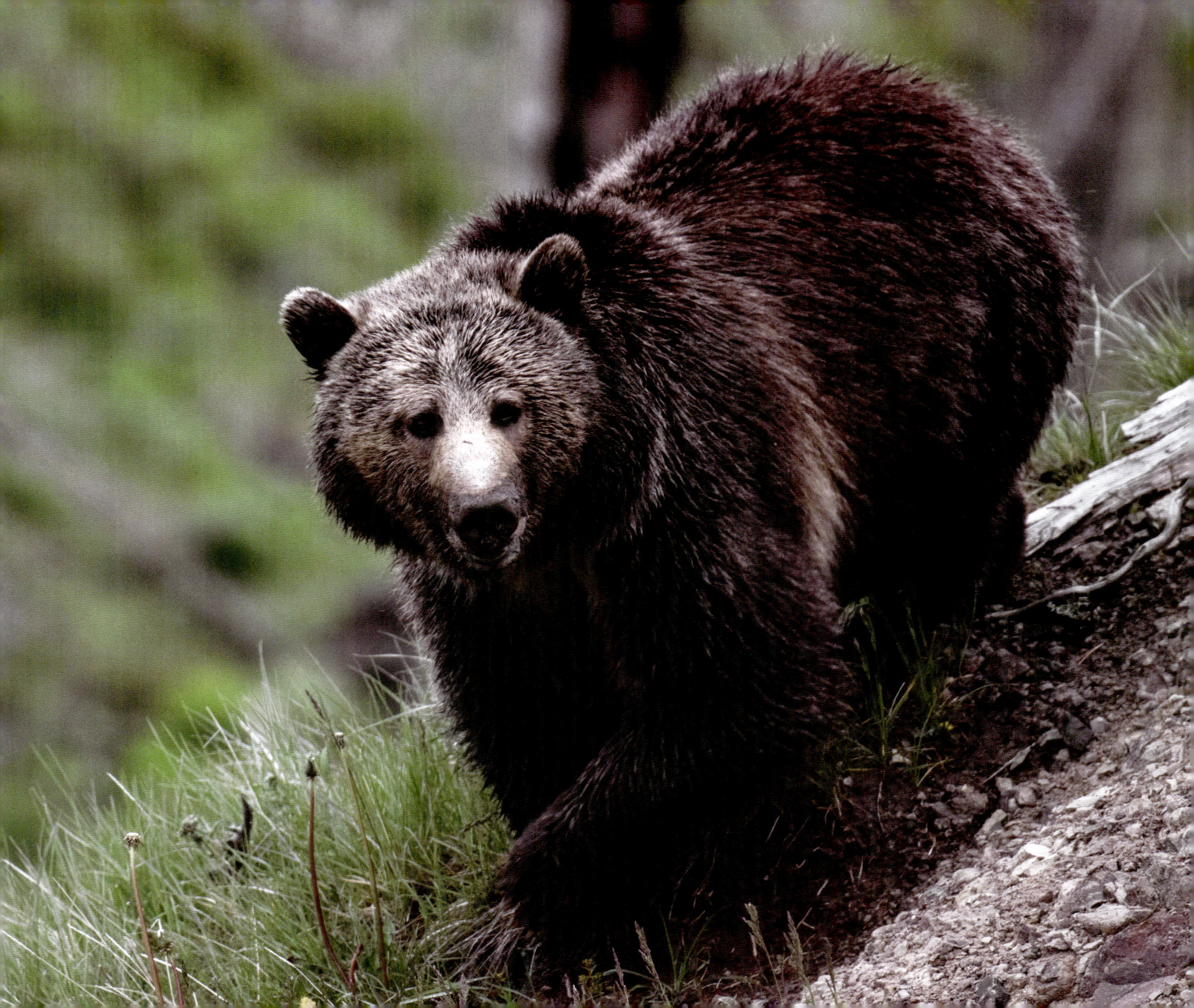

A BEAR *Called* RASPBERRY

IN 2007 ONE OF MY FAVORITE YELLOWSTONE GRIZZLIES WAS BORN. This grizzly is known as Raspberry, and she and her two siblings were born to a sow named Blaze. As far as Yellowstone grizzlies go, Blaze was a well-known bear, and a favorite to many.

Shortly after appearing out of the den, Blaze and her three COY (cubs-of-the-year) set out on a journey and headed toward Mary Bay. Unfortunately, not long after reaching the bay, one of the three cubs disappeared, leaving Raspberry with only one sibling. This sibling eventually became known as White Claws. Raspberry and White Claws grew up around Yellowstone Lake, and you could often find them alongside their mom in the areas of Mary Bay and Steamboat Point.

In an effort to protect their cubs, grizzly sows will often stay in a location close to the road. This is because the big male bears generally don't like the road, nor do they like the crowds of people. Most of the time the boars stay in the back country until it is mating season and then they venture out in search of a mate.

Blaze took good care of her cubs. They stayed in the lake area most of the time, and many park visitors were able to watch as Raspberry and White Claws grew into their teenage years.

By 2009 it was time for Blaze to mate once again. So, after being with their mother for two and a half years, Raspberry and White Claws were basically kicked out the door and sent into the world to start a life on their own. Even though the area around the lake was officially Blaze's territory, the two could still be seen there on a regular basis.

Not long after, it was said, that White Claws got into some trouble and had to be euthanized. Then in another sad turn of events, in 2015, a bear that might have been Blaze was also euthanized. The story is that a hiker came between her and her COY, and as all mama bears do, she protected them at all costs. I can't say for sure if it was actually Blaze or White Claws in either of these situations, but those are the stories I've heard.

Thankfully, after leaving her mom, Raspberry thrived. She managed to make it through her teenage years without getting hurt or getting into trouble, and she eventually grew into a beautiful adult grizzly. Just like Blaze, she made her home around the lake, and has delighted many, many visitors as they catch glimpses of her.

I have had the opportunity to watch her on many occasions. One of my favorite memories of Raspberry is from October of 2014. I, along with my mom, my son, and my young daughter, watched as Raspberry foraged along the hillside, crossed the road, walked along the shore of Yellowstone Lake, and then went for a swim. It was a beautiful sunny afternoon and the sun glistened on her coat as she moved in the water. When she got out, she shook the water off, approached and smelled a tree, and then stood on her hind legs and scratched her back on the trunk. She then continued down the shore, crossed back to the other side of the road, and disappeared over the hill in Mary Bay.

Not long after that particular sighting, it was time for Raspberry to head to her den. Many photographers speculated and wondered if she would show up with COY the next spring, or if once again she would be alone. Even though she had been seen mating earlier in the year, it is never a guarantee that a sow will show up with cubs.

The wait from winter to spring was a long one, and the wait for Raspberry to come out of her den was even longer. As the snow started to melt, many other bears appeared, but there was still no sign of Raspberry. People started to wonder if maybe she hadn't survived the winter, but finally, after what seemed like forever, Raspberry showed up and she was not alone! She had two beautiful COY in tow. One was dark in color and the other was very light. This duo became known as Rocky and Snow.

Seeing Raspberry with cubs of her own was a highlight many of us had waited a long time for. It was as if she had finally come full circle and had overcome all of the hardships that grizzly bears face in the wild.

Unfortunately, Rocky only survived a short time. He (it was assumed this cub was a male) just disappeared and no one can say for sure what happened to him.

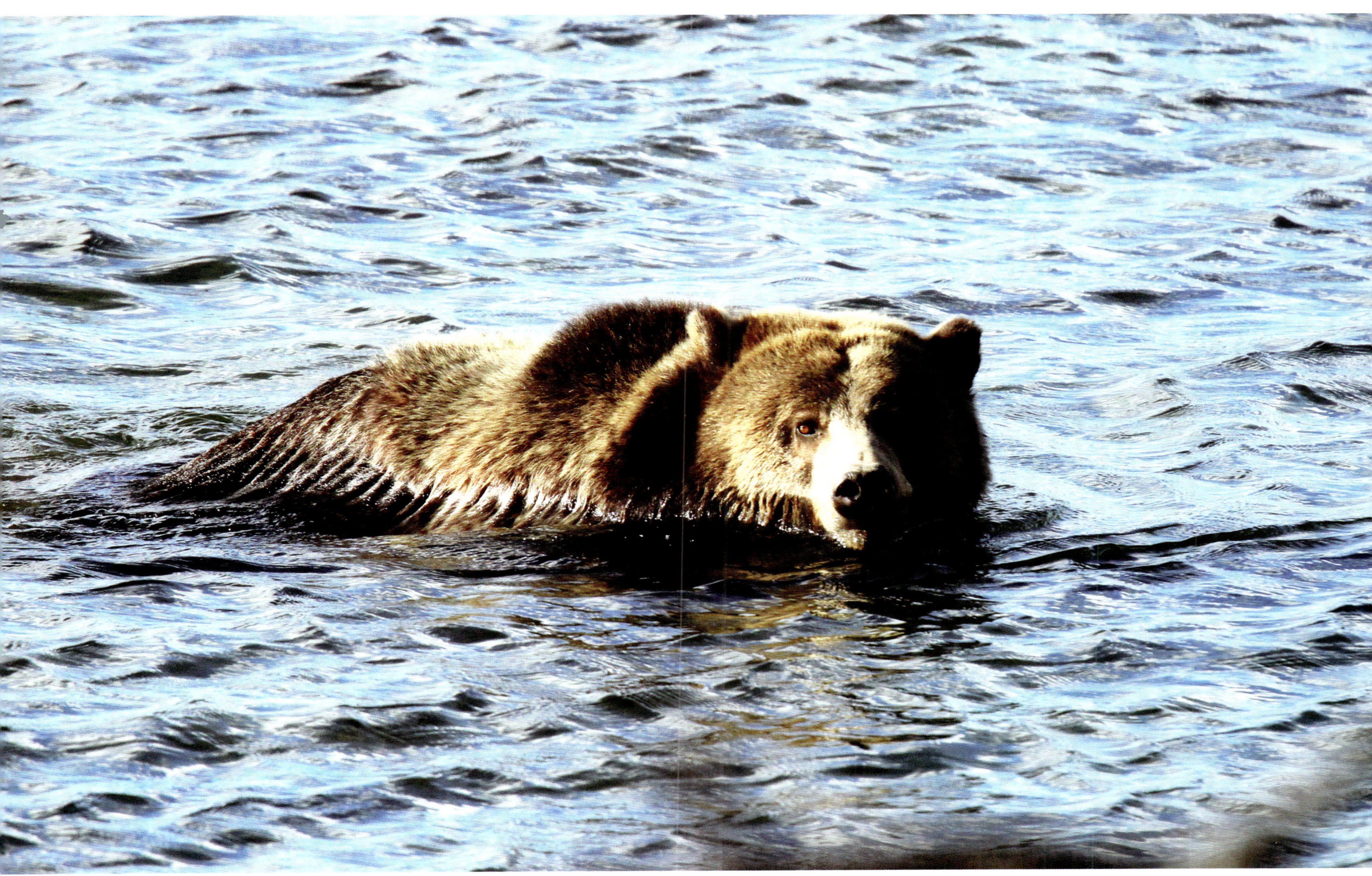

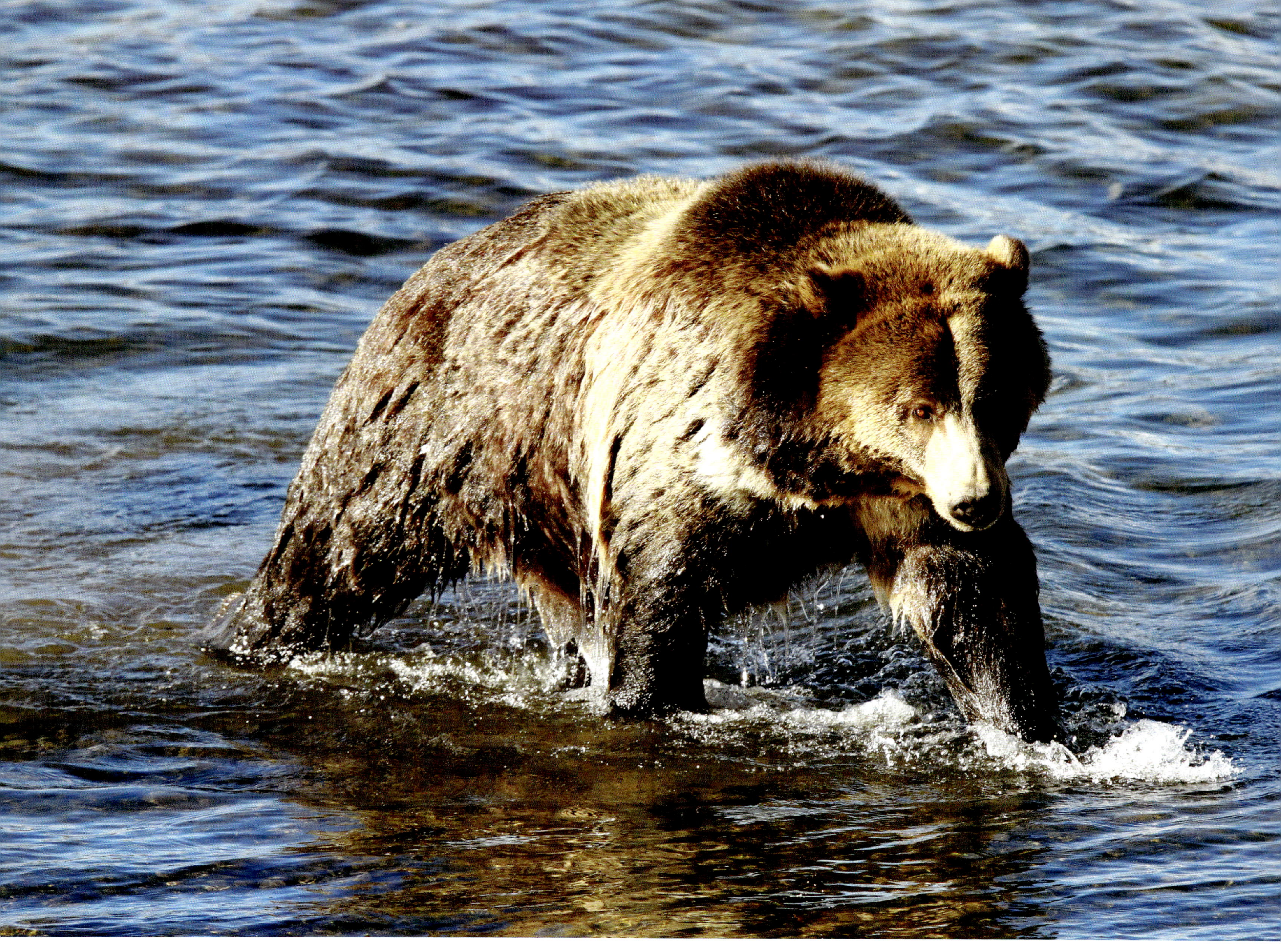

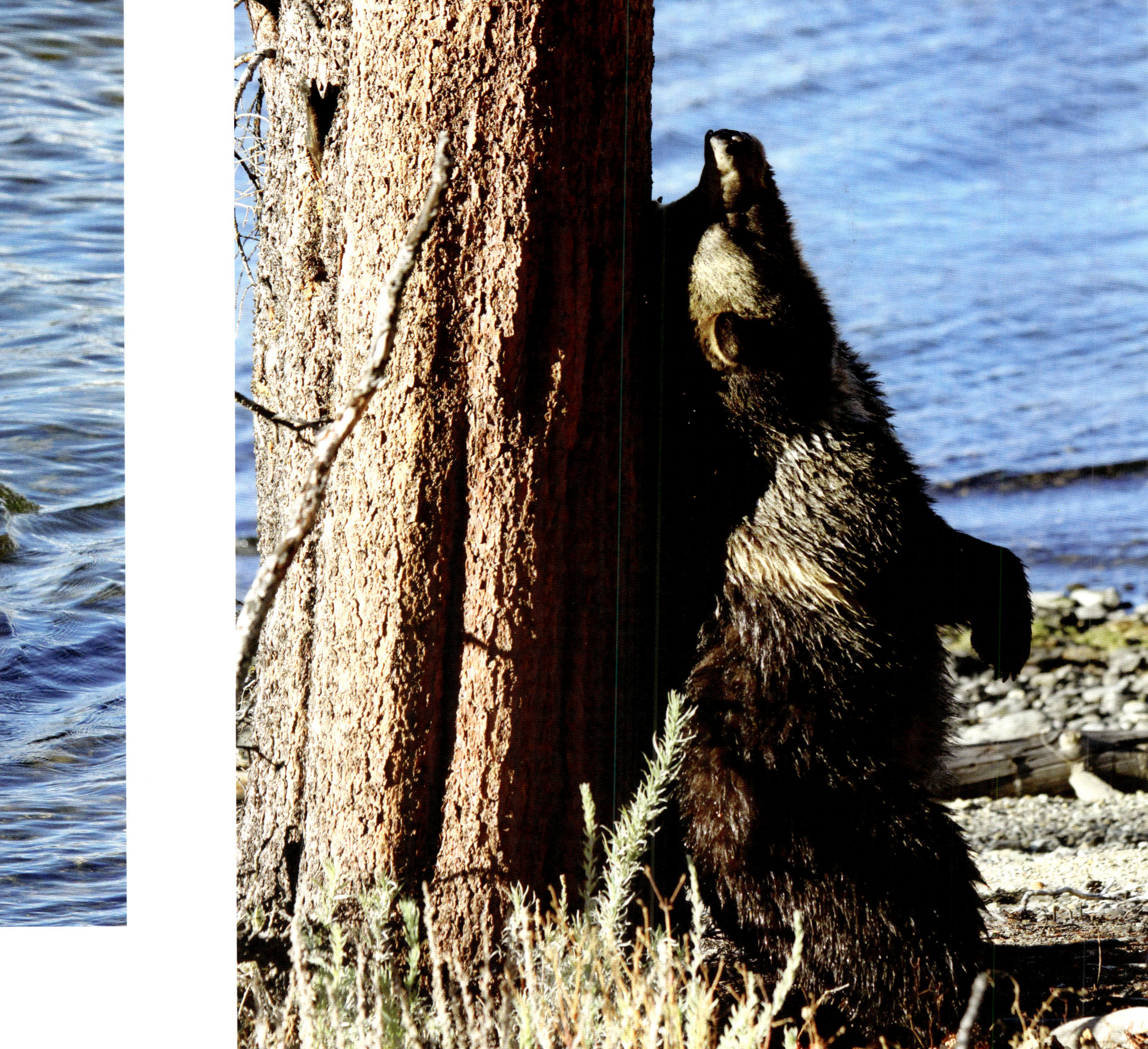

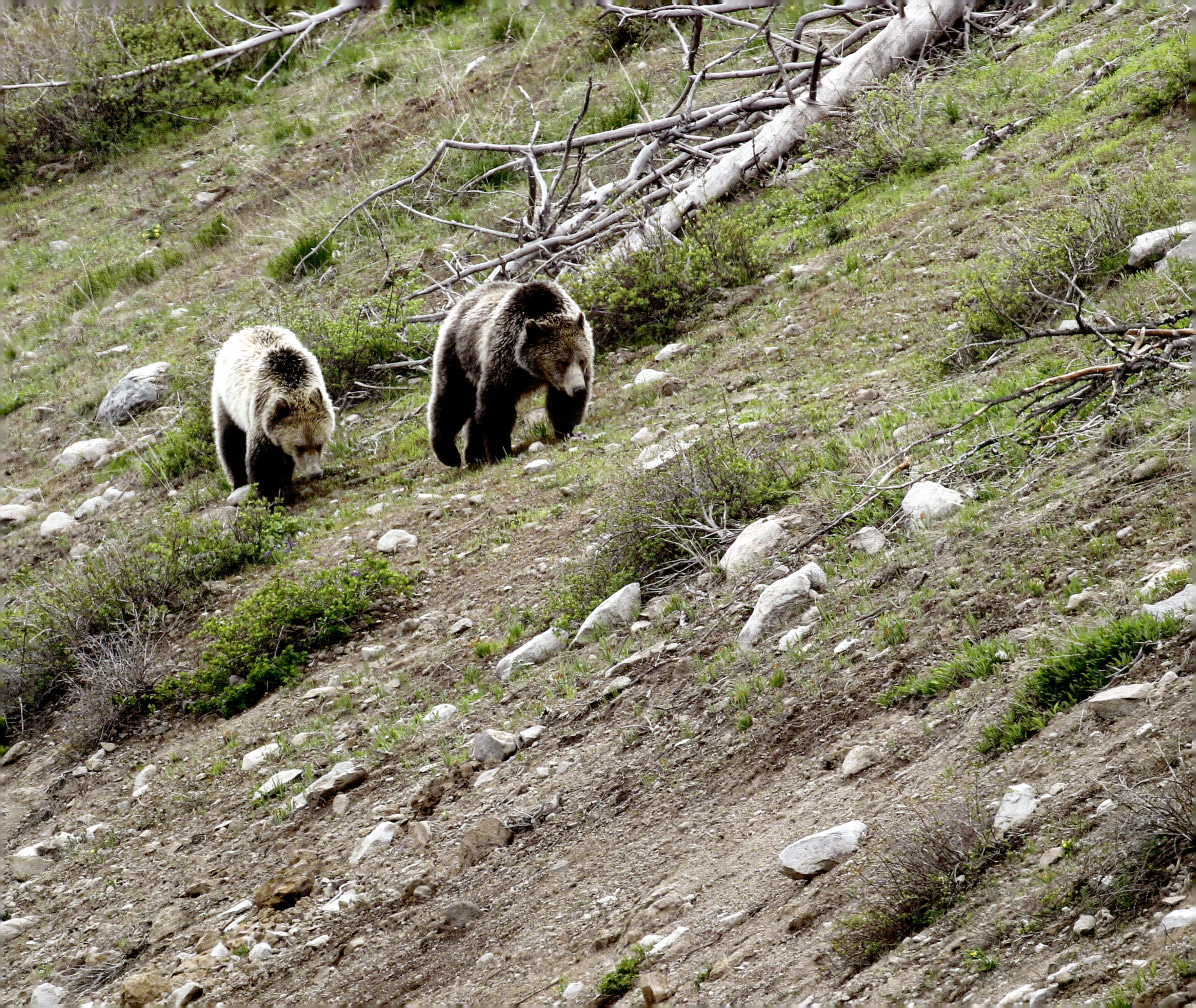

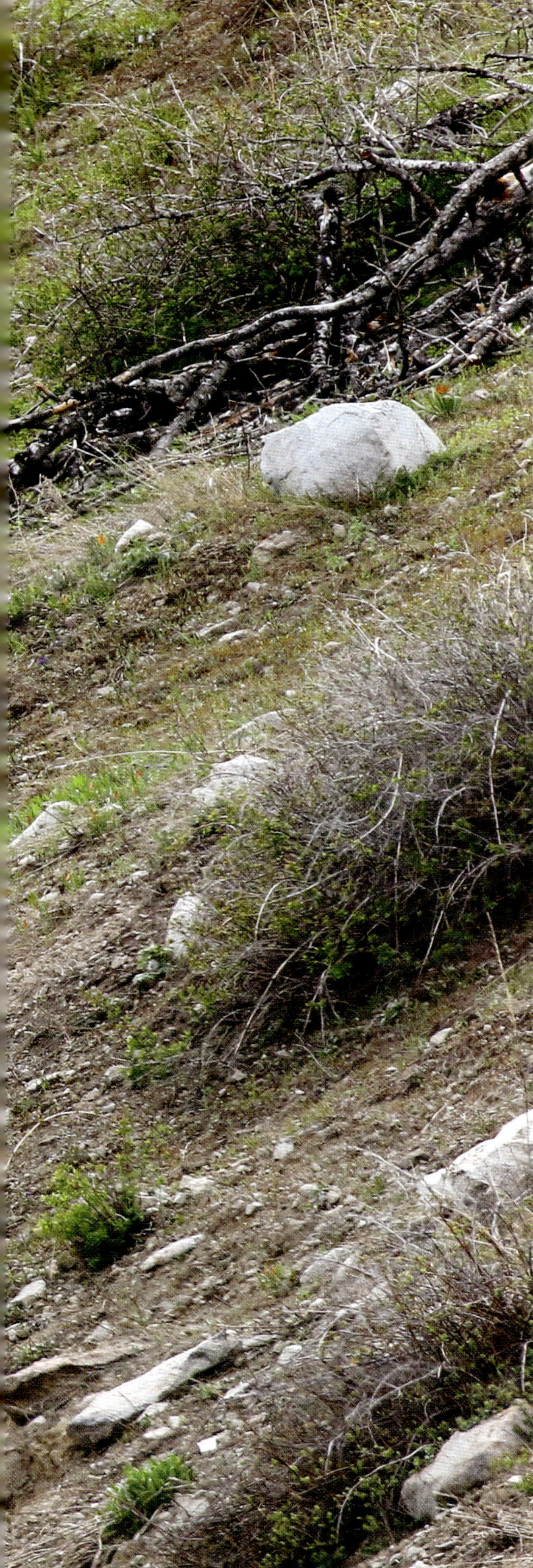

Snow on the other hand is an entirely different story. She became the apple of mom's eye, and it seemed as though Raspberry never let her get too far out of sight. The bond that developed between these two bears was stronger than any other mother and cub I have ever seen. So strong in fact that when the usual time frame of two and a half years came along, and it was time for Snow to be on her own, Raspberry avoided all of the big males and kept Snow around for another whole year. This is not something that is unheard of, but it rarely happens.

Eventually in the spring of 2018, Raspberry sent Snow out on her own. As sad as it was to watch this happen, it was definitely time. Raspberry had taught Snow all of the important lessons of life and given her the tools she would need, and it was now up to Snow to use them.

Raspberry was courted by many bears that spring. There were times that I almost felt sorry for her. Eventually her mating period ended and she disappeared to higher terrain for the summer. When she reappeared that fall, I was a little bit worried. I didn't think she looked very good. She didn't appear to be her normal fall weight and something about her just didn't seem right. Normally she would hang out for a while at that time of year, but for some reason she was only seen for a short time. She disappeared quickly, and it seemed she had headed to her den very early.

Thankfully she showed up that following spring, and she looked great. When she appeared, she had no COY with her. I wasn't surprised at all considering how she had looked the previous fall, and I was honestly thankful that she didn't have any cubs. This would give her plenty of time to herself, and I believed she needed it.

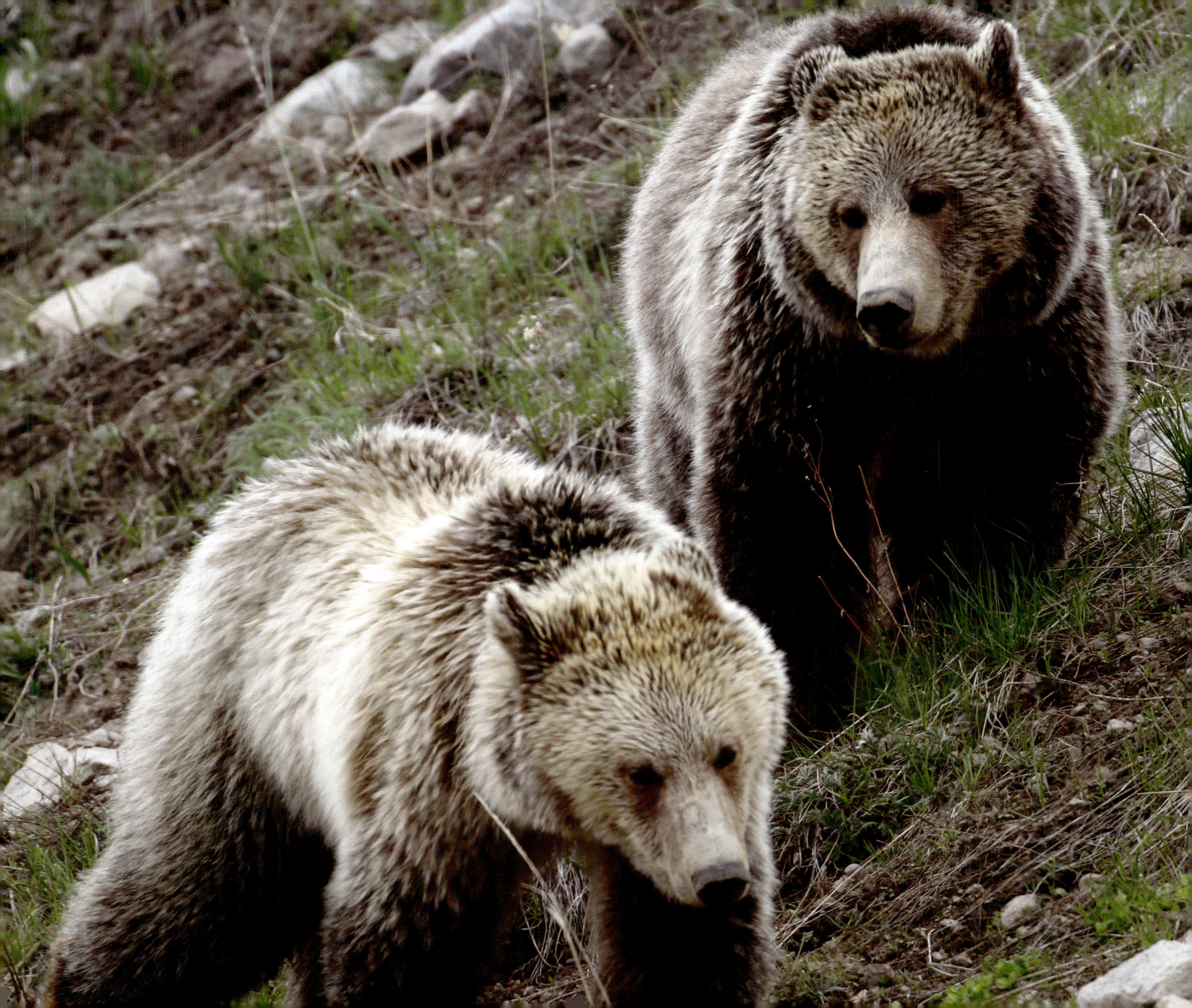

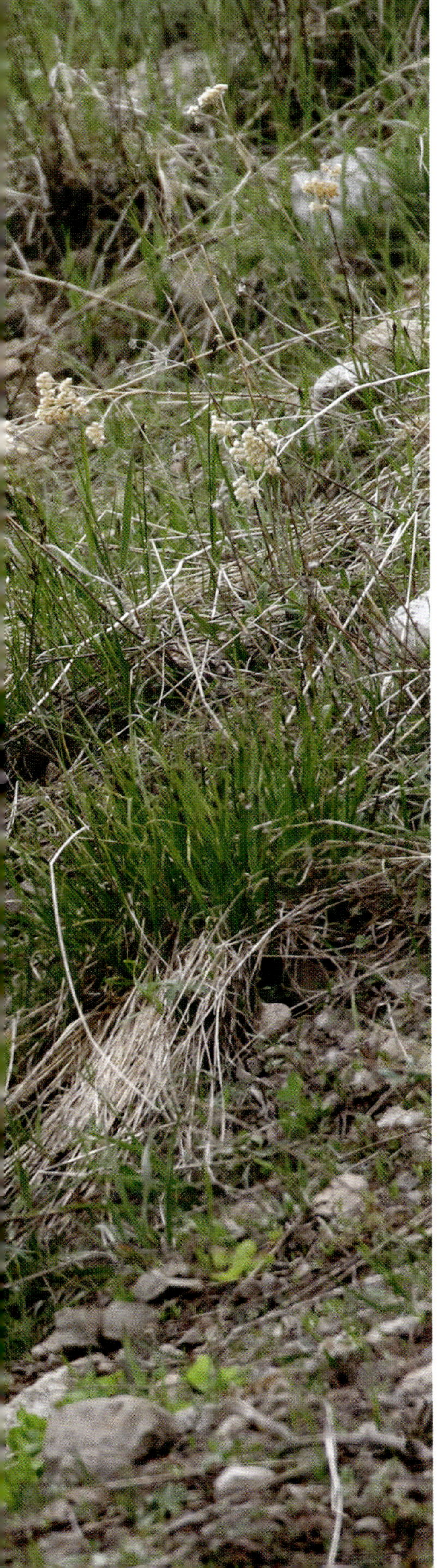

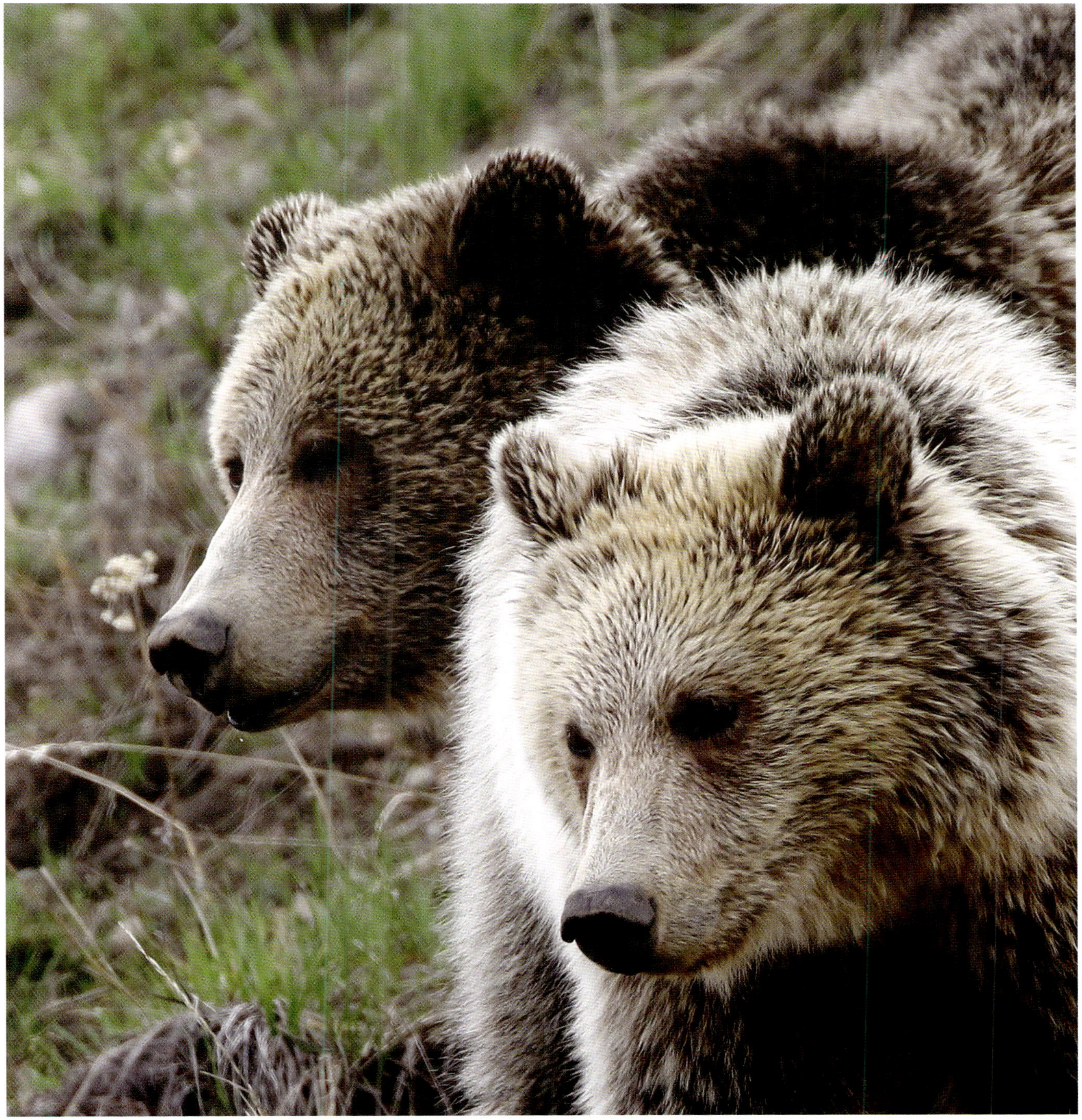

That year seemed to fly by. Raspberry mated and went to higher ground for the summer. When she returned in the fall, she looked incredible. She hung out for a while in her usual areas, and eventually headed to her den.

In the spring of 2020, my daughter and I were driving along a road in Yellowstone and we noticed a truck pulled over and the people were watching something high up on the mountain. Initially I had driven past the vehicle, but at the last minute I decided to turn around and see what they were looking at on the hillside. I drove in the pullout, parked, and got out of my vehicle. I then walked up to the truck and asked what they were seeing. They said they had spotted a bear way up high, almost to the top of the mountain. I immediately went back to my vehicle, got out my scope, and pointed it in the direction they had shown me.

At that particular time, there had been no sightings of Raspberry. I knew the chances of that bear being her were very slim, and I assumed because of the location that it was going to be a big male. To my surprise, I was wrong. There on that snow-covered hill, way up high, was Raspberry, and she wasn't alone. Directly behind her, racing around in the snow, was one cub.

For me, that moment was surreal. I was instantly overcome with happiness and my eyes filled with tears. An old friend had survived another winter and brought another miracle into this world.

After a few days, Raspberry and her cub made it down the mountain and were seen in the usual spots. The love between them was obvious, and it warmed my heart watching them wrestle around and play together. It was another one of those moments with Raspberry that I will never forget.

As for that little cub, it looked just like her. And although there were many names being tossed around as to what it should be called, in the end, it became known to many as Jam. Raspberry's Jam.

Of all the bears I have watched over the years, Raspberry is the one who will be embedded in my heart forever. She has taught me so many things about my life, and also about the life of a grizzly. Without her, I would have never known that the love between a mama grizzly and her cub is as strong as my love is for my own children. We would move mountains for them and protect them at all costs. She has taught me about strength when being alone, and about enjoying the simple things in life like sleeping on logs, early morning walks on the beach, and those much-needed back scratches.

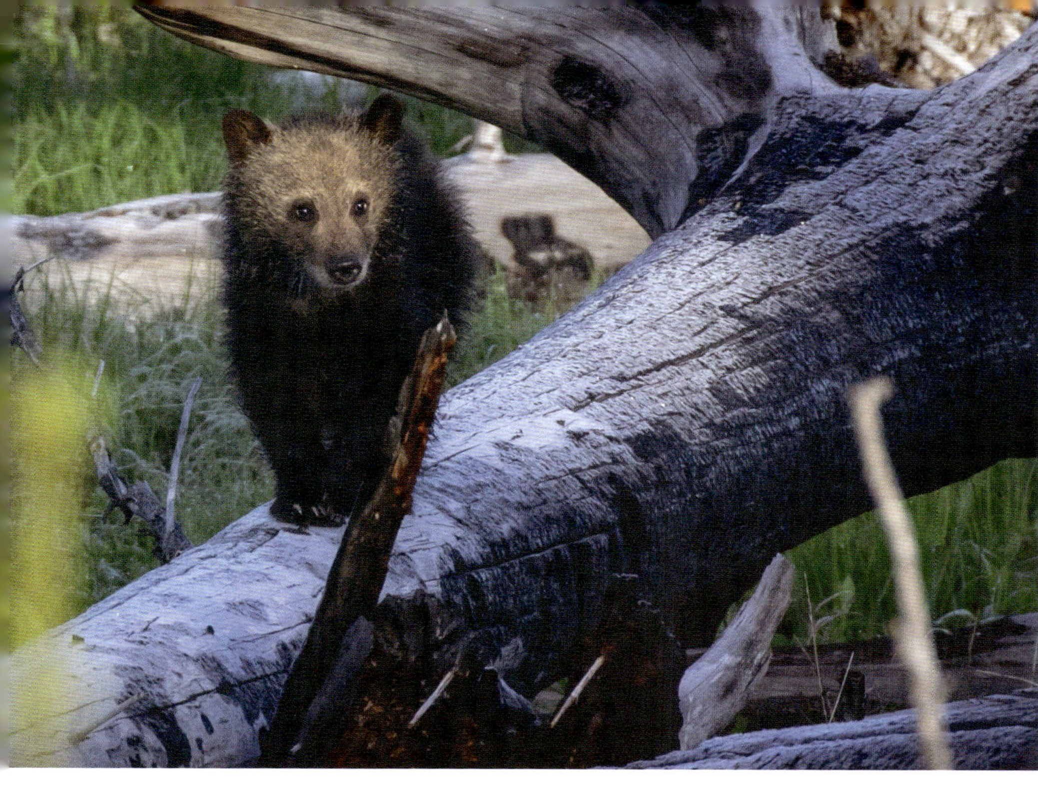
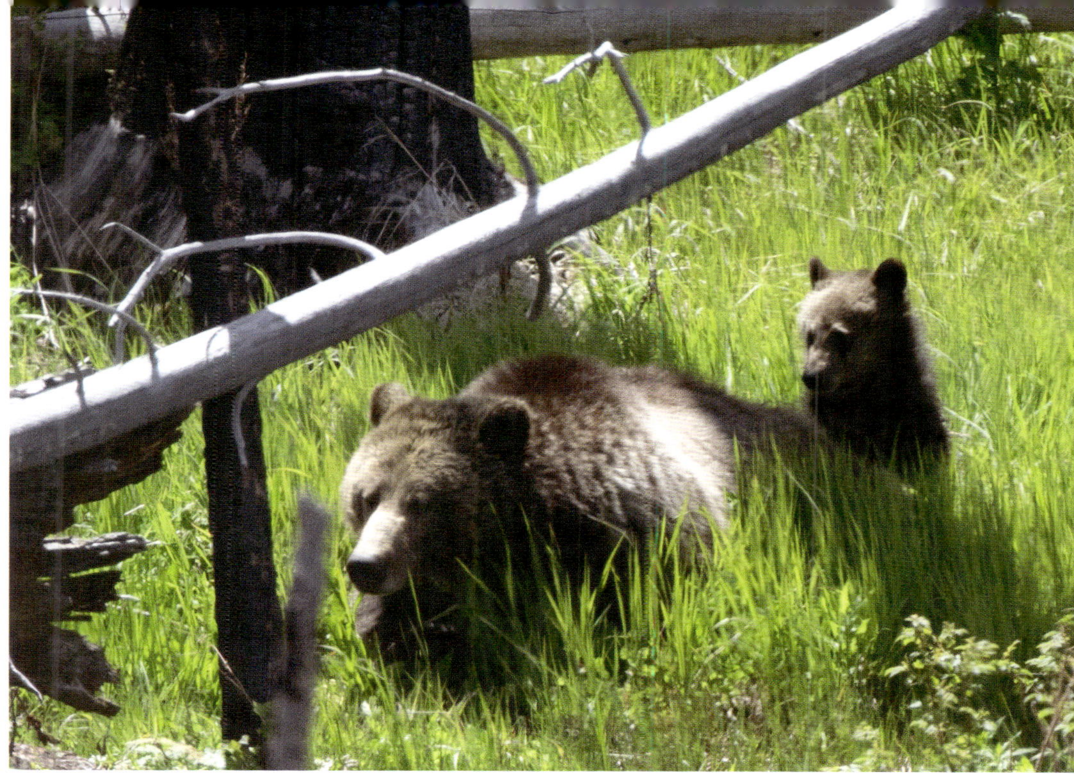
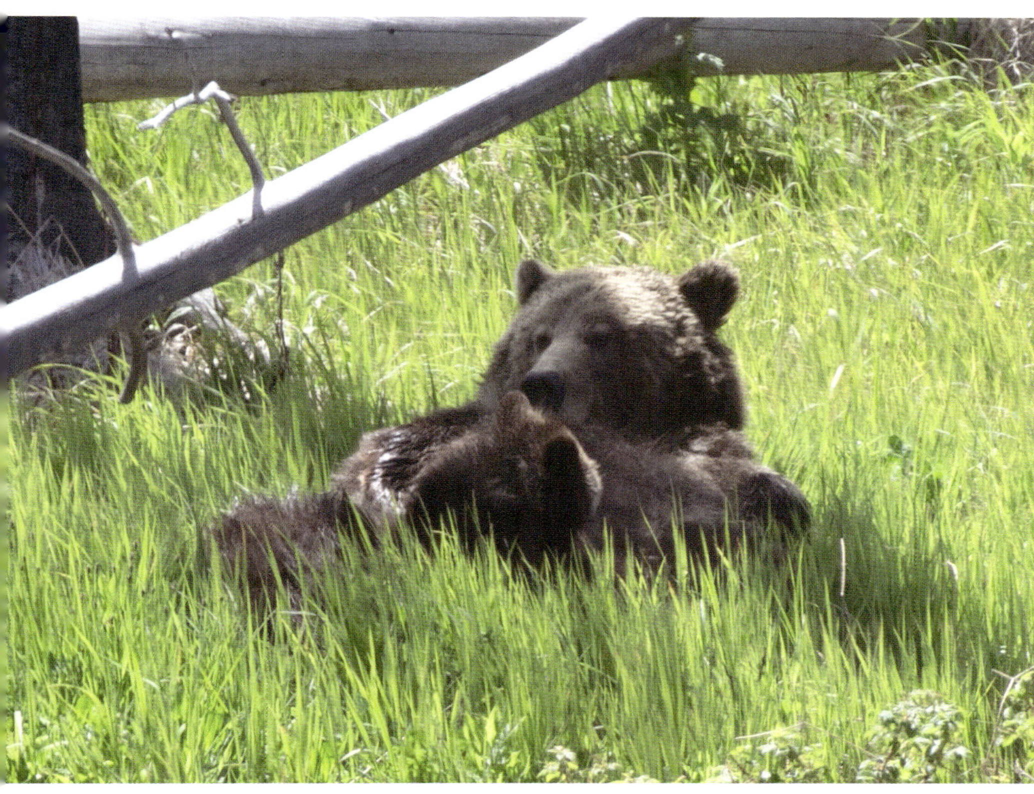
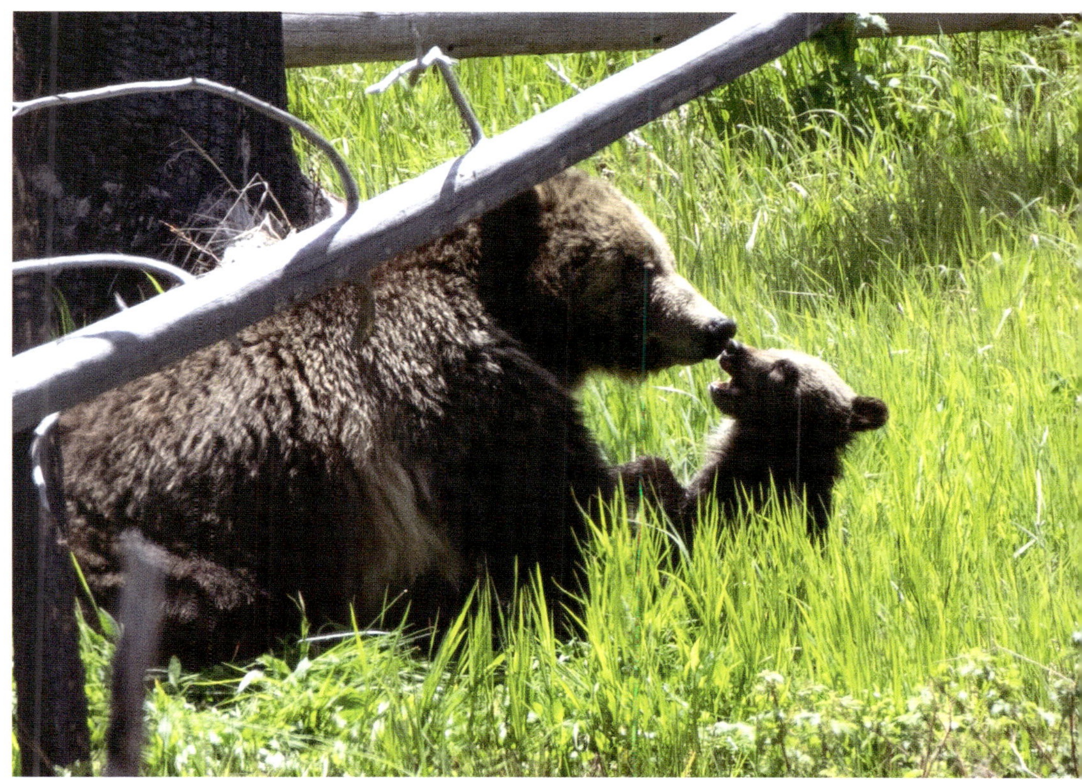

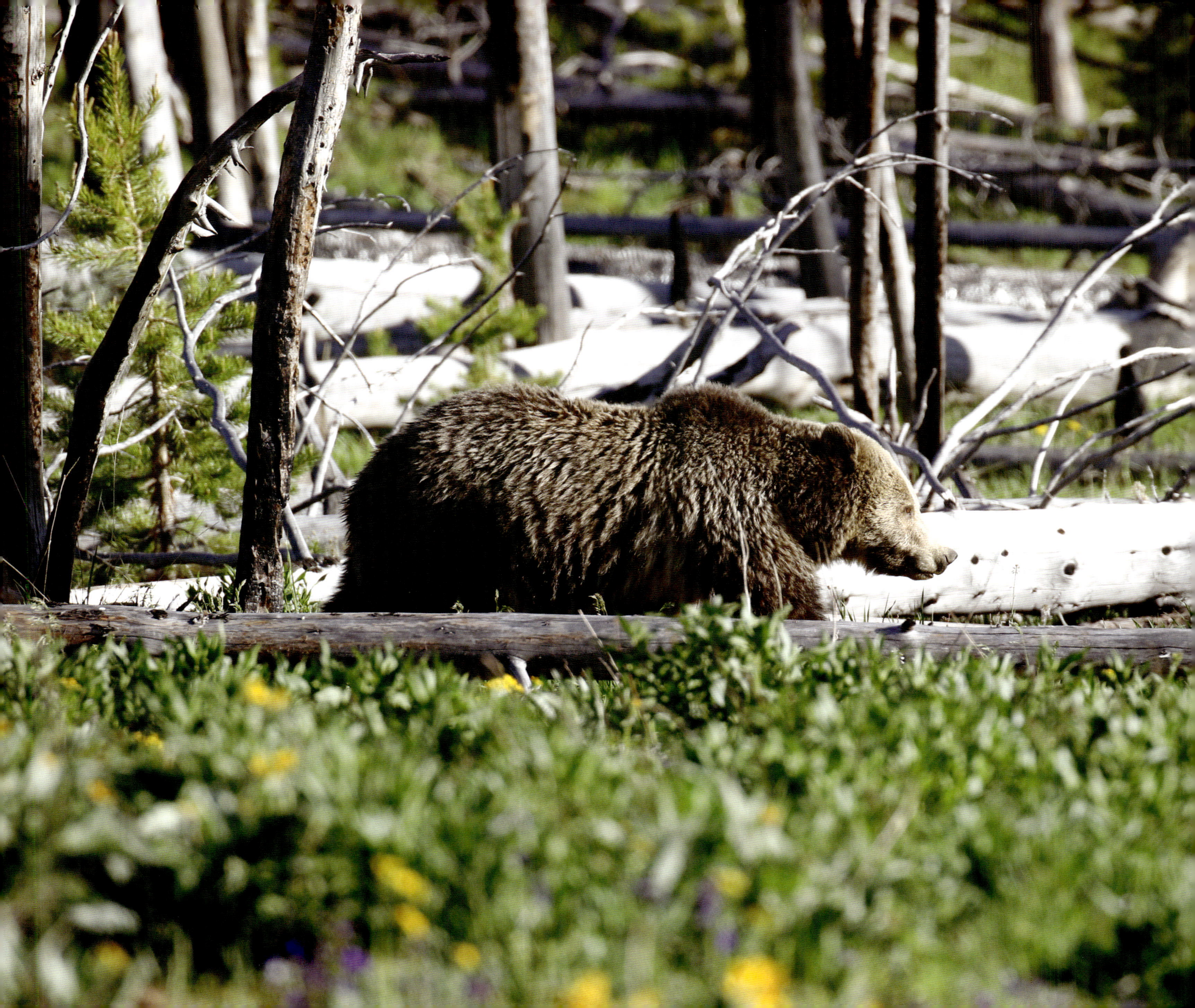

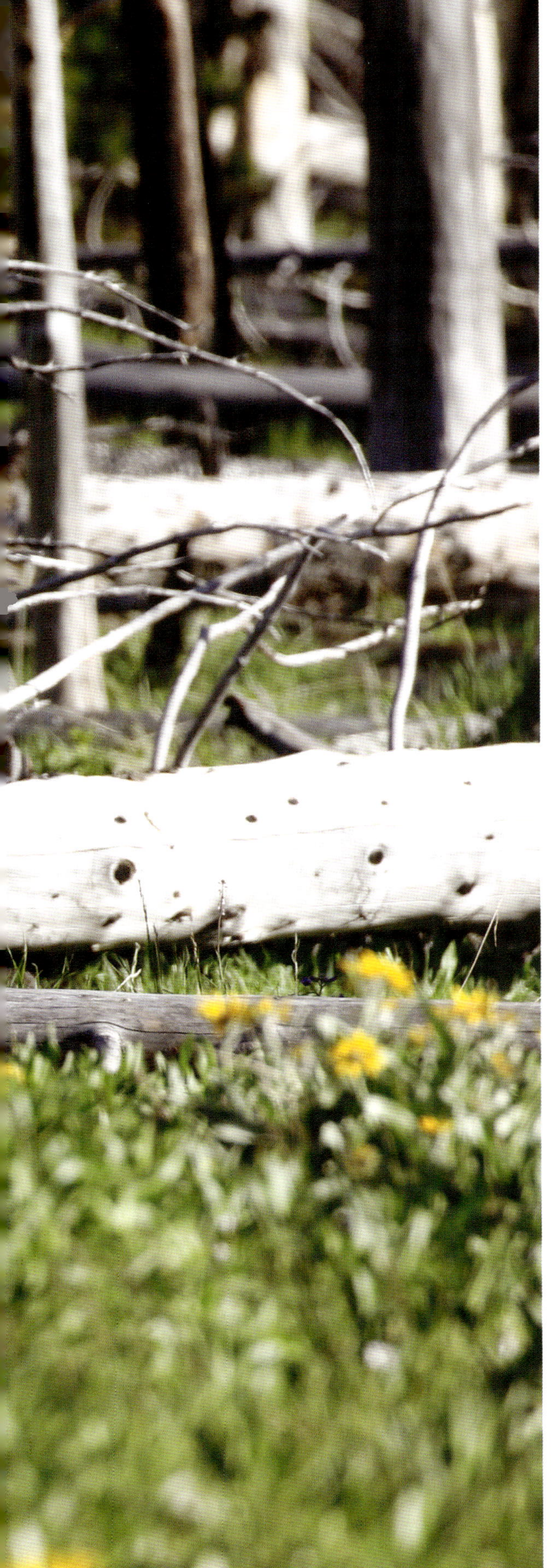

I am so thankful for the time I have been able to spend watching her, but most of all I am thankful for the gifts she has given me. Without even knowing it, she has brought peace and comfort, laughter and tears, and has made my heart smile on so many occasions. Perhaps the most important gift of all is the gift of knowledge—a very powerful tool when educating others about this incredibly amazing species.

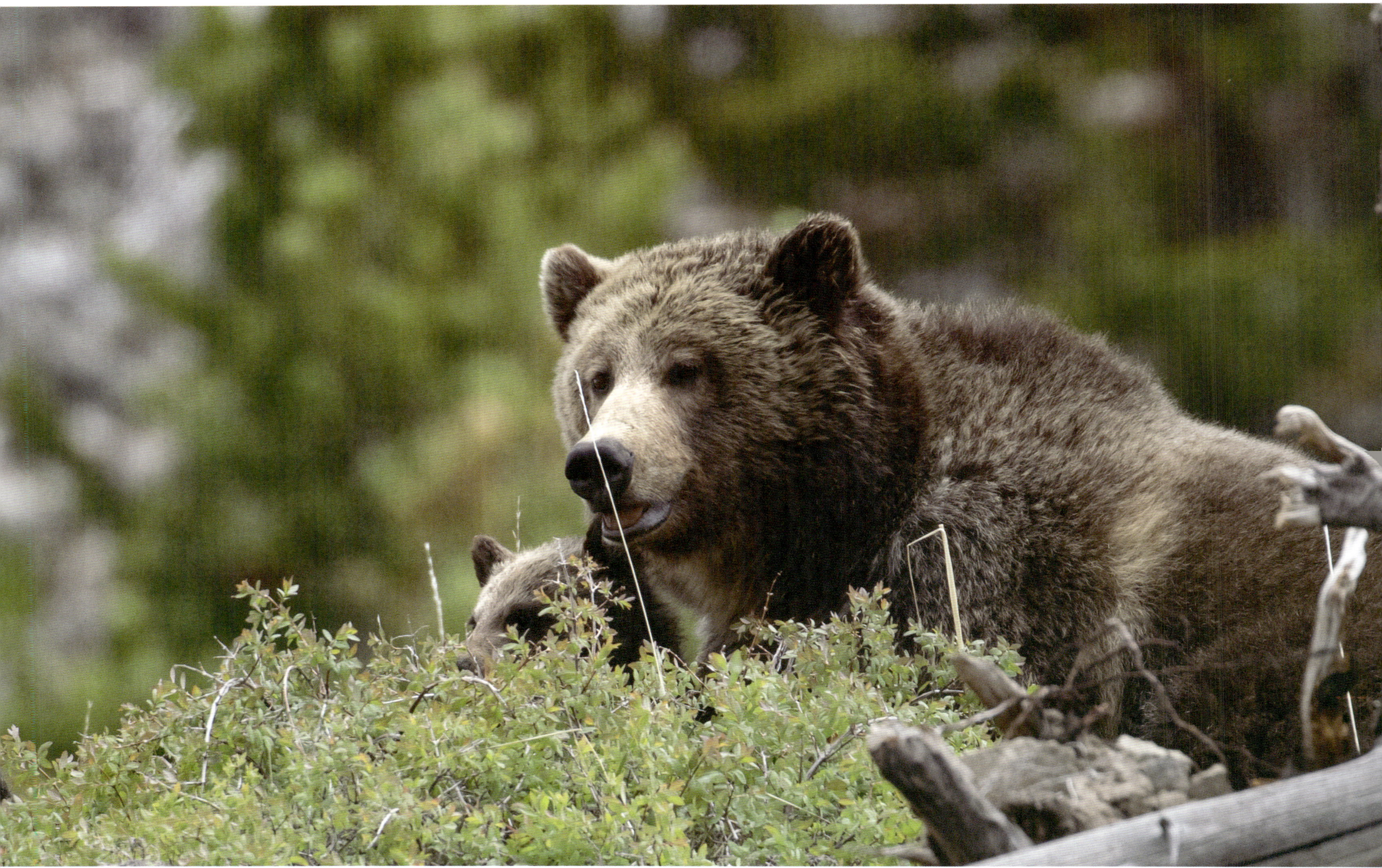

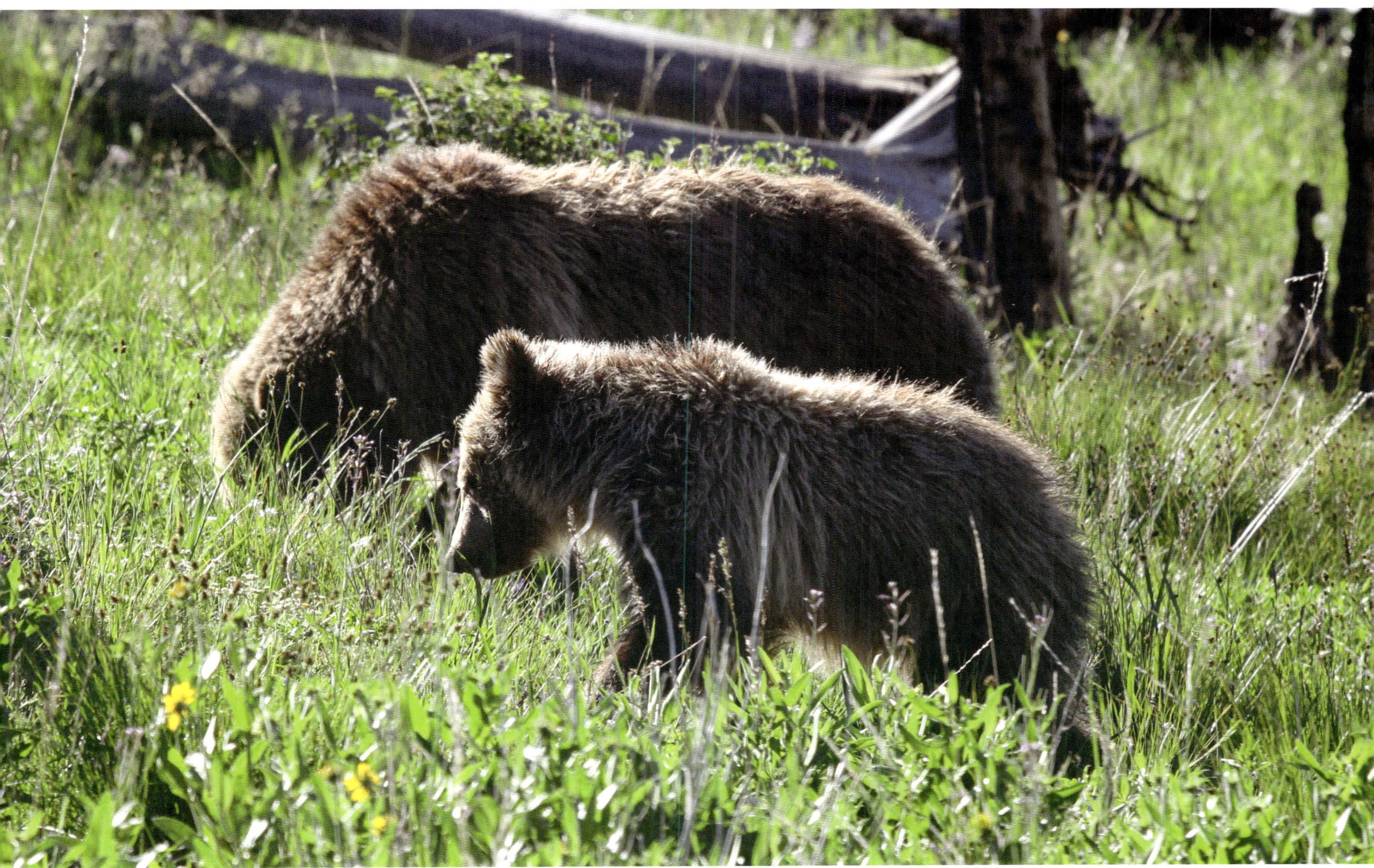

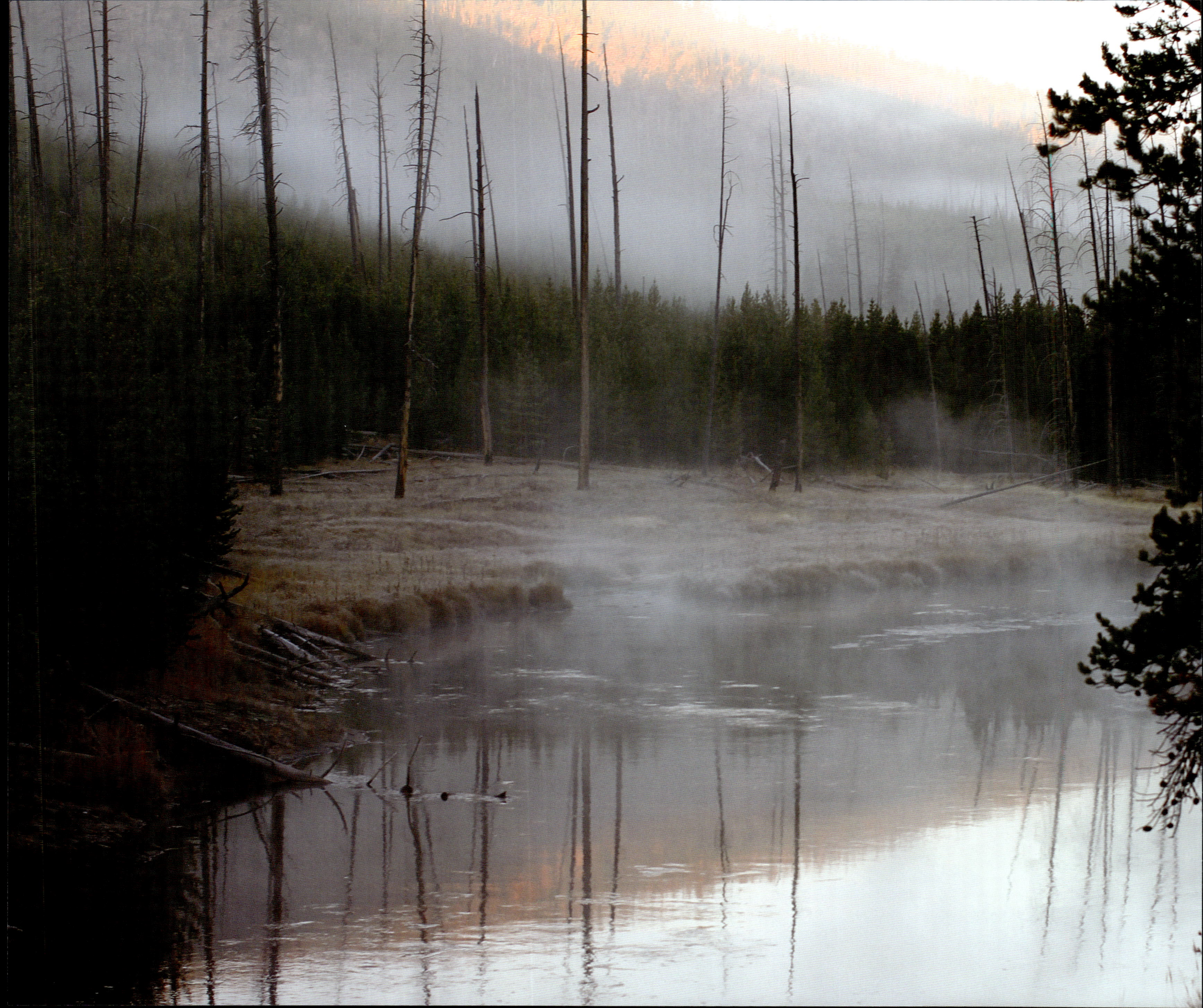

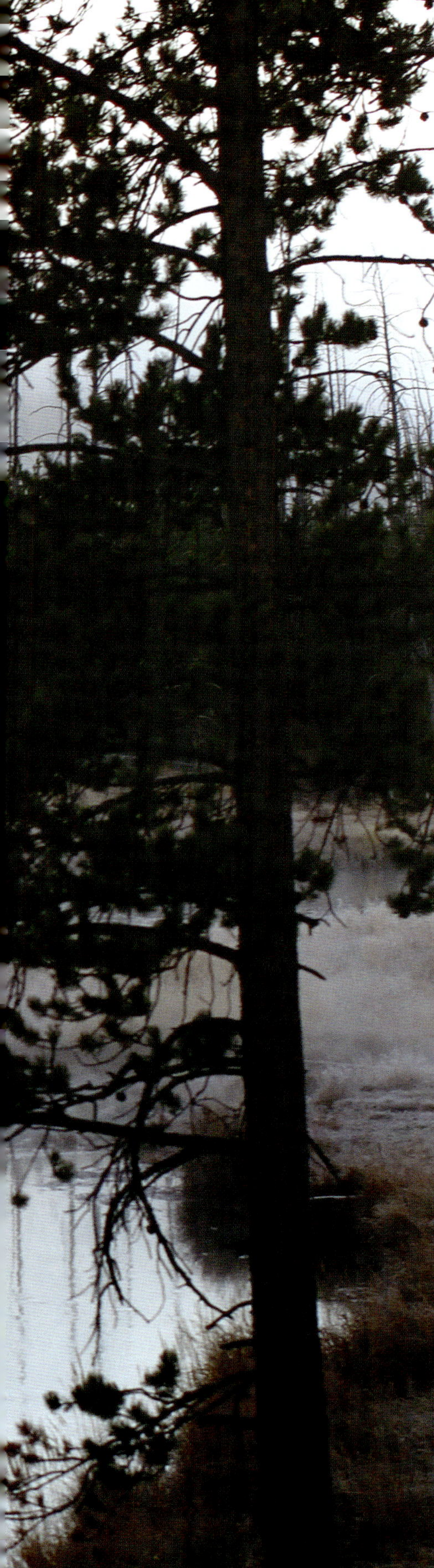

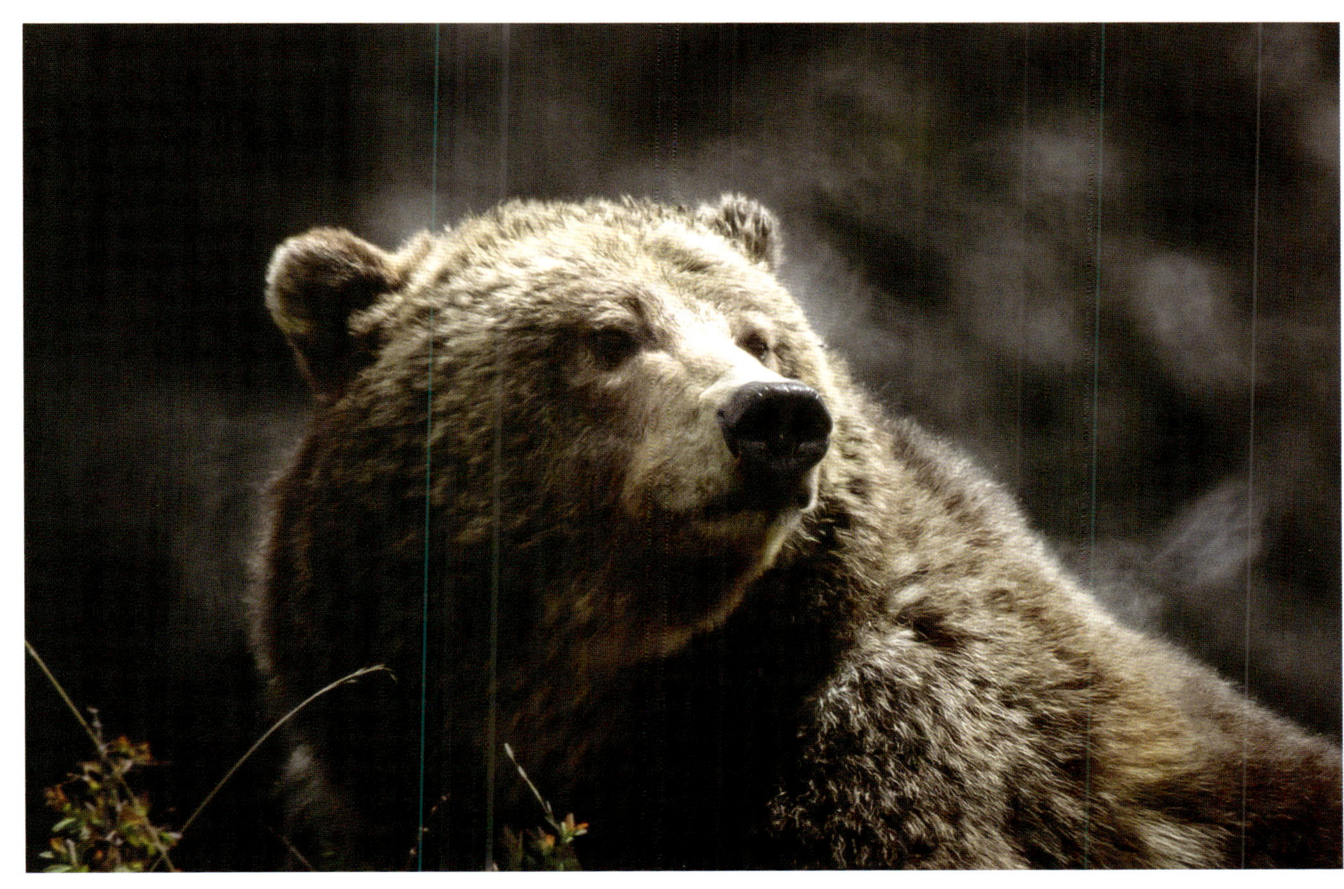

If you're going to be a bear, be a grizzly.

—*Mahatma Gandhi*

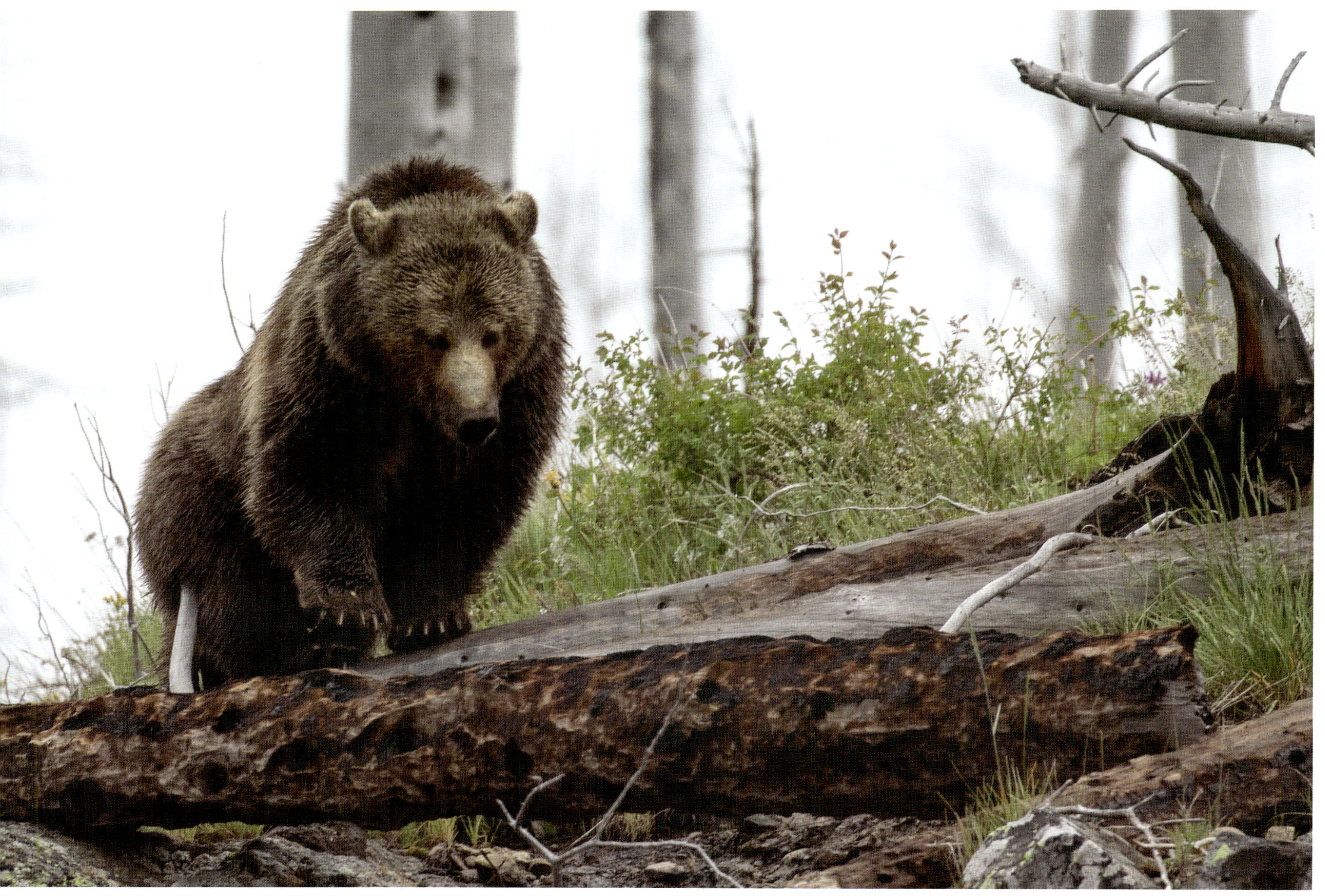

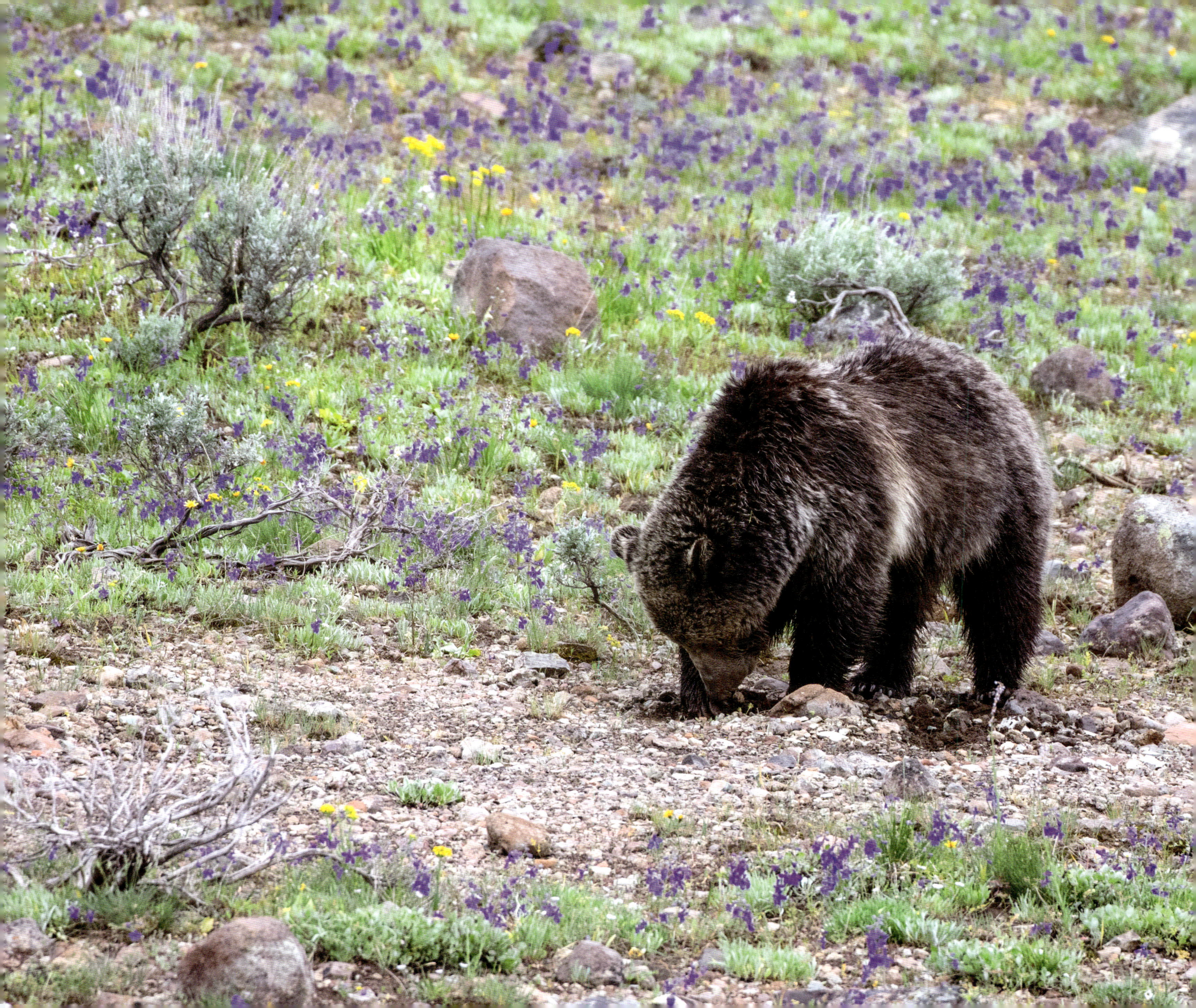

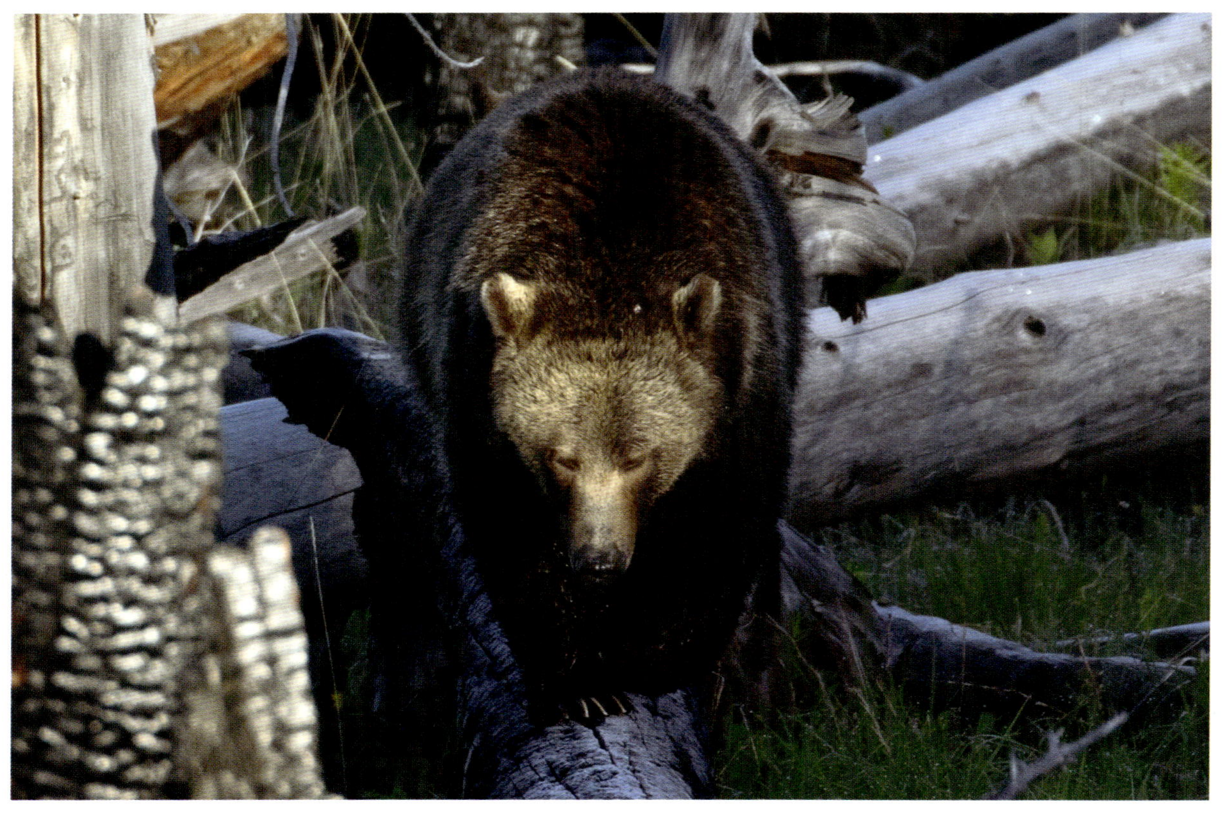

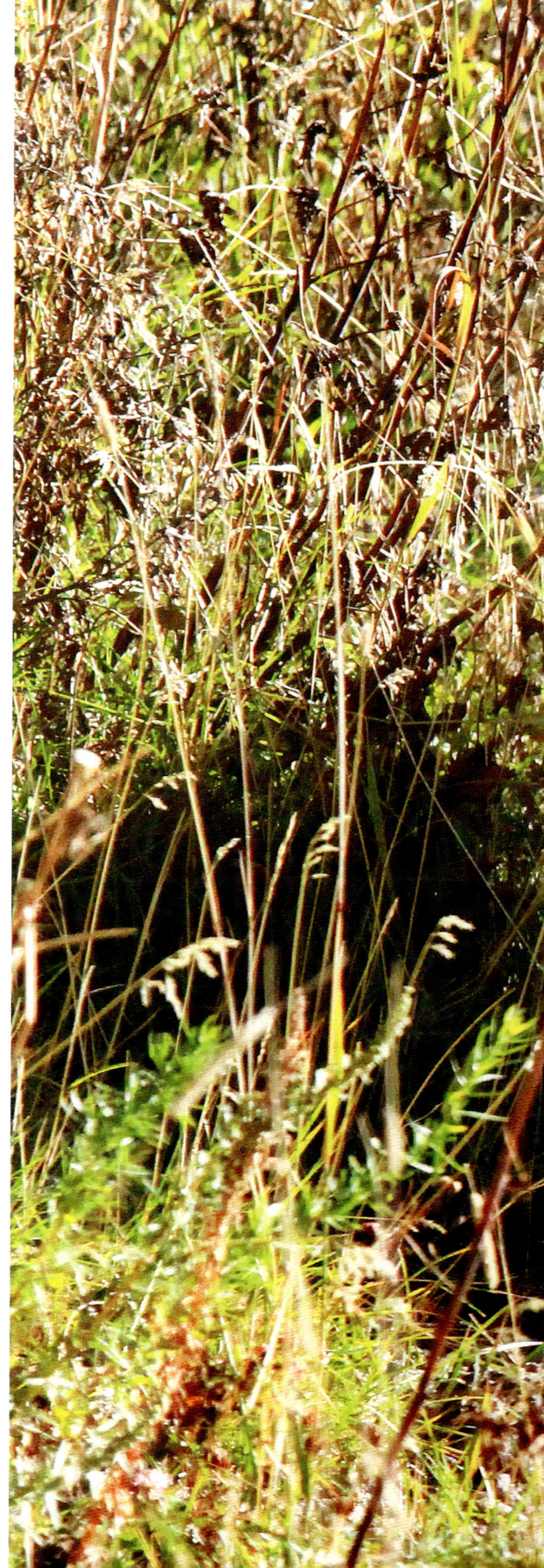

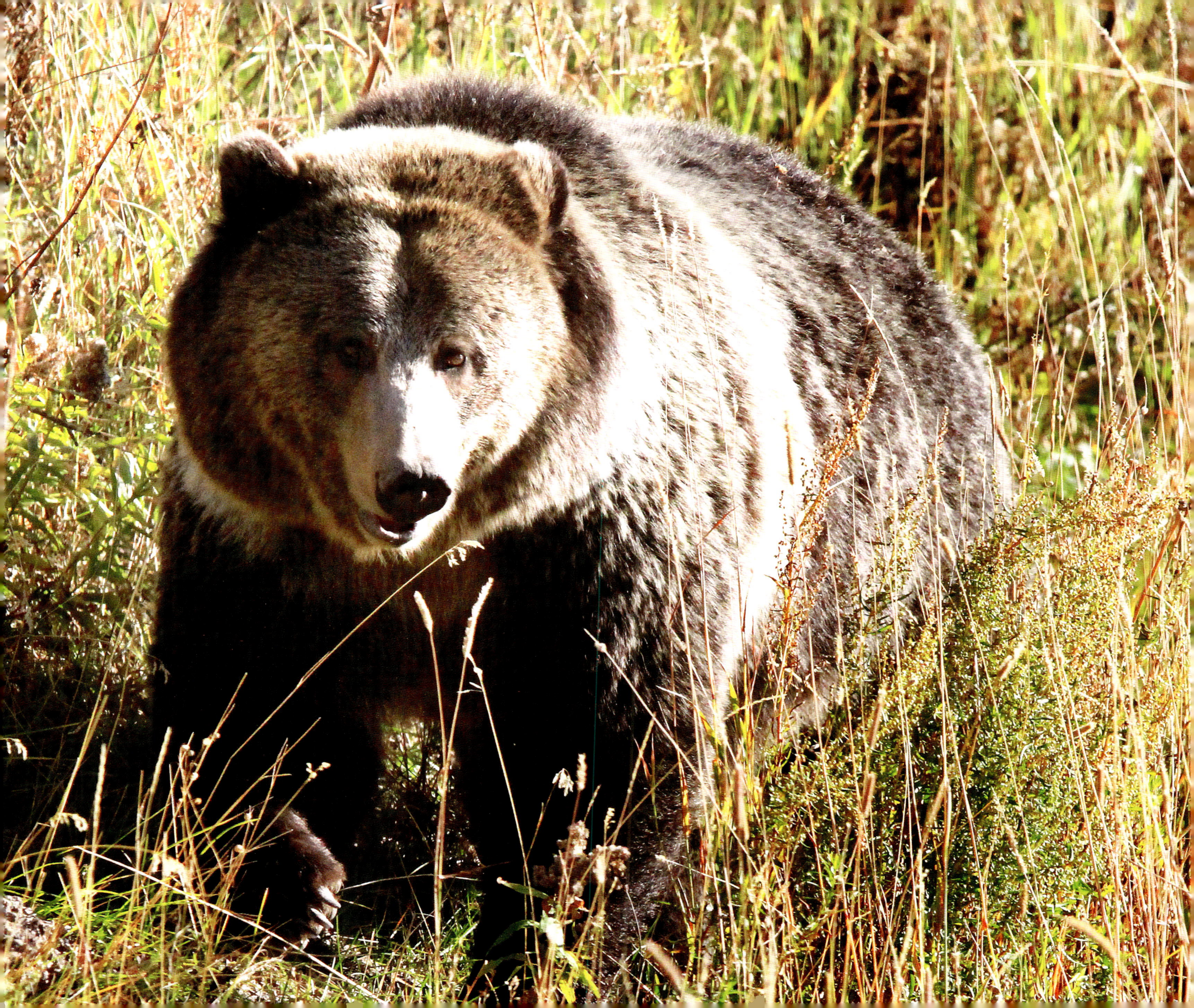

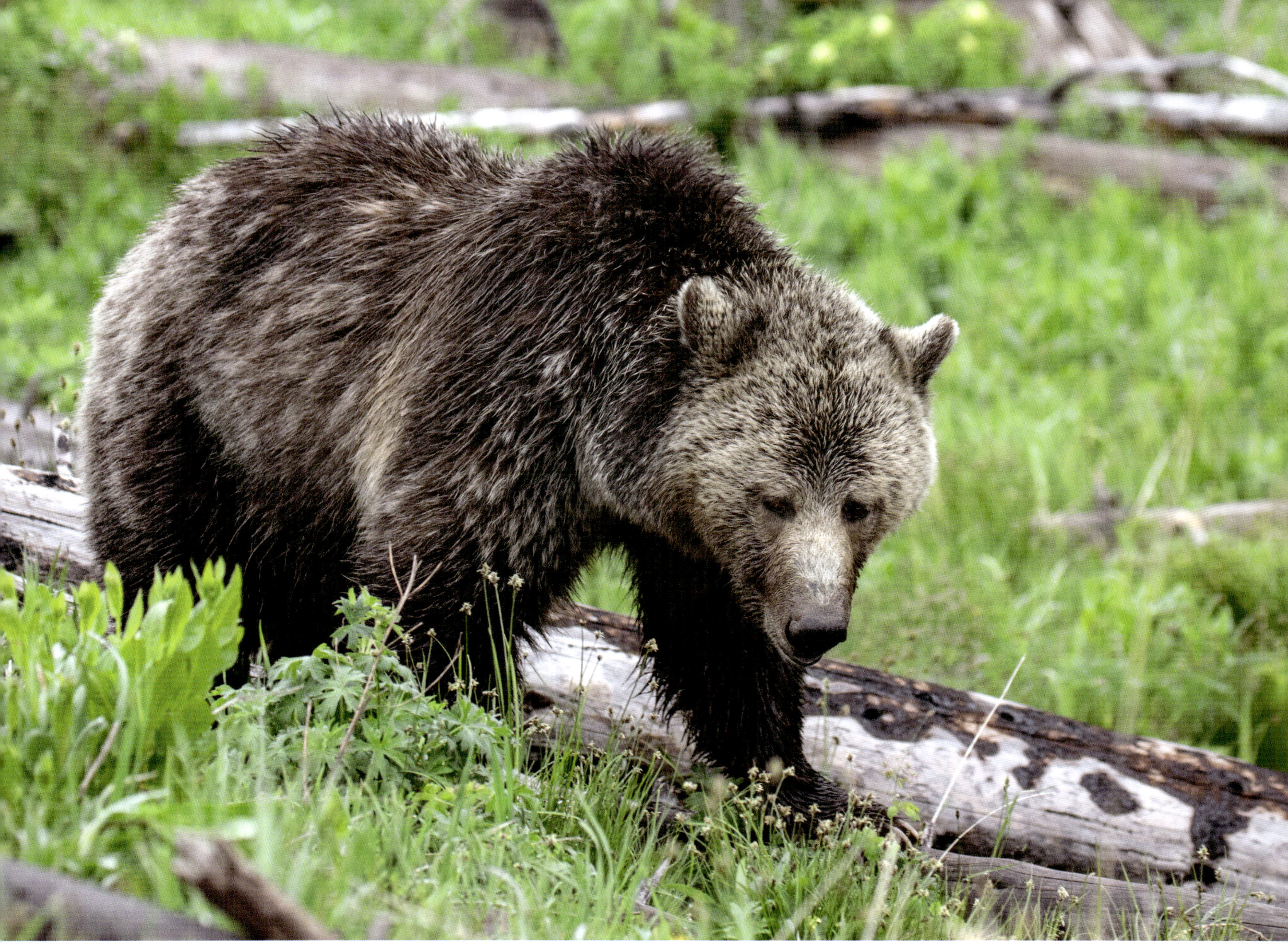

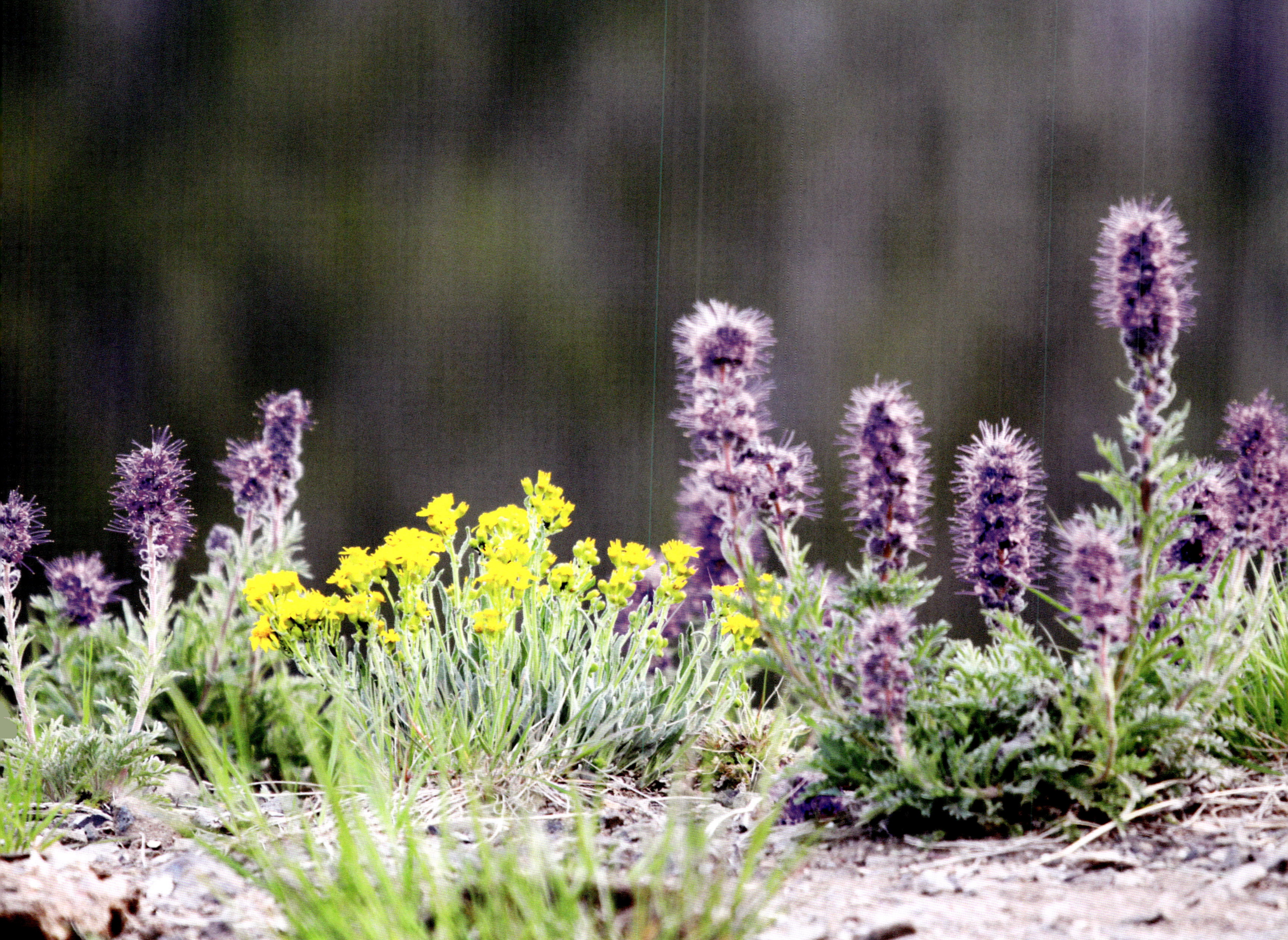

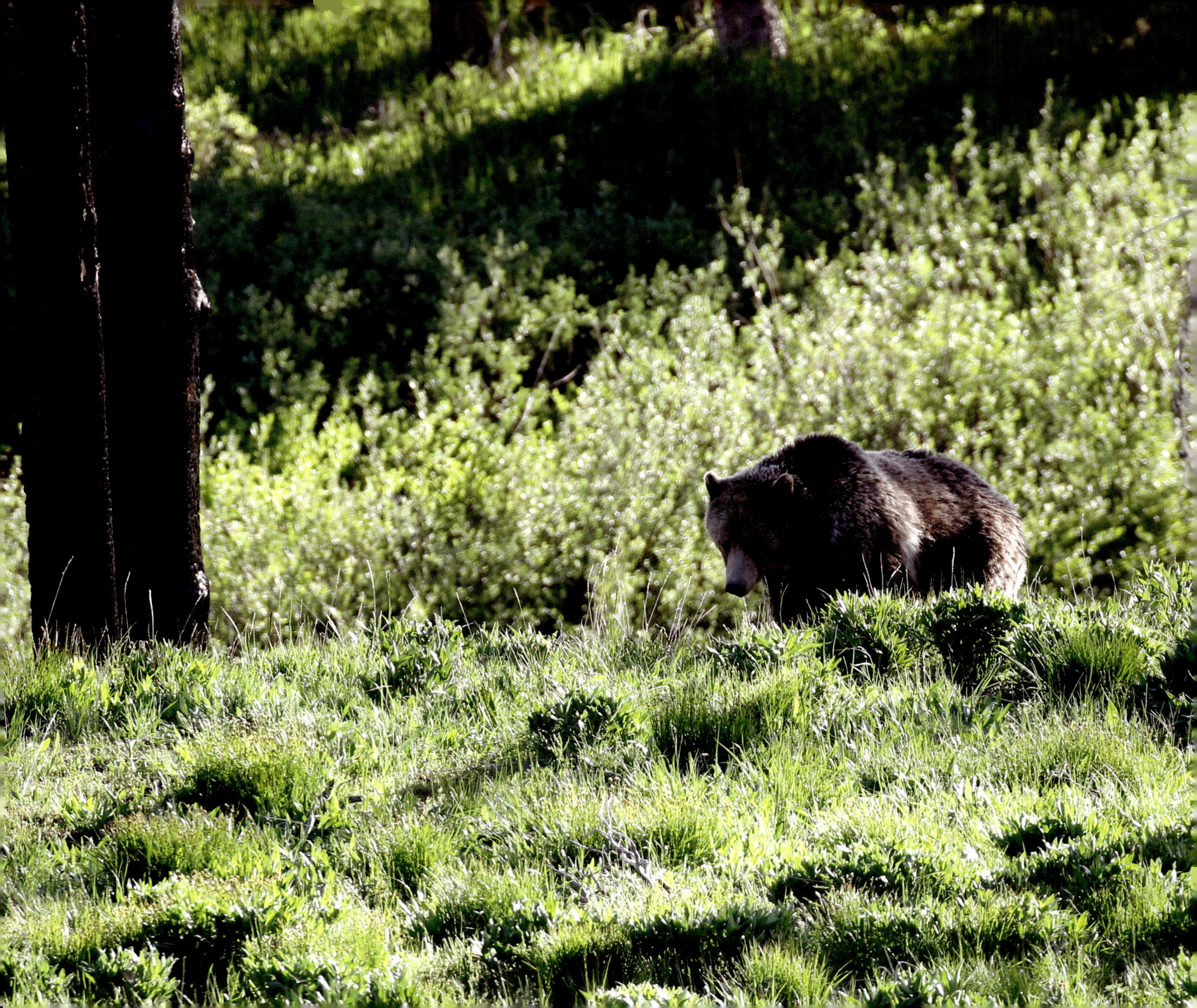

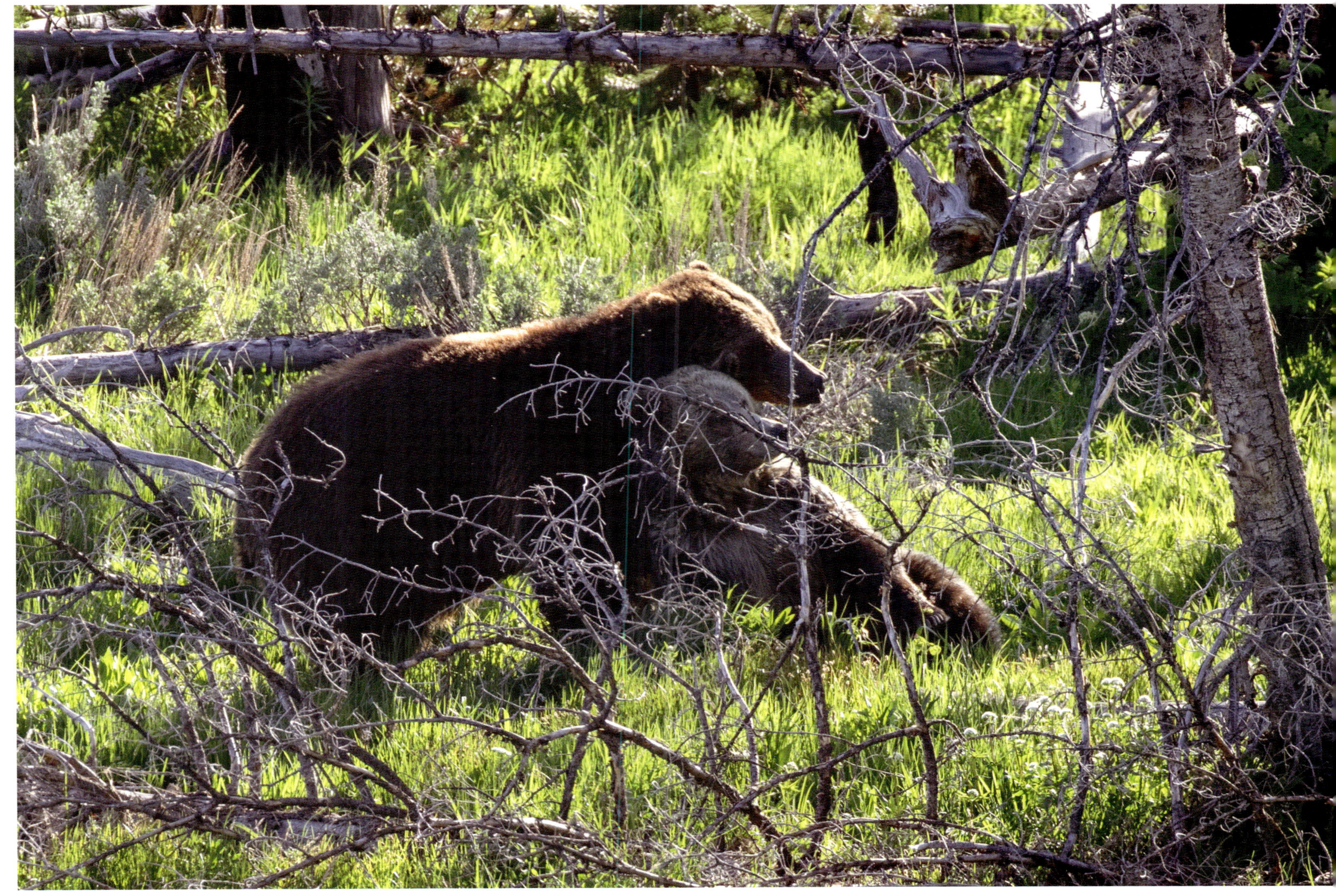

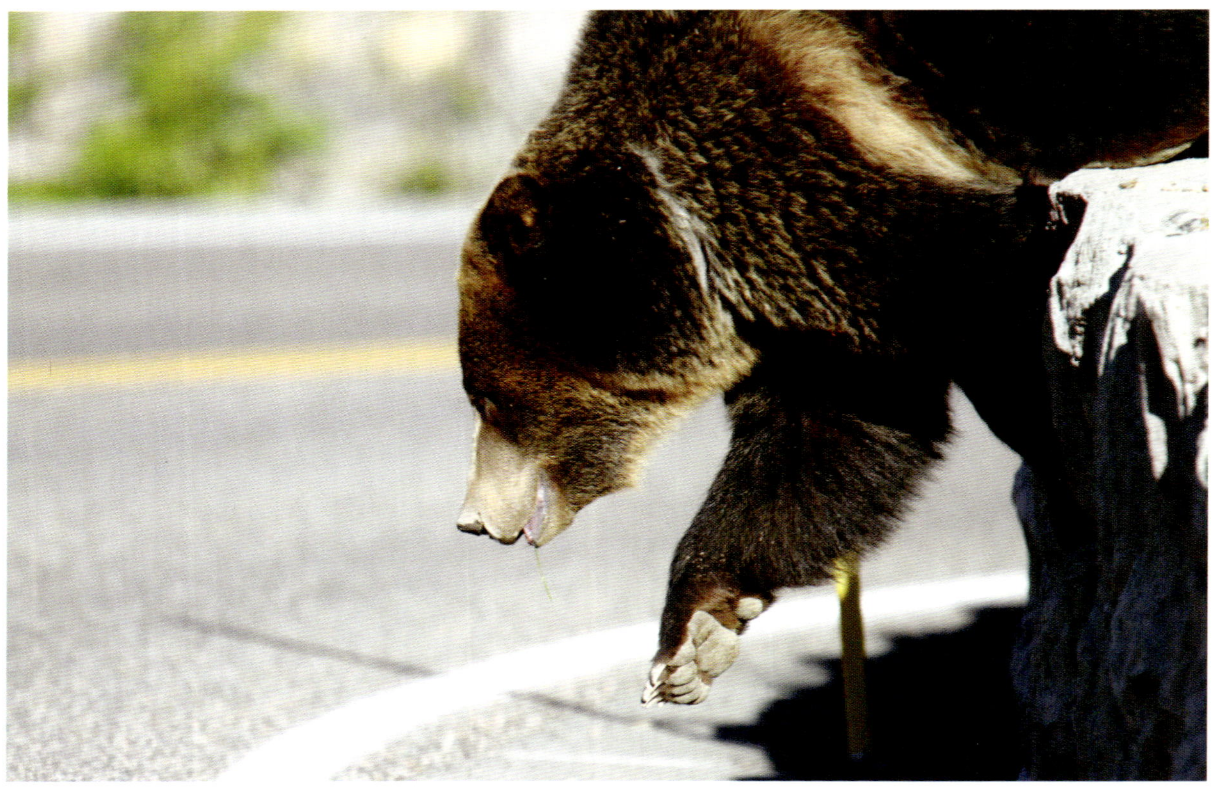

In an effort to protect their cubs, grizzly sows will often stay in a location close to the road.

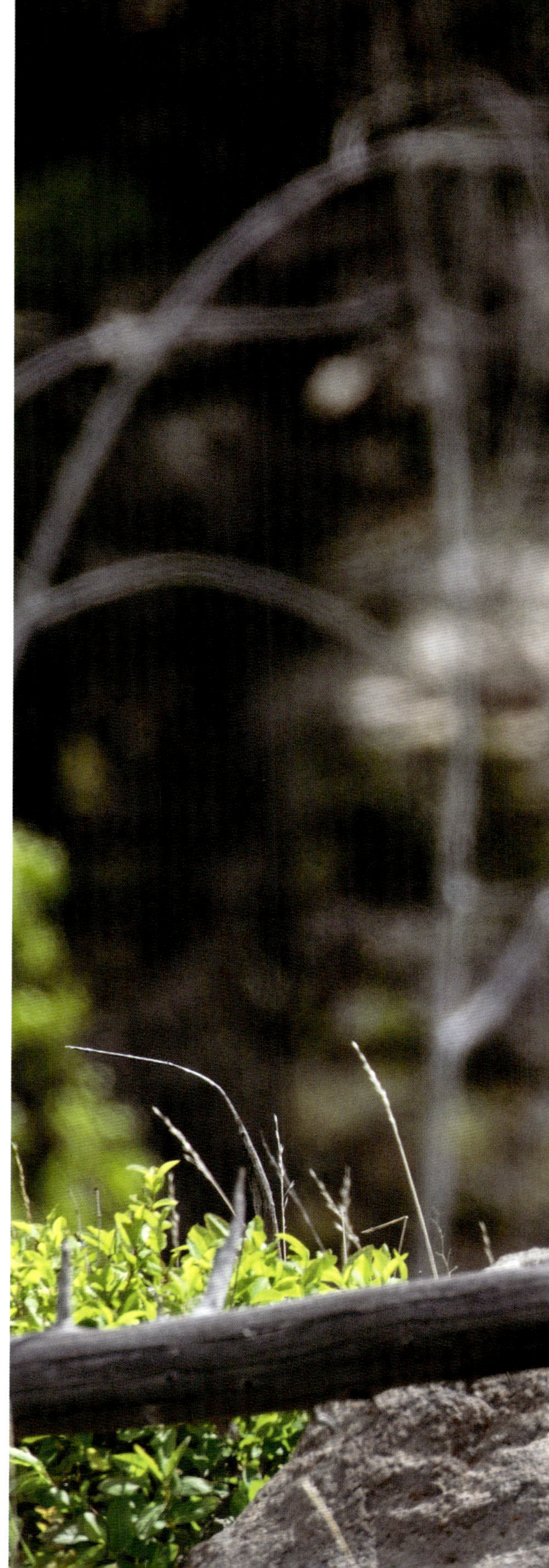

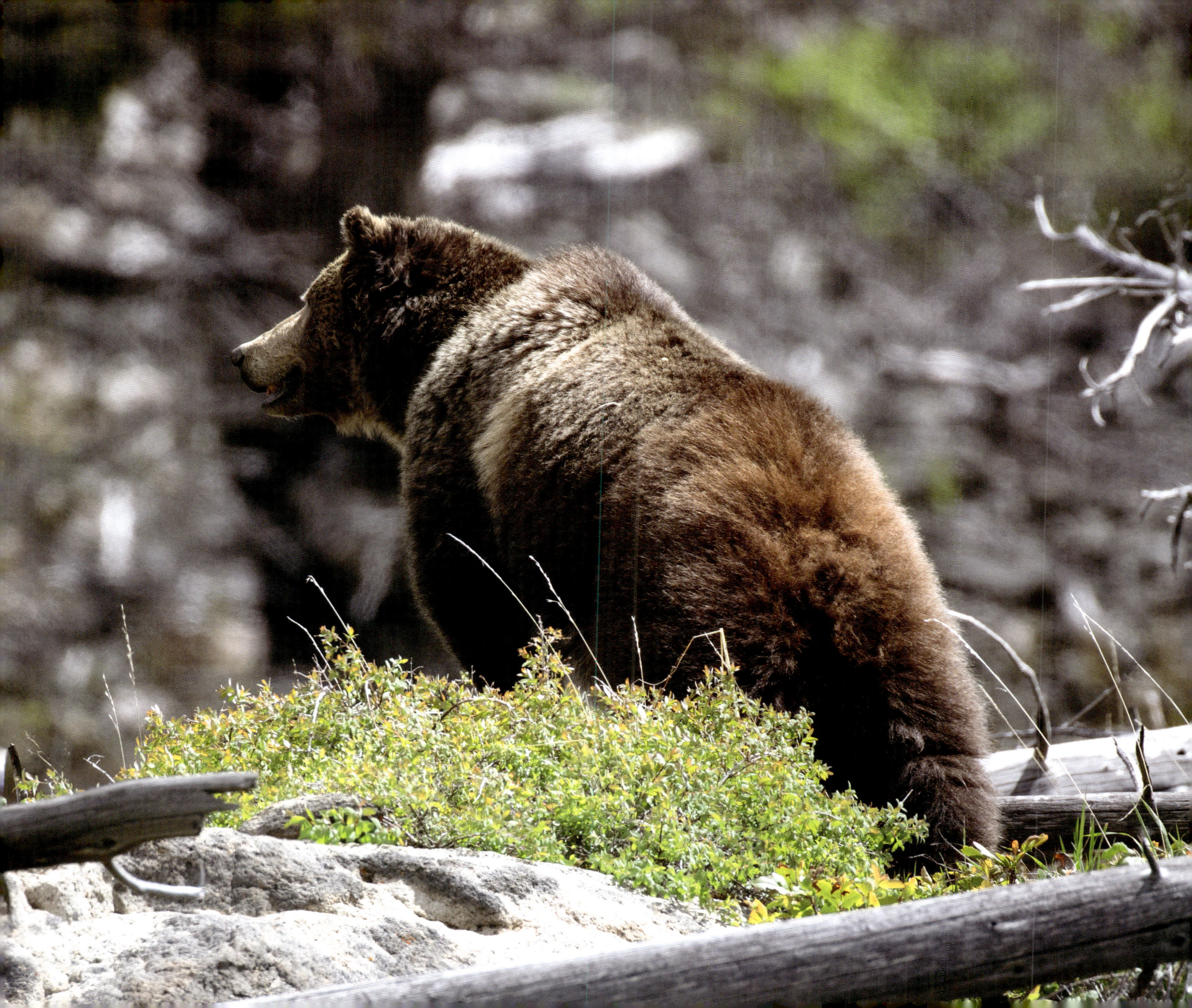

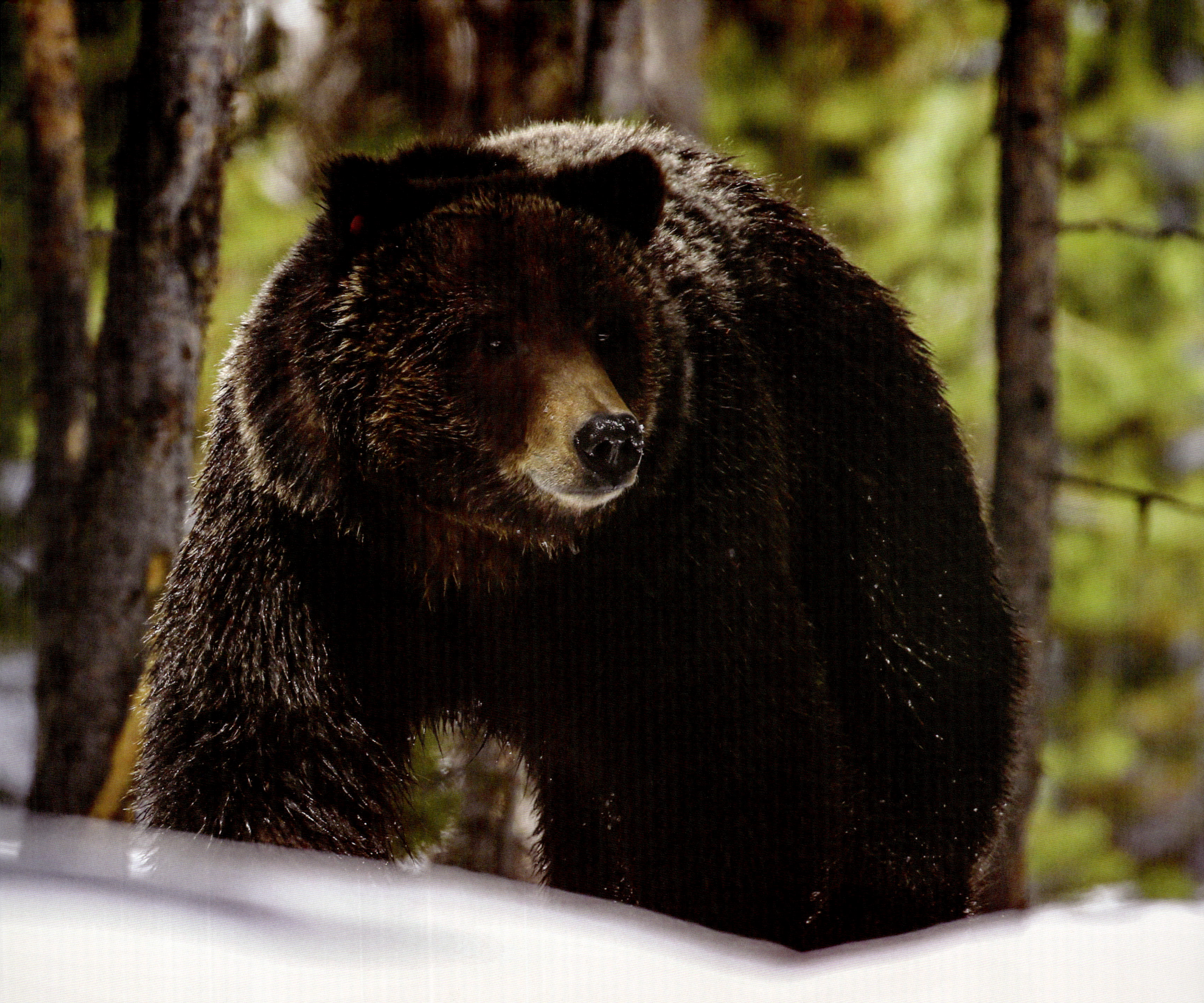

GRIZZLY 791

791 WAS BORN IN 2011. Before he came to Yellowstone, he had been in trouble for chasing cattle and was pretty much on his last chance before possibly being euthanized. Once in Yellowstone, he never left, and has since made the area his home. At eight years old, he weighed in at 500 pounds. Male grizzlies in the Greater Yellowstone Ecosystem average between 200–700 pounds. So, 791 is a big bear. He has been collared twice, but at the time of writing this, he does not currently have a collar. He also has tags in both ears.

Grizzly 791 will ultimately go down in history as the bear who, in the fall of 2020, took down a giant six-point bull elk in the Yellowstone River in Hayden Valley. While I do have many memories of 791, this is one that will top the charts forever.

I was driving along at first light on a beautiful fall morning. A light fog covered the valley floor and there was a slight chill in the air. At the time, even though I love grizzly bears, the only thing on my mind was finding wolves. I pulled over in a pullout, parked, got out, and waited for the morning light to spread across the valley. As I was standing there, I looked back to where I just came from, and saw cars starting to pull over. Suddenly someone pulled up and mentioned the word "bear."

I immediately jumped in my vehicle and headed back that direction. When I pulled up, I saw a grizzly charging into the water with an incredible bull elk a few feet ahead of it. By the time I got out of my car, the grizzly had jumped on the elk's back and was trying to take it down. The elk was thrashing his antlers and splashing around in the water trying to get the bear off its back. Unfortunately, the elk wasn't having any luck. The bear then moved to the side of the elk and started spinning it, back to belly, and within a few seconds, the elk had drowned.

Still not believing what I had just witnessed, I looked up and realized that the current was pulling the bear and the elk toward my side of the river. I slowly started to back up, and within a few seconds, I was looking directly into the eyes of this enormous grizzly. I took a deep breath, and waited to see what happened next. Thankfully, the bear grabbed the elk with his teeth and started swimming with it to the opposite side of the river.

By the time 791 got about ten feet from shore, he was exhausted. He moved forward about another six feet and then stopped. He turned the elk on its back, and ever so delicately, opened it up and ate a few bites of this well-deserved meal. He then pulled it to the bank of the river and sat down while slowly looking around at the audience that had gathered.

During the next few days, 791 cached the carcass, had visits from a wolf pack, and also had another bear come in and take the carcass away from him. Even though 791 was much larger than the other bear, he must have realized that bear was considerably older and wiser than he.

What an incredible bear, and what an amazing experience.

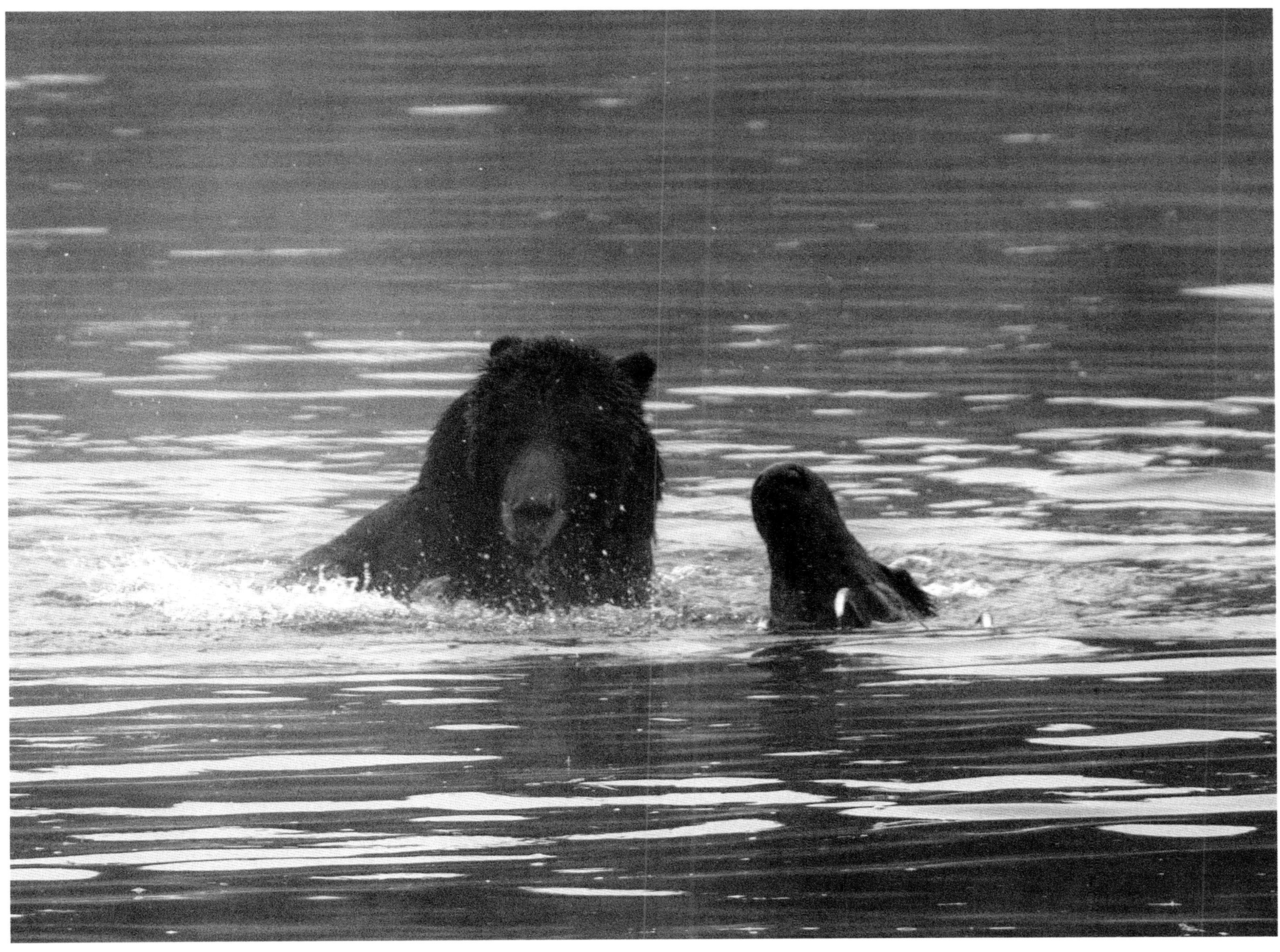

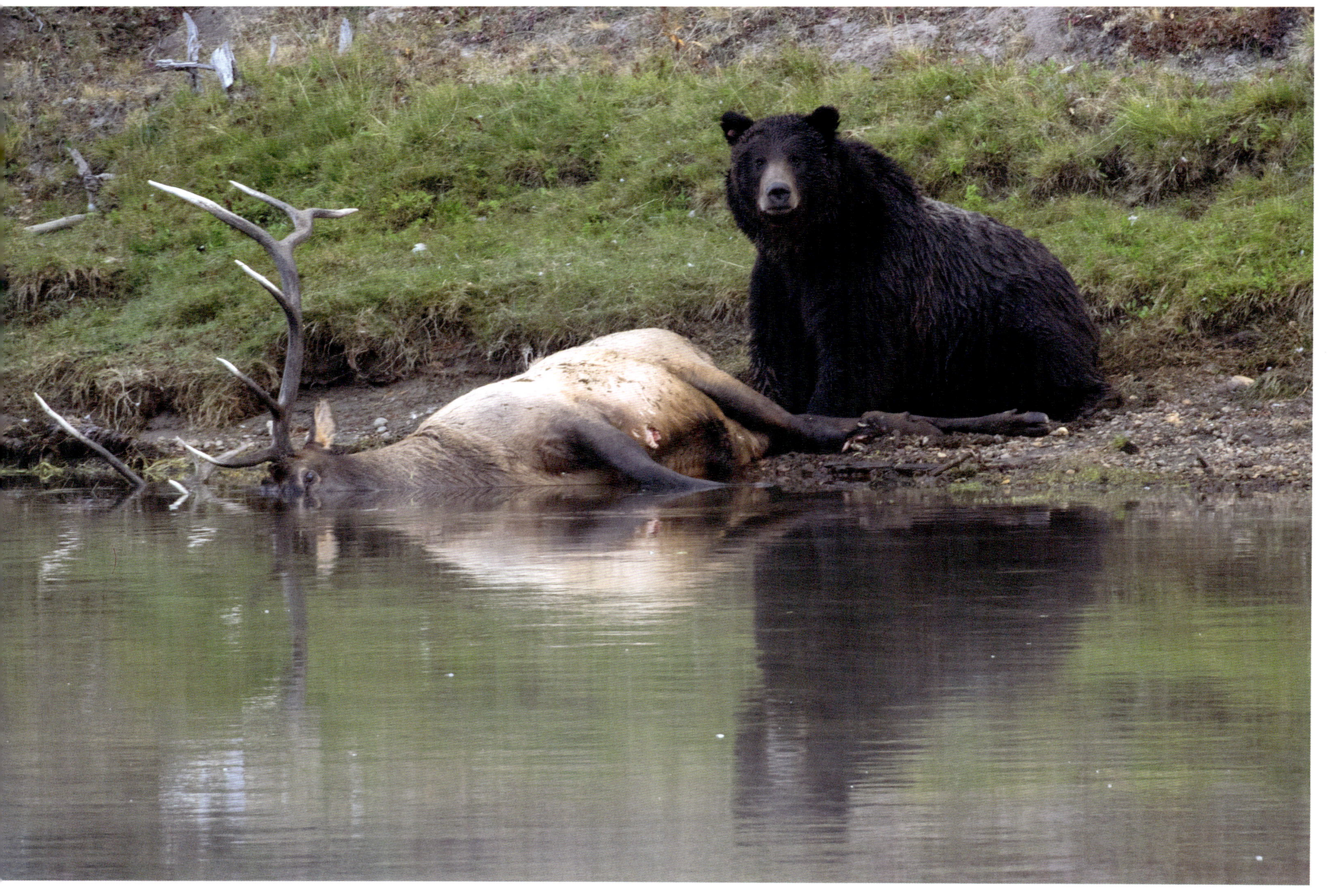

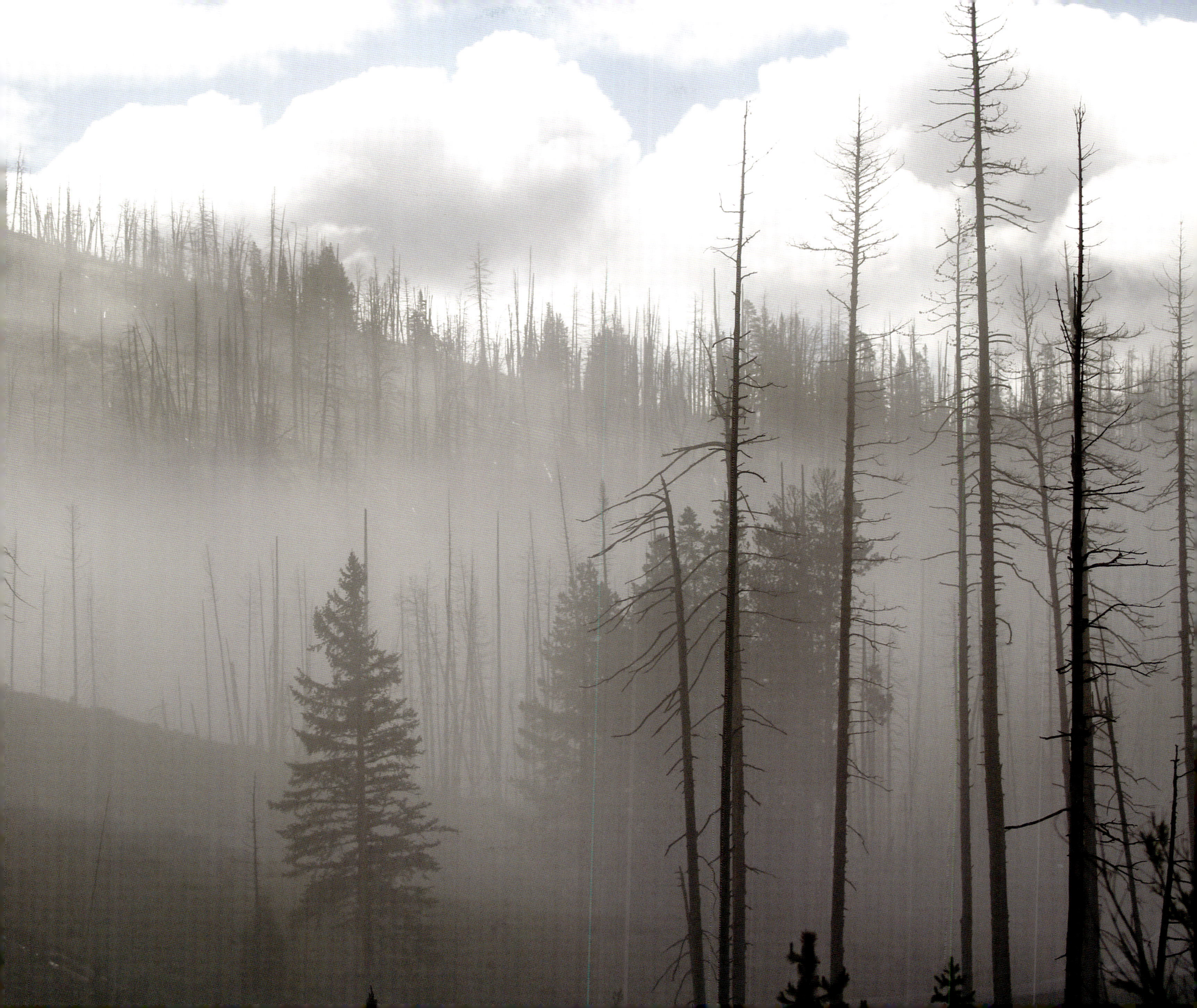

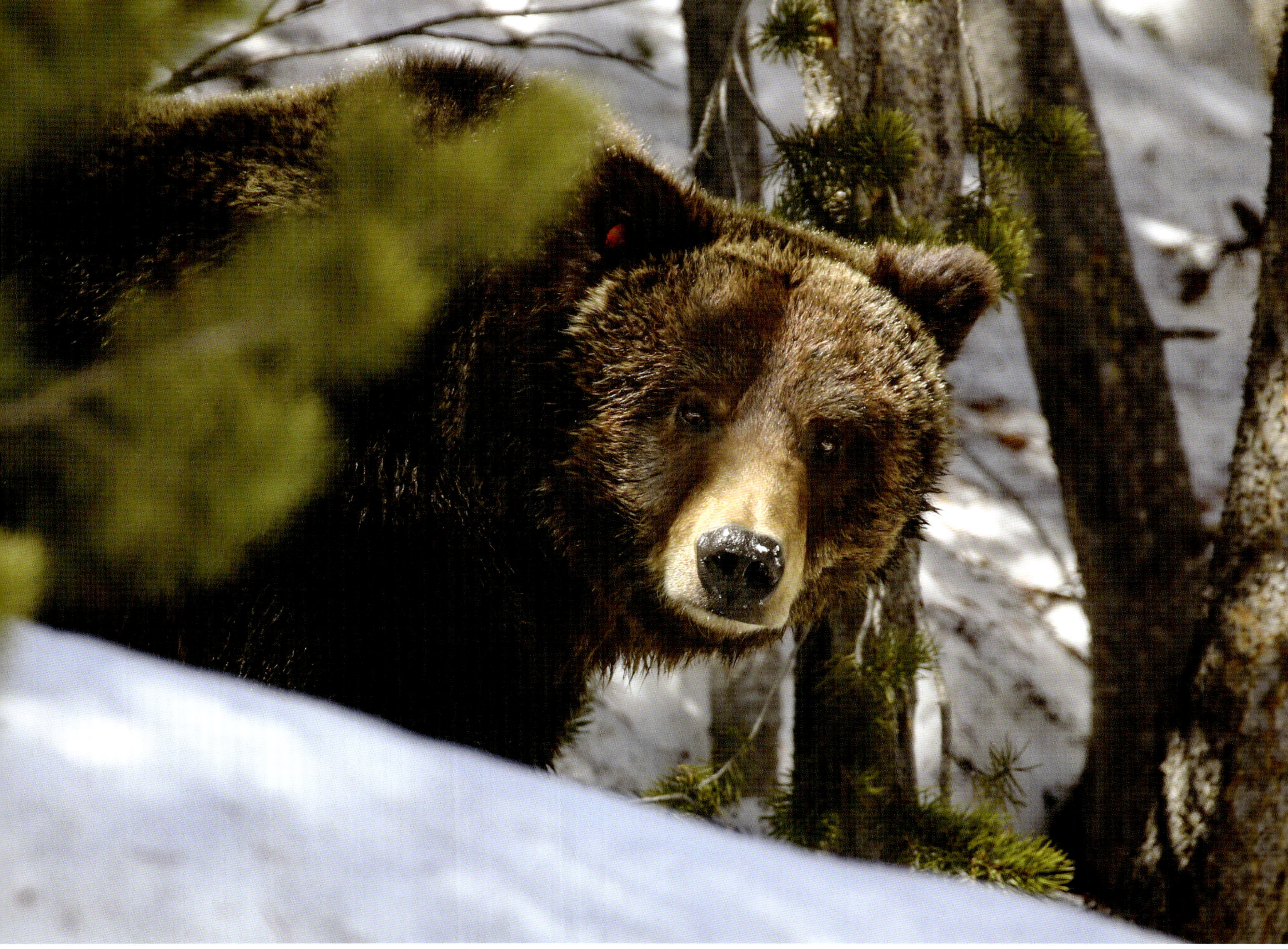

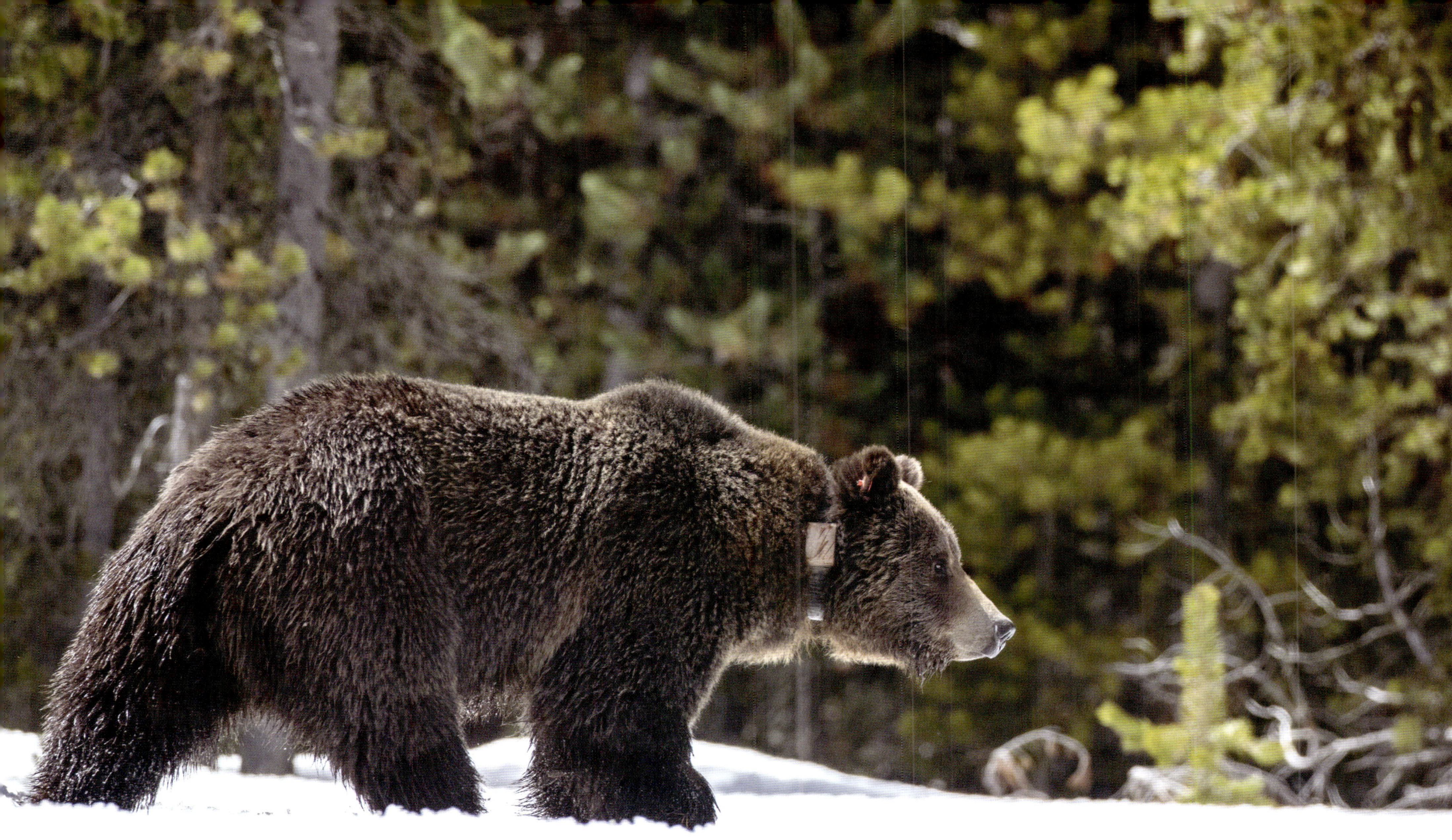

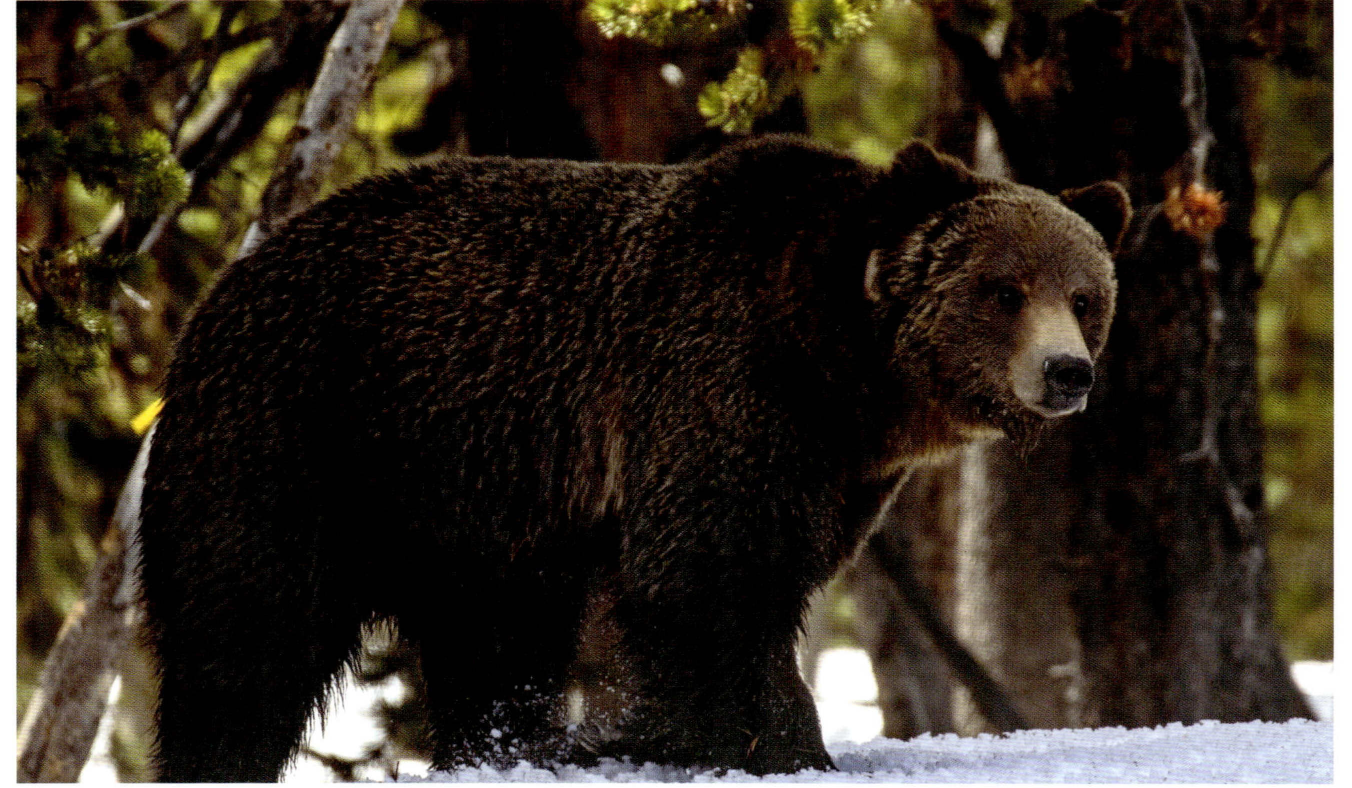

Not long before taking the bull elk down in the river, 791 took a bison carcass away from a pack of wolves. For a couple of days he lay on top of the bison and never let the wolves have a bite. He is definitely a force to be reckoned with.

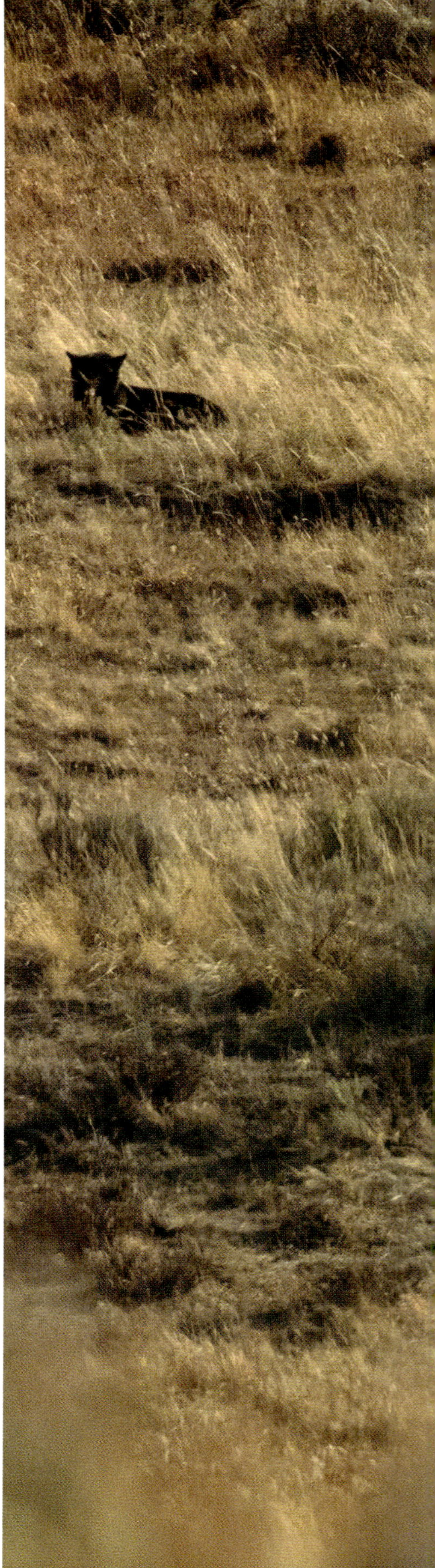

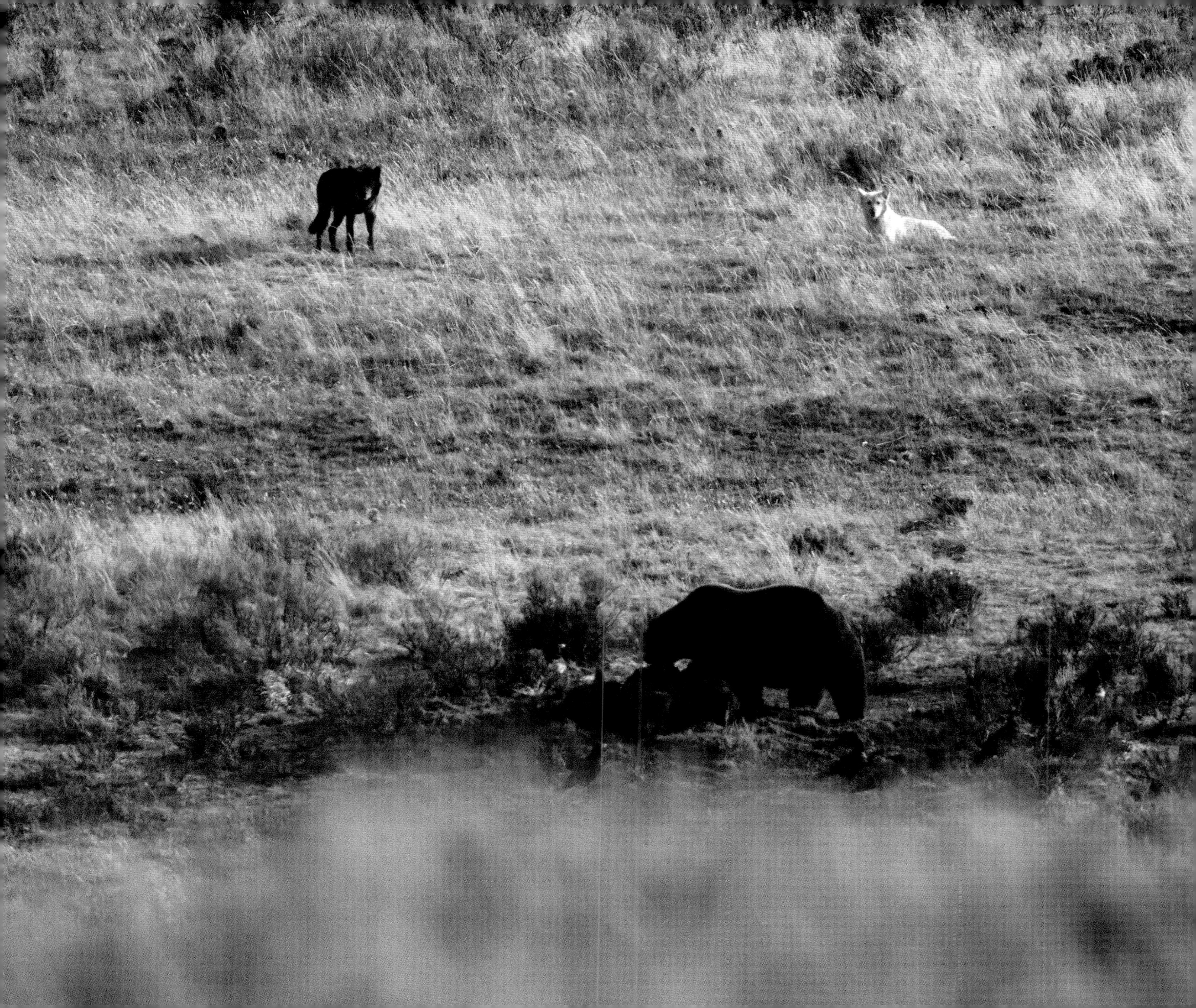

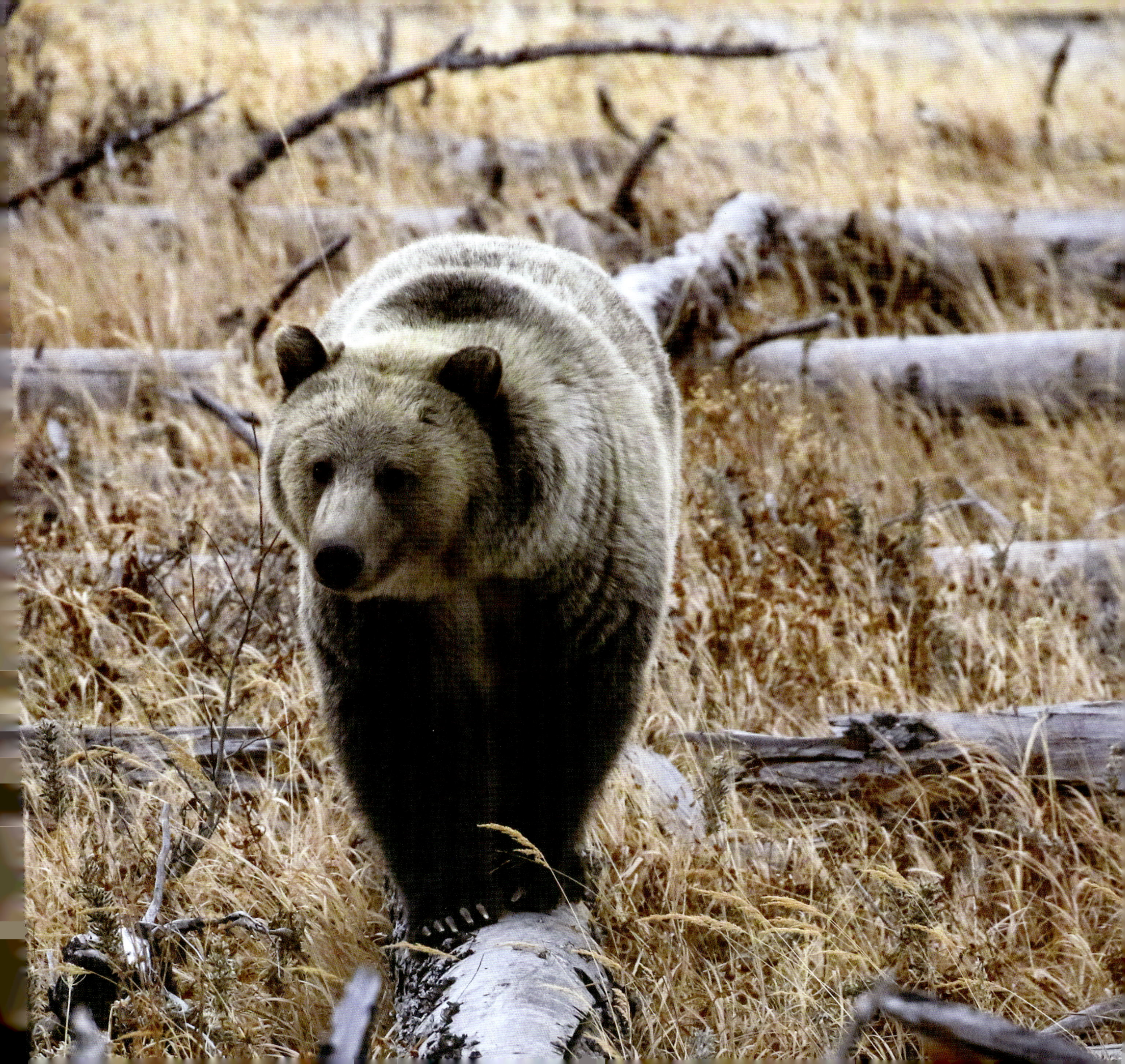

ALONG *Came* SNOW

ON A SPRING DAY IN JUNE OF 2015, the cutest little blonde grizzly cub made its way down a mountain in Yellowstone. This cub was accompanied by its mom, Raspberry, and one sibling. At that time, no one knew this little blonde bear would become one of the most loved, and most photographed, bears of its time.

The light coloring of this cub was very unique, and while park officials shy away from naming bears, because of the coloring, this cub was given the name Snow. I'm not sure whether it was a photographer or a visitor who came up with the name, but it fit perfectly.

Just like all young grizzlies, this little one was mischievous, playful, and full of trouble. After its sibling, who had been called Rocky, disappeared, Raspberry became its new best friend. Grizzly sows are very protective of their cubs and they are on high alert watching for danger at all times. Because of this, it's not very often that you will see them playing with their cubs. Raspberry, however, was different. She loved to play, and no one played with their cub the way that she did.

These two would spend hours playing together. They would wrestle and roll, bite, full out swat each other, climb over logs, and Raspberry would even grab Snow by the rear, with her teeth, and toss her in the long grass. This was all in the name of fun. Shortly after playtime was over, Raspberry would lie on her back, Snow would nurse, and the two would then snuggle in close for a nap.

Given the area they were in, you could often find them near Lake Butte Overlook, and watch as they scratched on their favorite rock or on the visitor sign that is located there. At times they would even wrestle on the concrete, giving the people who happened to be there an incredible show.

They could also be seen at the 9-mile pullout, sleeping on logs right next to each other. Watching little Snow try to balance on a log while sleeping always brought a smile to my face.

Their walks along the shore of Yellowstone Lake are some of my favorite memories. Something about the love they had for each other, the water, and the beautiful setting made these times magical. For three and a half years we got to watch as these two bears lived life to the fullest. They captivated our hearts, and taught us things we never would have known about the life of a grizzly sow and her cub without this experience.

When the time finally came for Snow to set out on her own, it was one of the saddest mornings I have ever had in the park. The days leading up to that moment were full of those last lessons that only a sow could give her cub. I remember watching just a couple of days prior Snow going out on her own as the two of them very roughly played on a snow patch on the back side of Lake Butte Overlook. Neither was holding back, it was the strength and power of two grizzlies fighting, and then playing with the softness that only comes from such an incredible bond. It was one of the most fascinating things I had ever seen.

At that time, there had been a boar in the area, and he was pursuing Raspberry, but she had not given in yet. A short few days later, just as the sun was setting, the boar was close and Raspberry started chasing Snow away.

The next morning, I was headed toward the Overlook and I got a message from friends letting me know that Raspberry and her boyfriend were further up the road, and that they had just located Snow. When I reached the area, there on a log, looking very depressed and lonely, was Snow. Just looking at her you could tell what she was feeling. My heart was broken.

The next few days were rough. Snow wandered aimlessly in the area and seemed very sad and confused. Often times she would walk up to cars and look at the people inside as if she were looking for a long-lost friend. She then started scent-trailing Raspberry, and on a several occasions she got very close. Raspberry would once again have to chase her away. I don't think Snow ever realized that her mom chasing her away was for her own good. If that big guy would have wanted to, he could have ended her life in a second.

After the courting period for Raspberry was over, many bear watchers wondered if she and Snow would end up back together. That never happened, and I think it took the entire summer for Snow to realize that she was no longer allowed to be close to mom and that she needed to live life on her own.

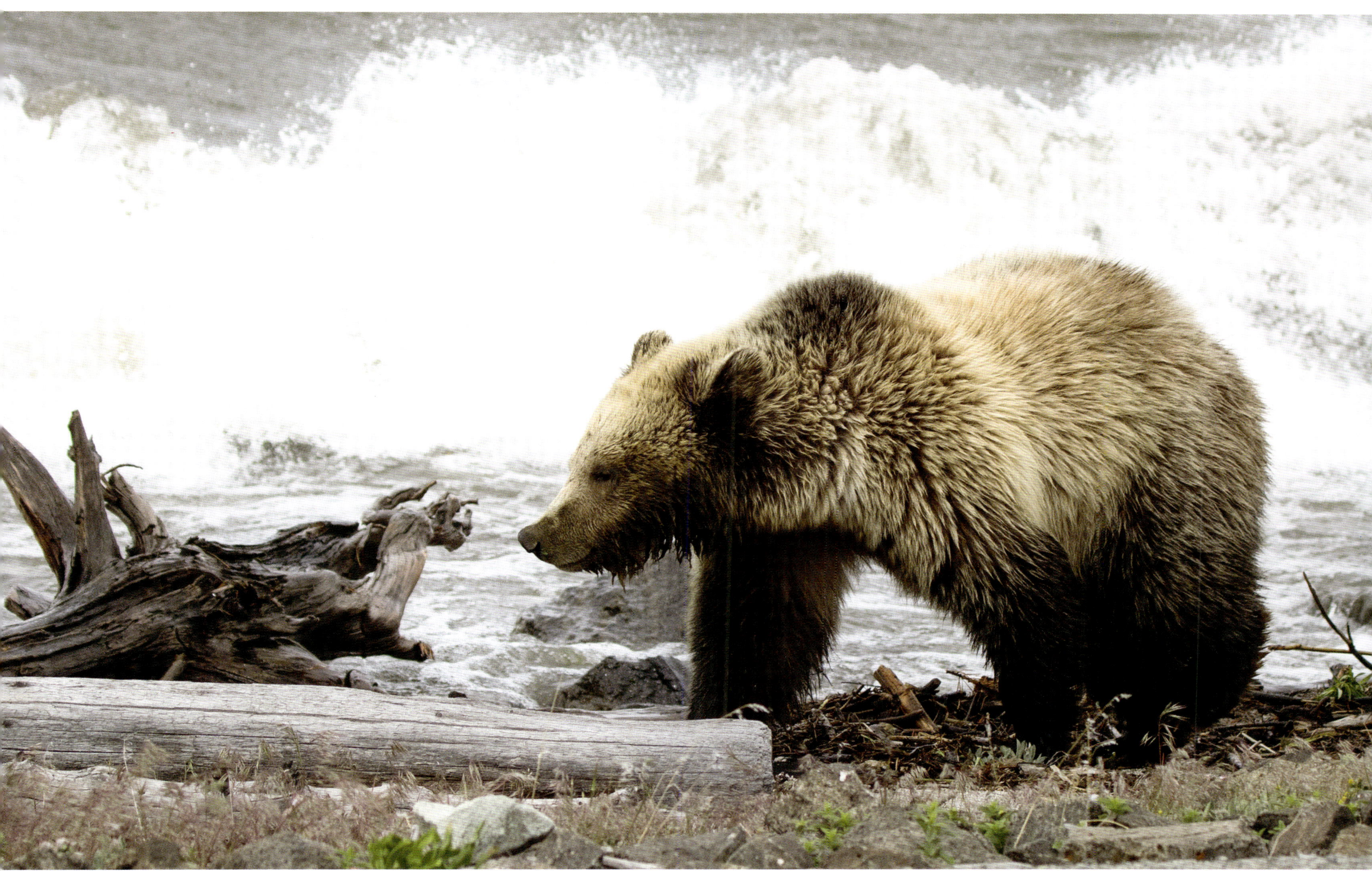

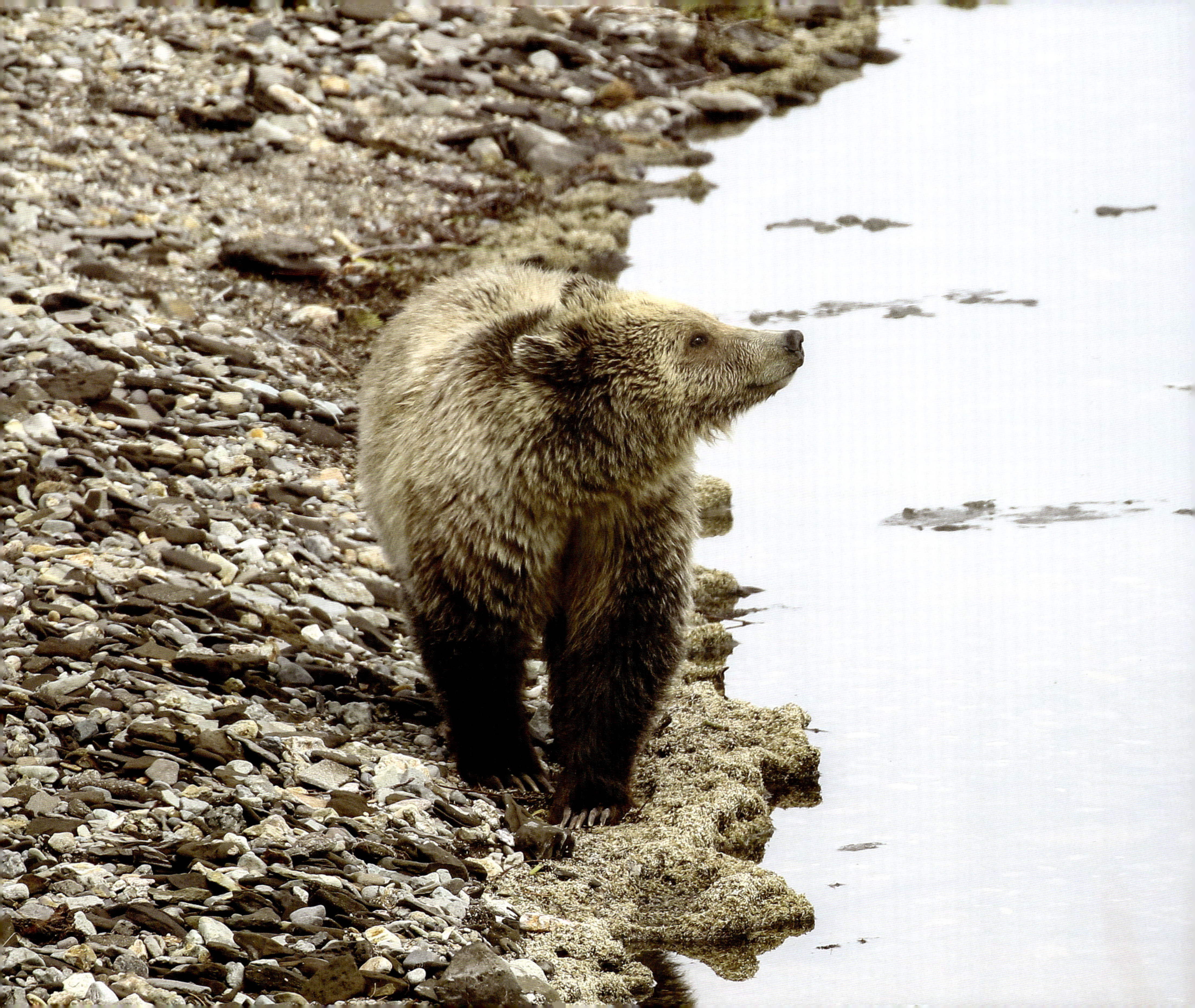

After those sad days passed, Snow once again began to thrive. She played in the water and took long walks along the shore of Yellowstone Lake. When she took her first elk calf by herself, people around the world cheered even if they weren't in the park to see it. She survived her first time denning alone, and she even survived taking some porcupine quills to the face. Raspberry had taught her well, and she was definitely a survivor.

One incredible thing that happened shortly after Snow was chased off by Raspberry, was when I was able to learn, without a doubt, that she was a actually a female. Though we had speculated over the years, no proof had ever been captured until that moment. I was snapping pictures of her in Mary Bay, and she stopped to go to the bathroom. After looking back through the photos several times, there was no doubt, Snow was a girl.

The next two years were full of many sightings, and although Raspberry and Snow crossed paths on several occasions, Snow had settled in to life on her own and was once again the smiling bear we had all come to love.

In 2019 Snow went on a couple of long walk abouts, and was seen in different areas of the park. It was assumed she was trying to find a territory of her own, but she always ended up back near the lake. This area apparently had captured her heart.

In the spring of 2020, she once again emerged from the den. She was now five and a half years old. What a blessing it was to have watched her grow from that little cub into a beautiful adult grizzly. That year she was seen in the company of a couple of male grizzlies. I can't wait for the day she shows up with a couple of COY of her own. Just the thought makes my heart skip a beat. What a wonderful life.

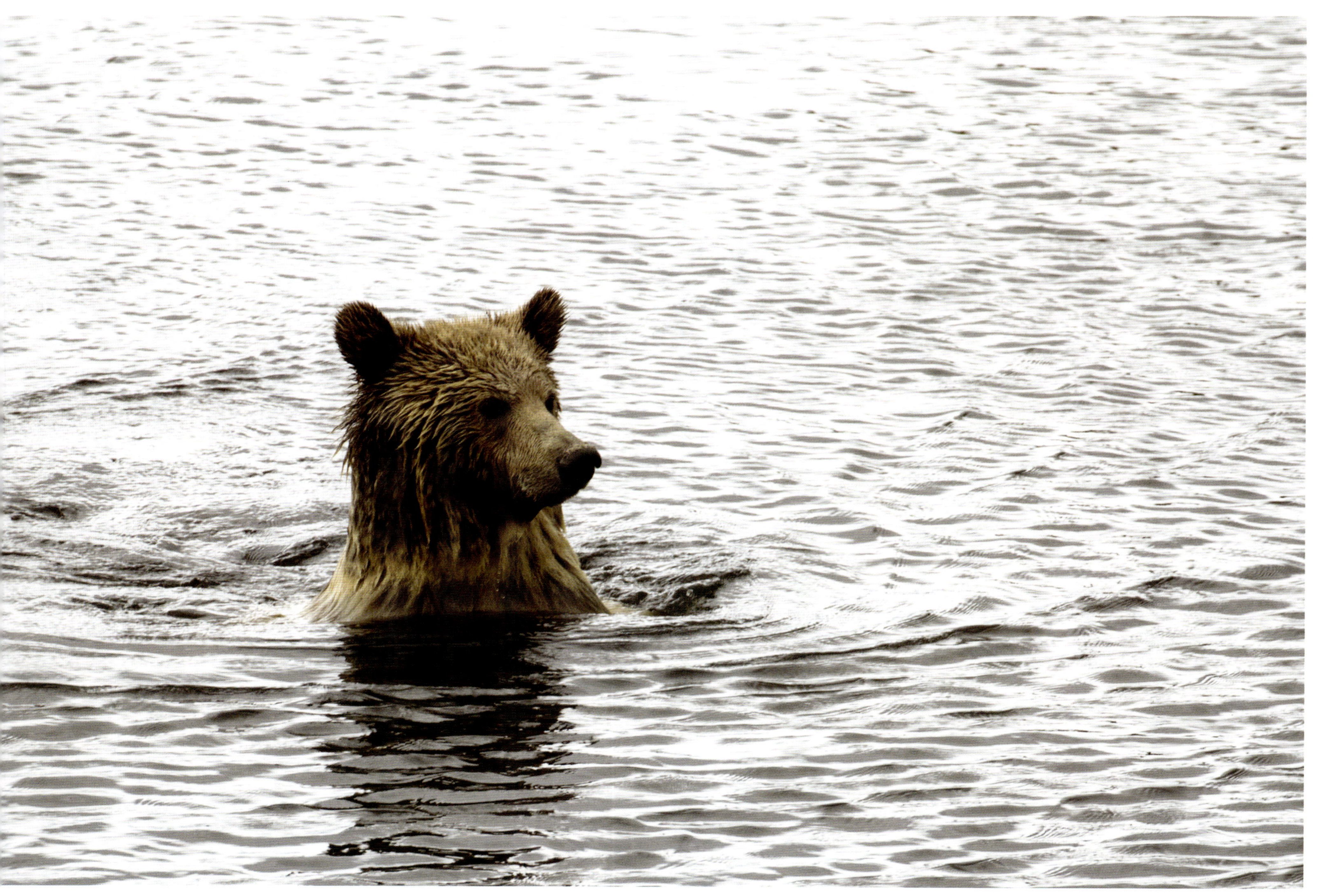

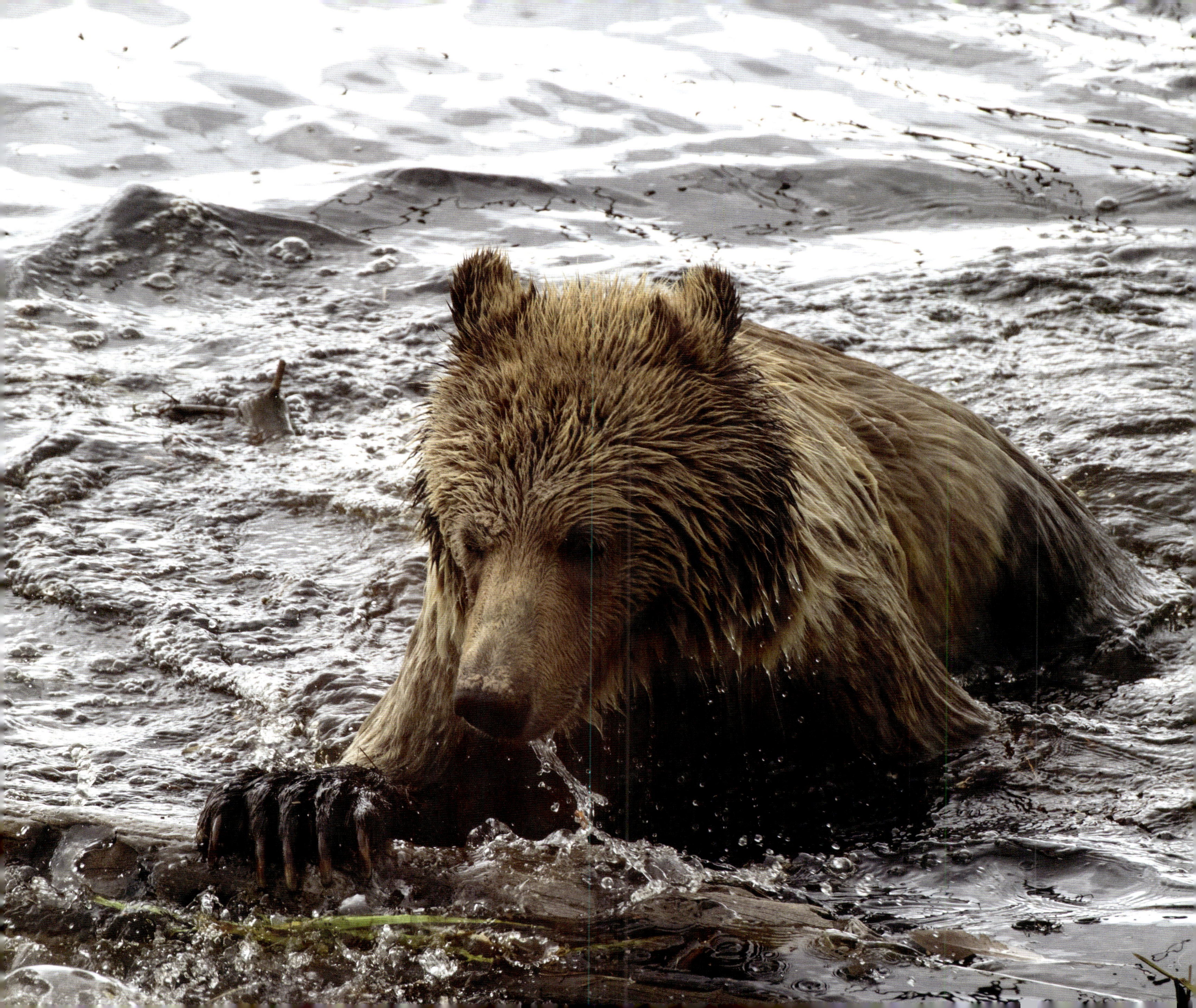

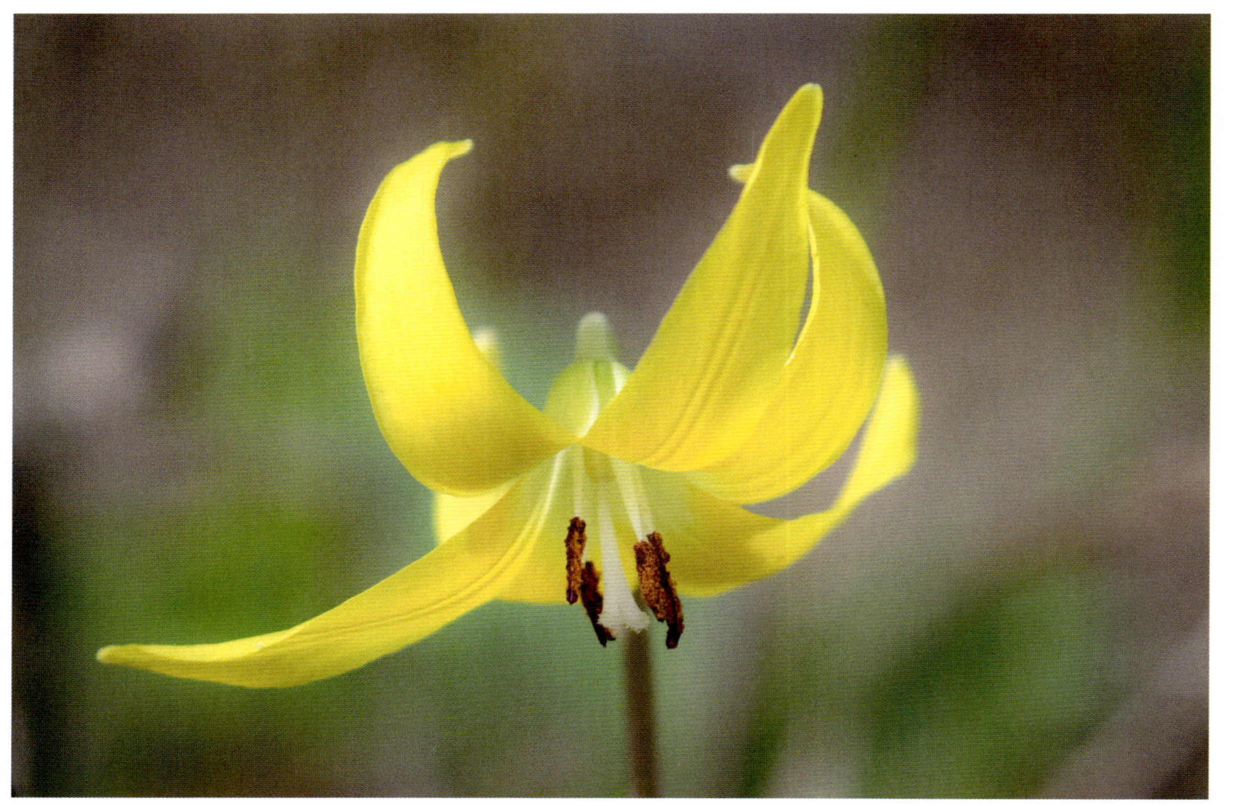

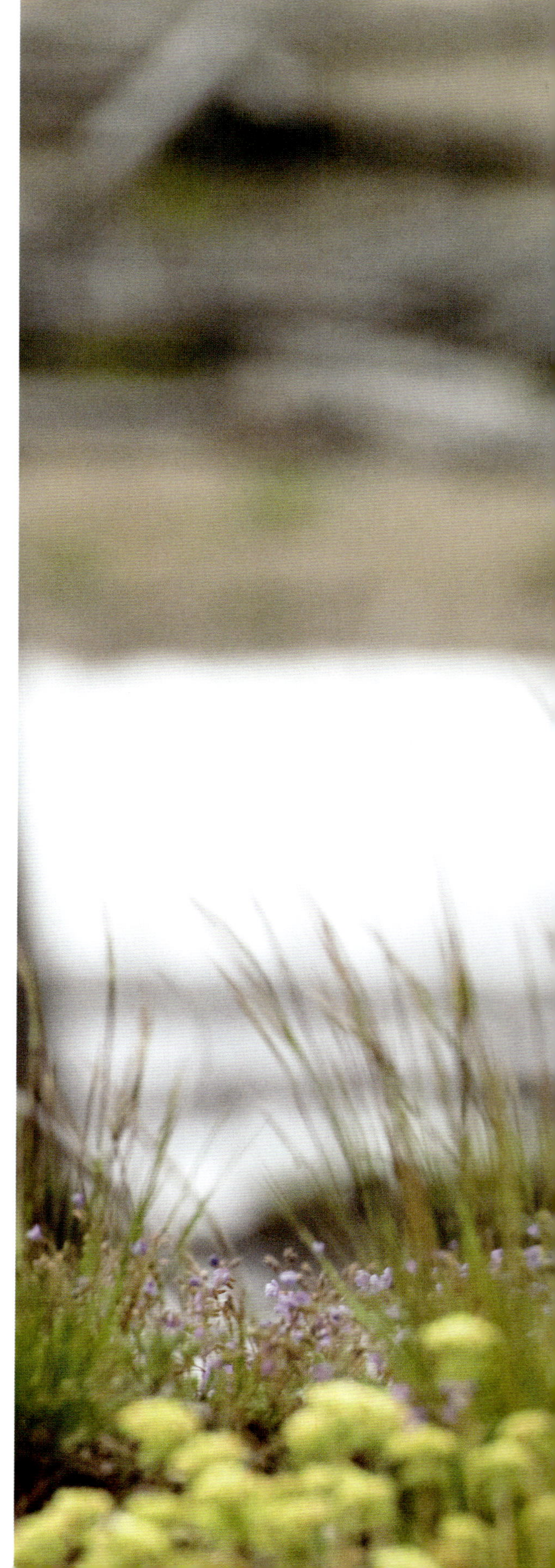

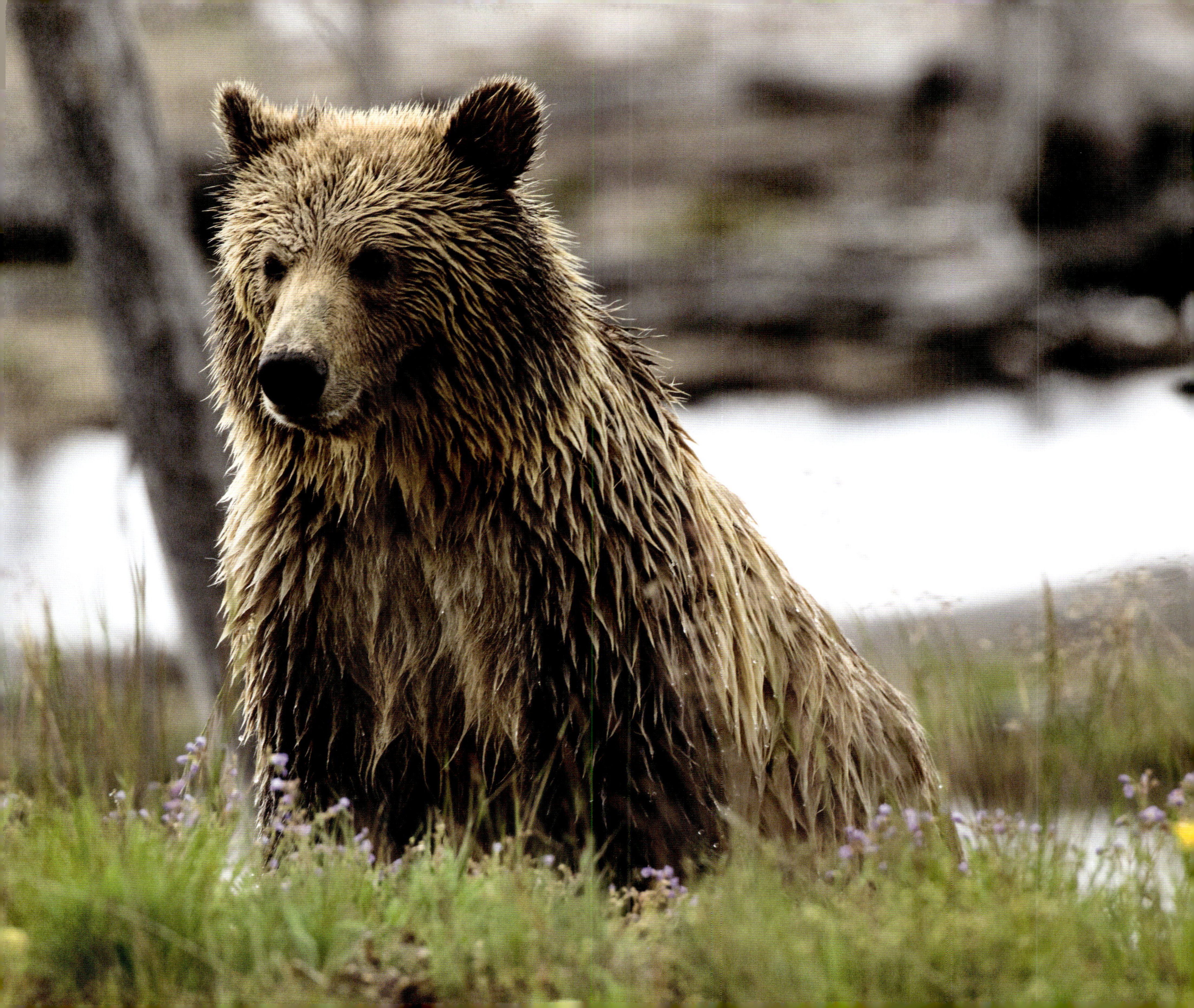

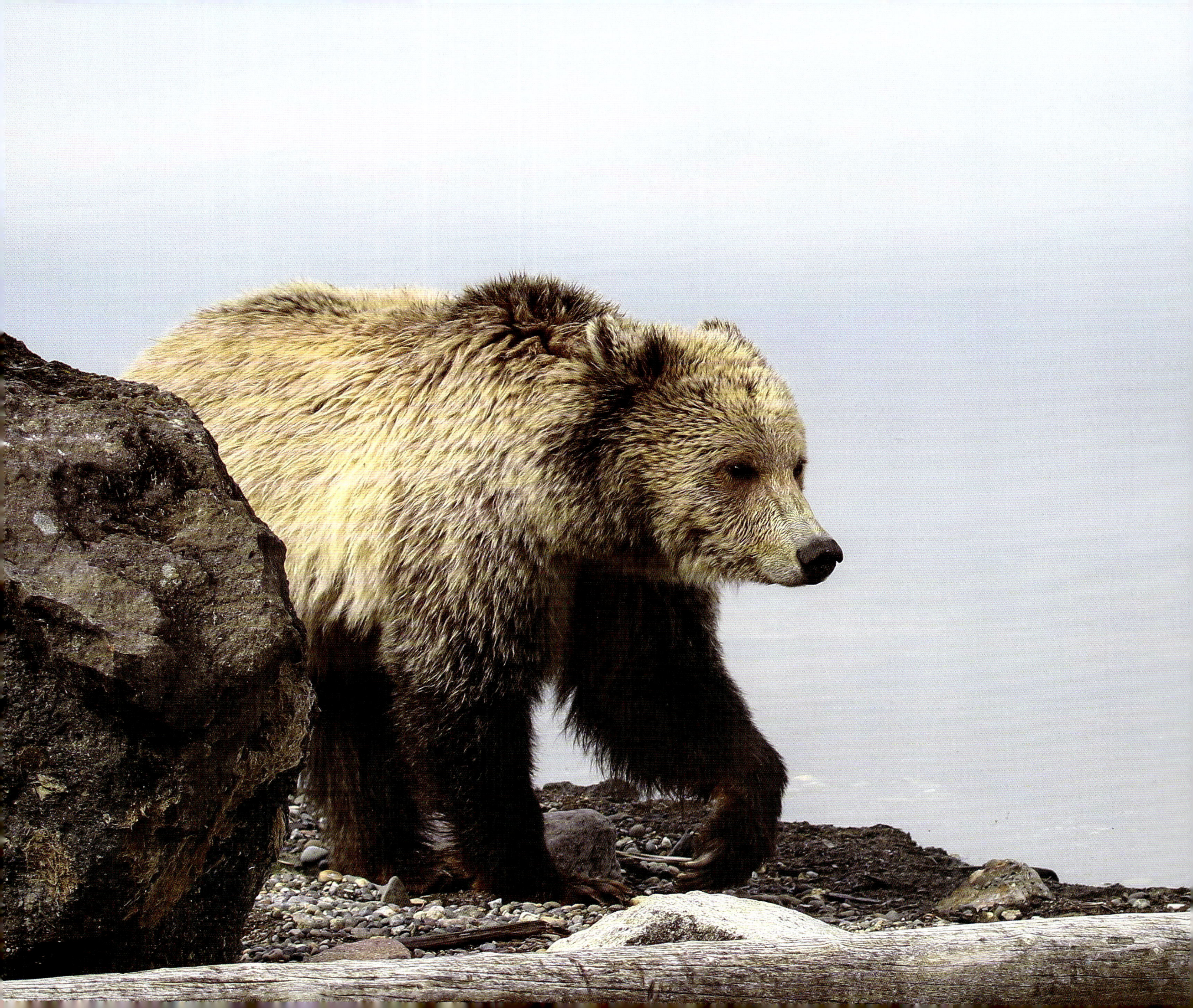

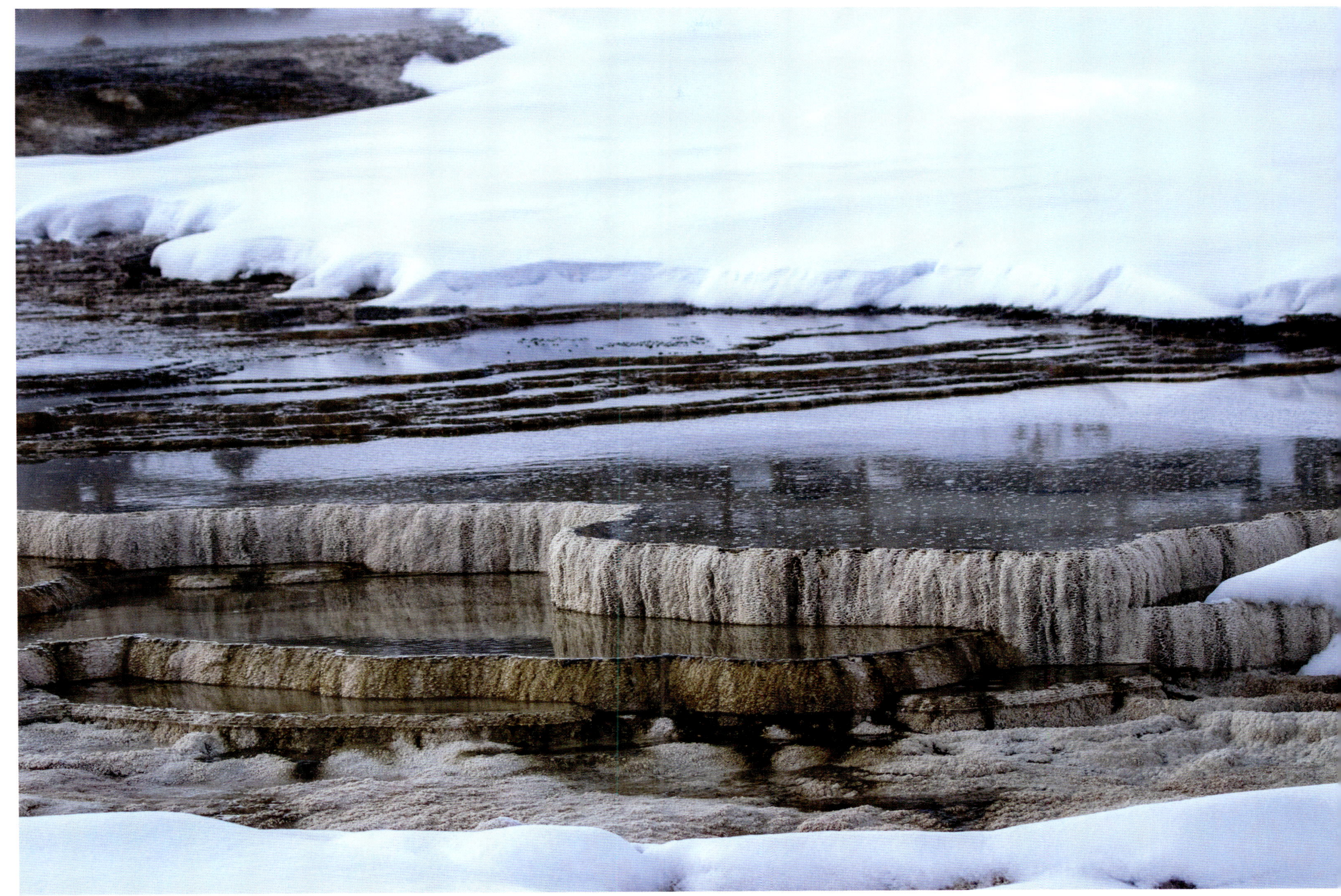

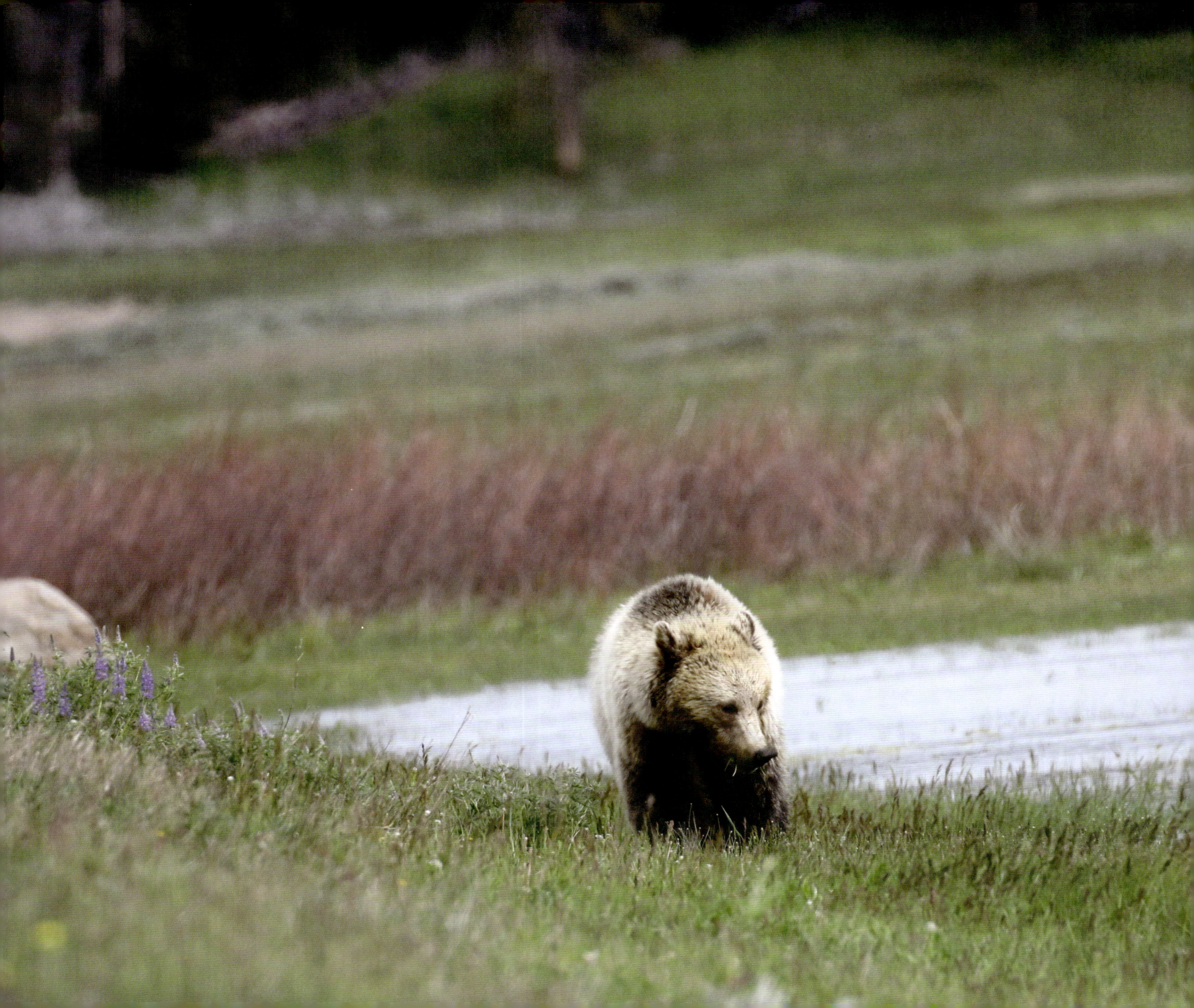

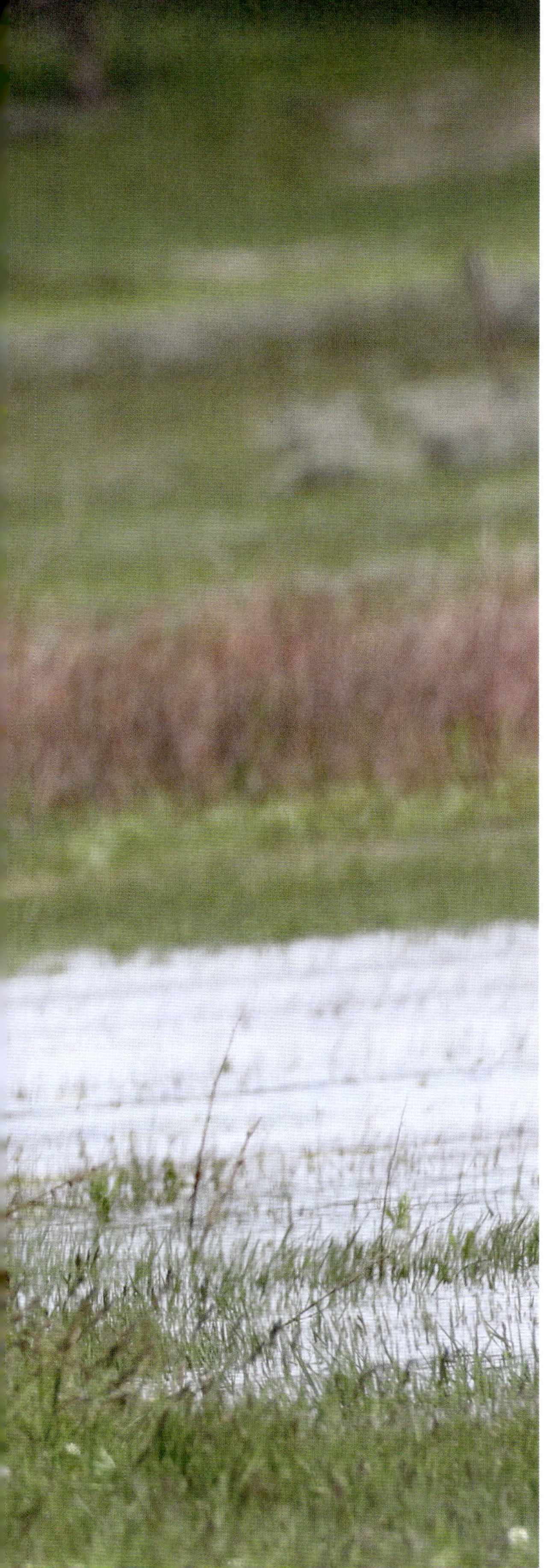

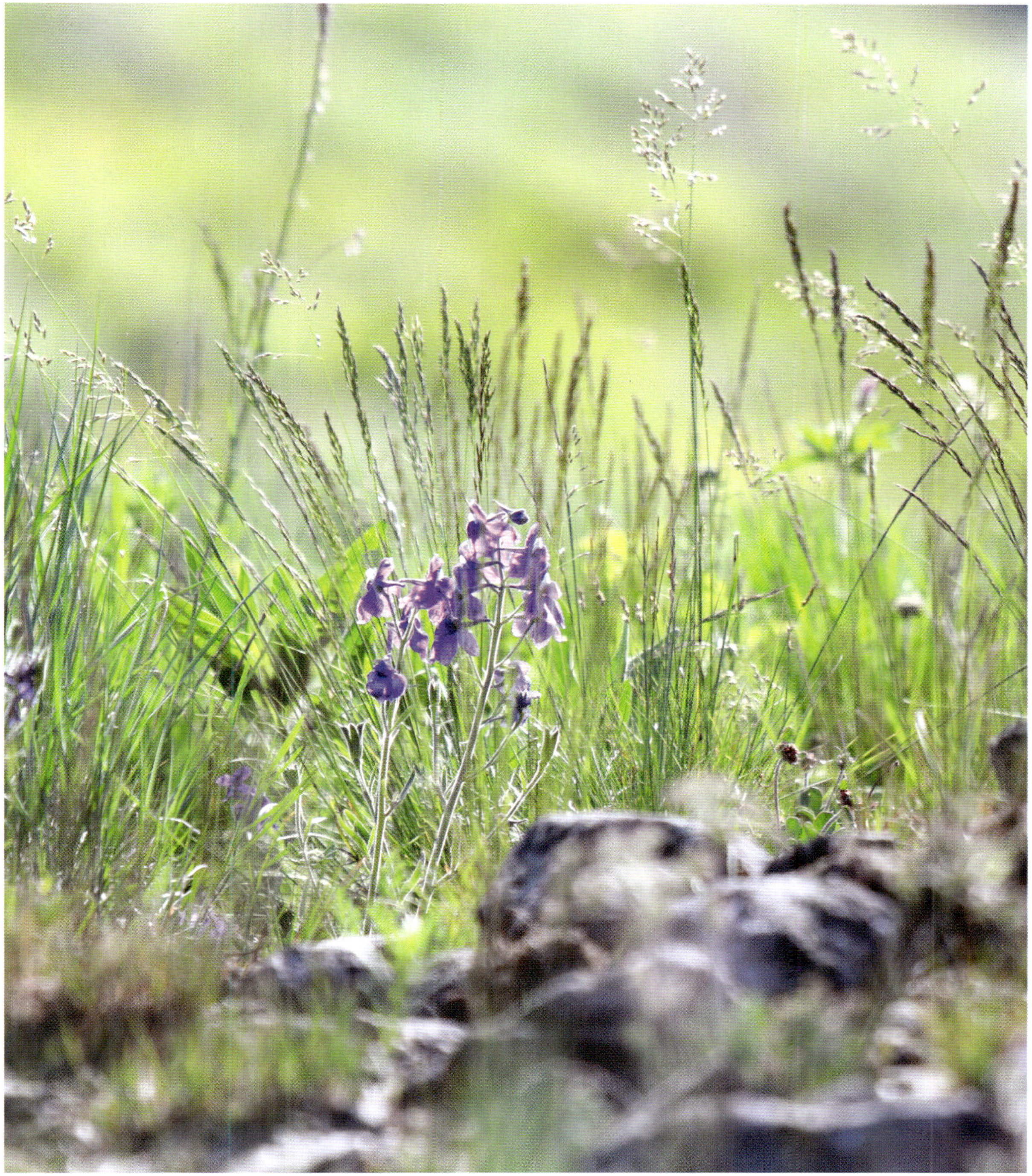

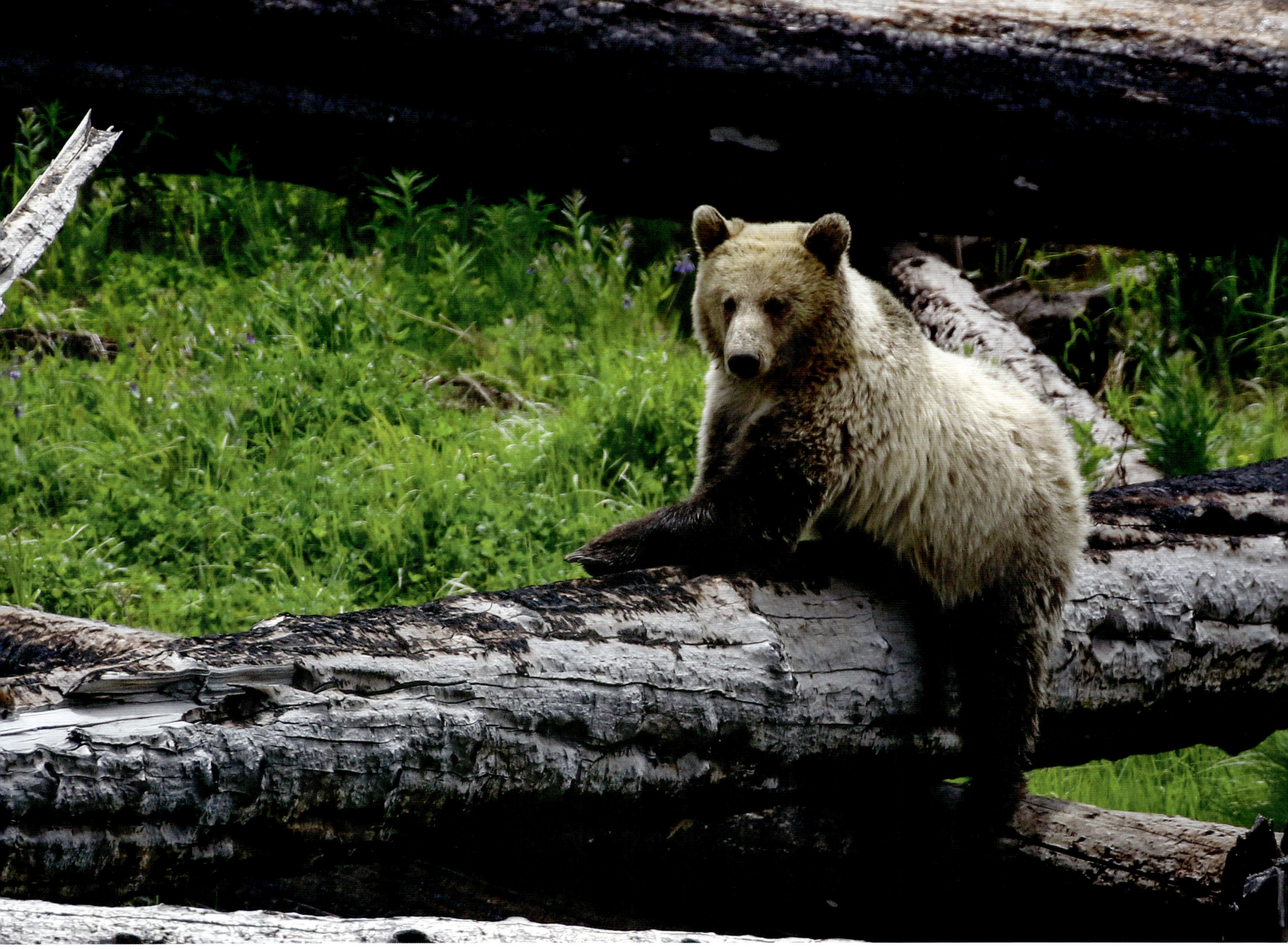

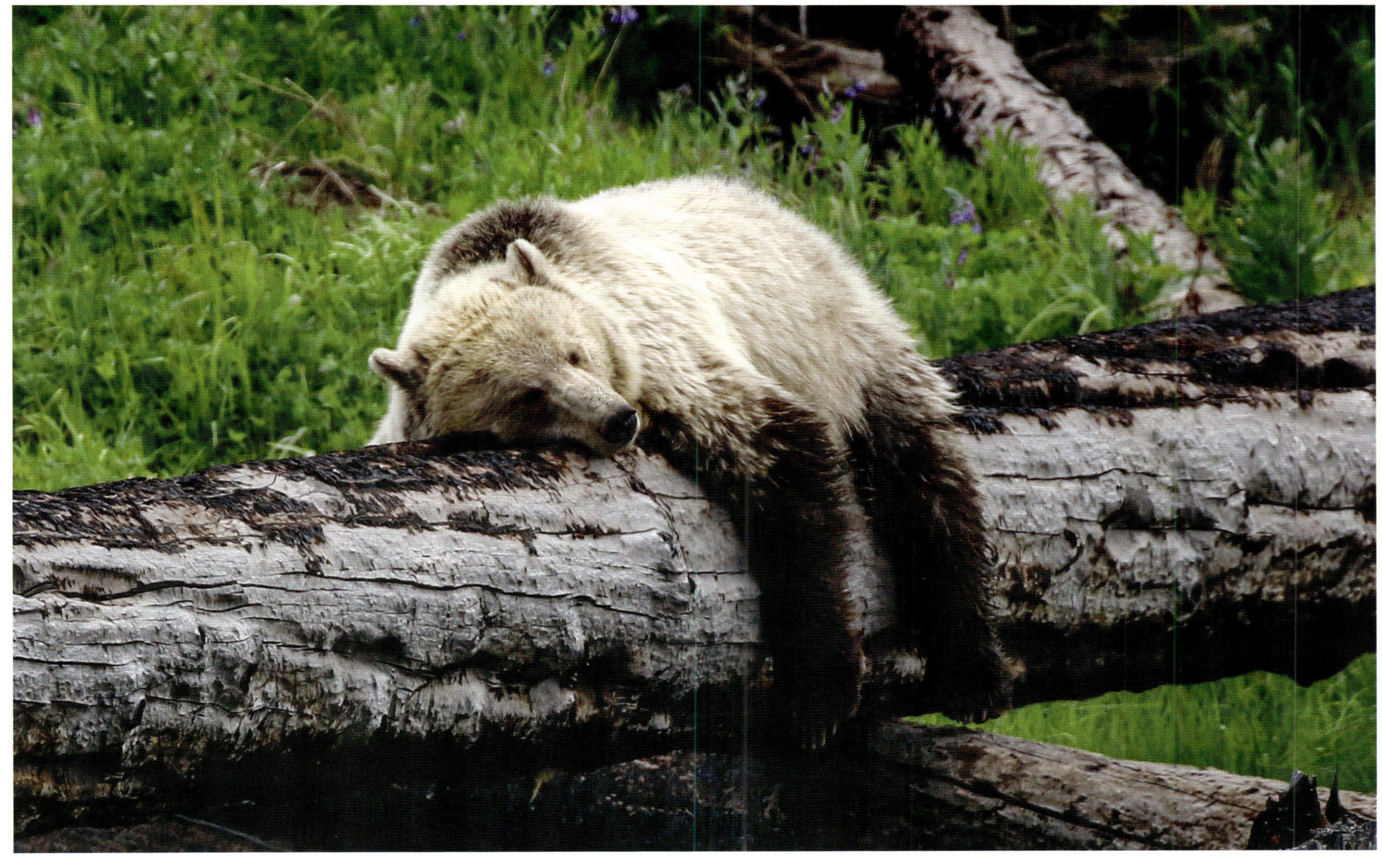

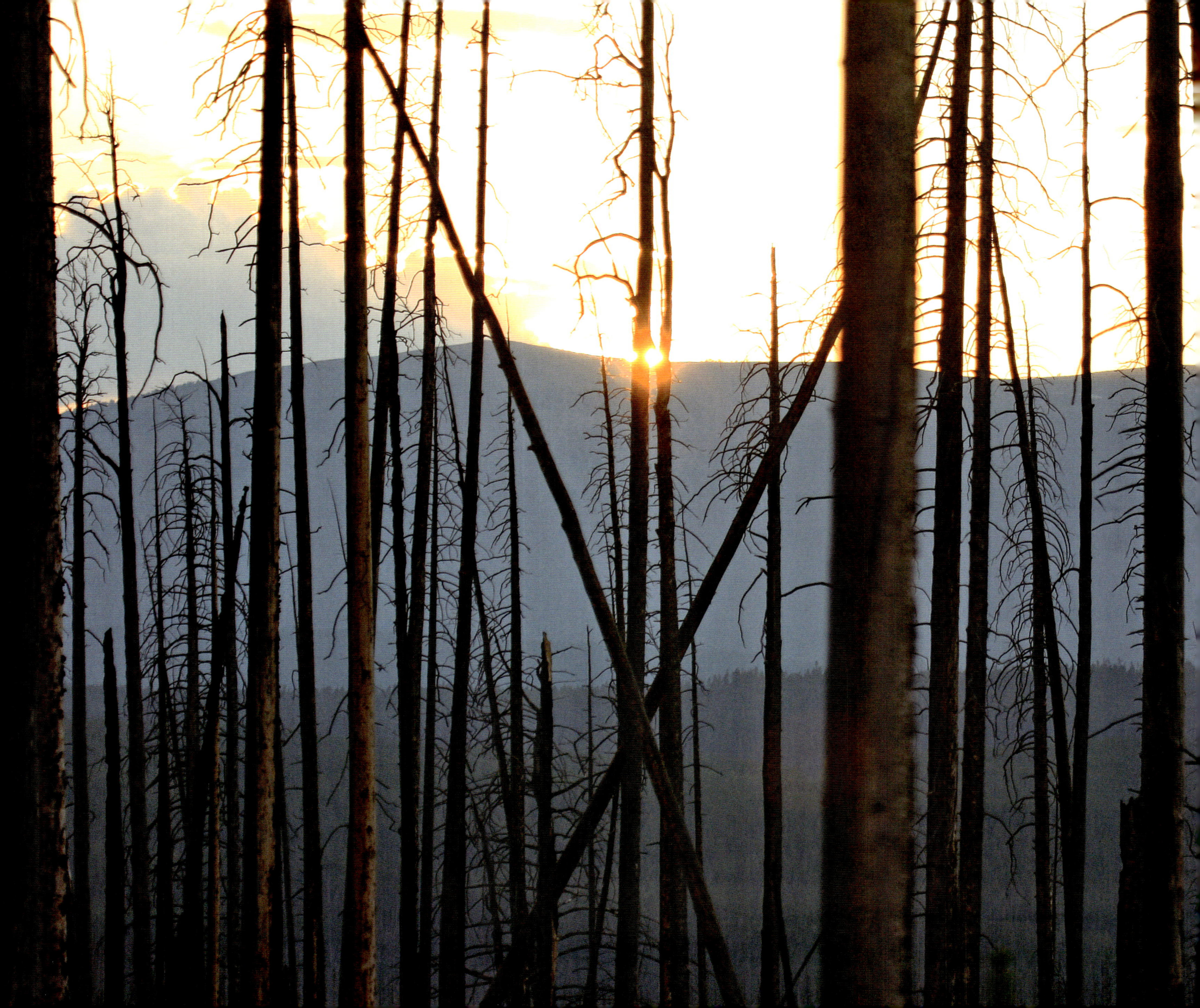

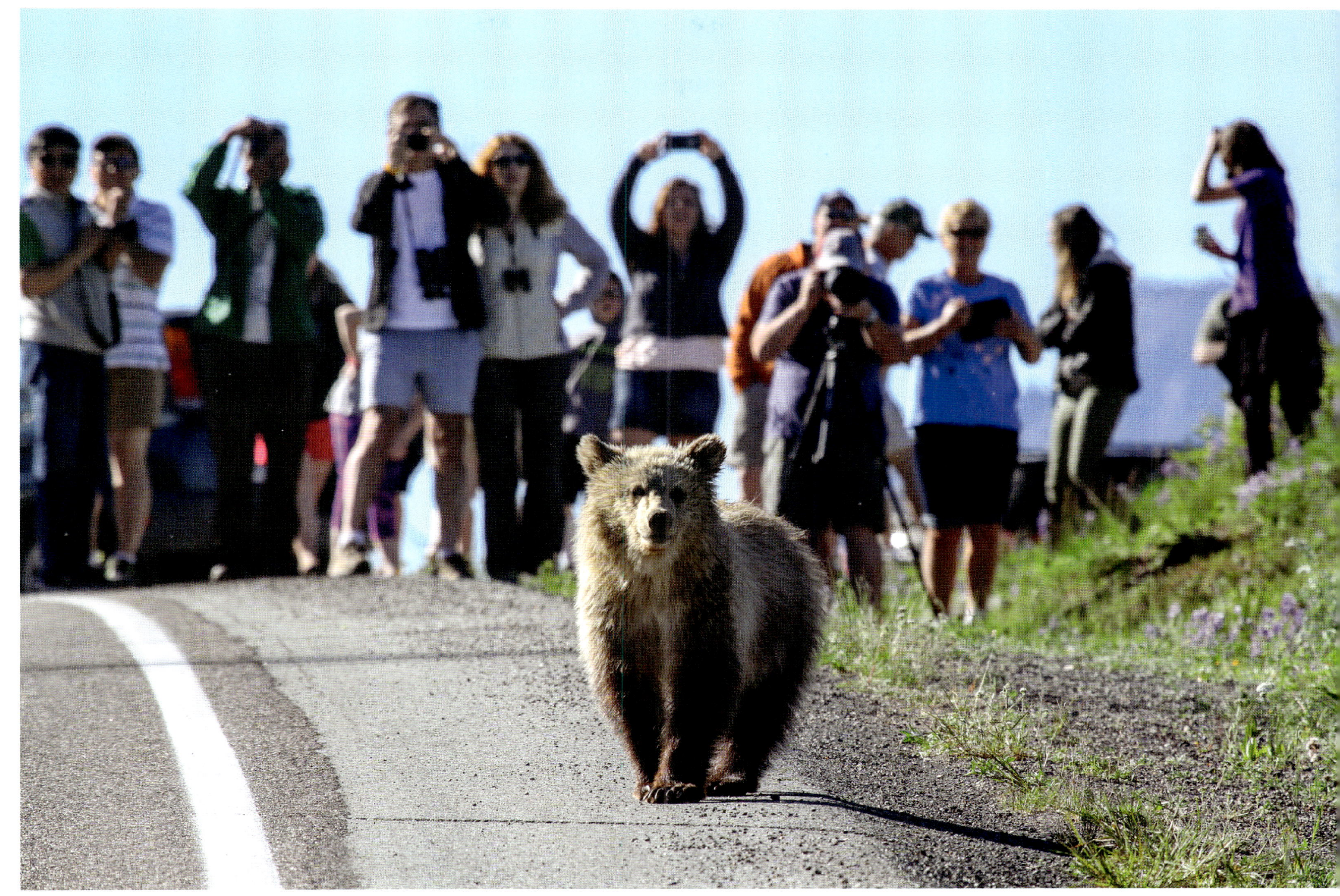

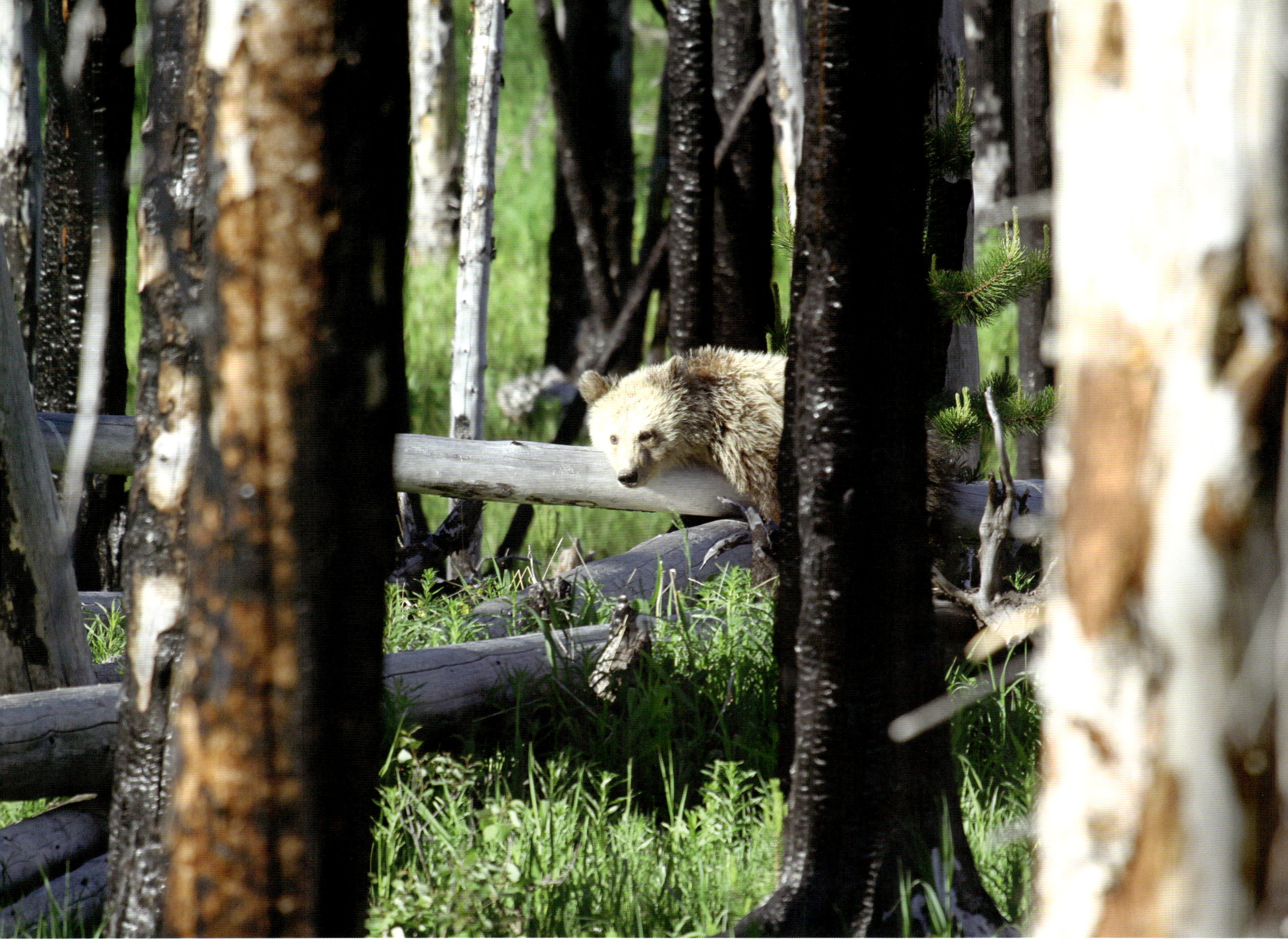

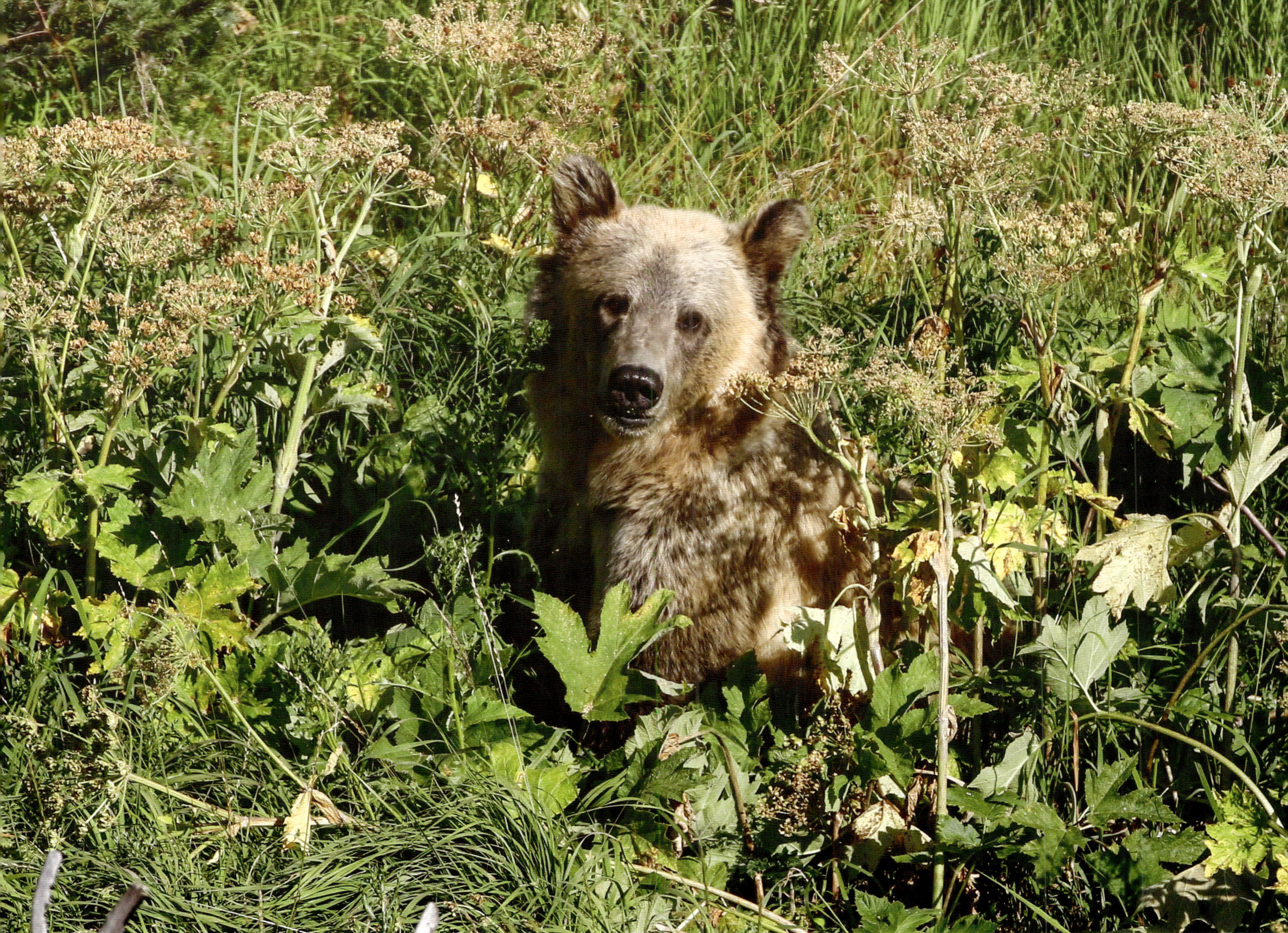

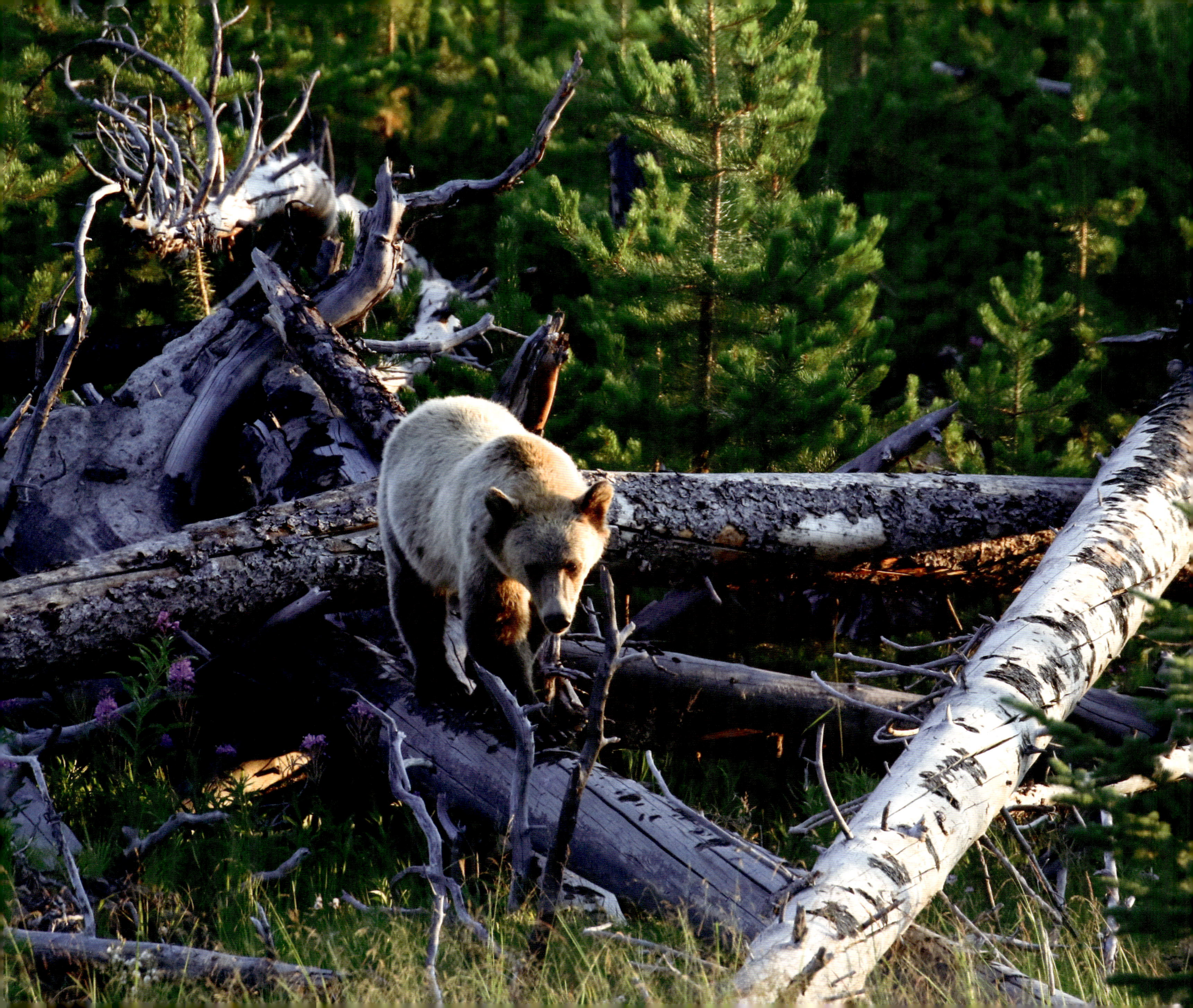

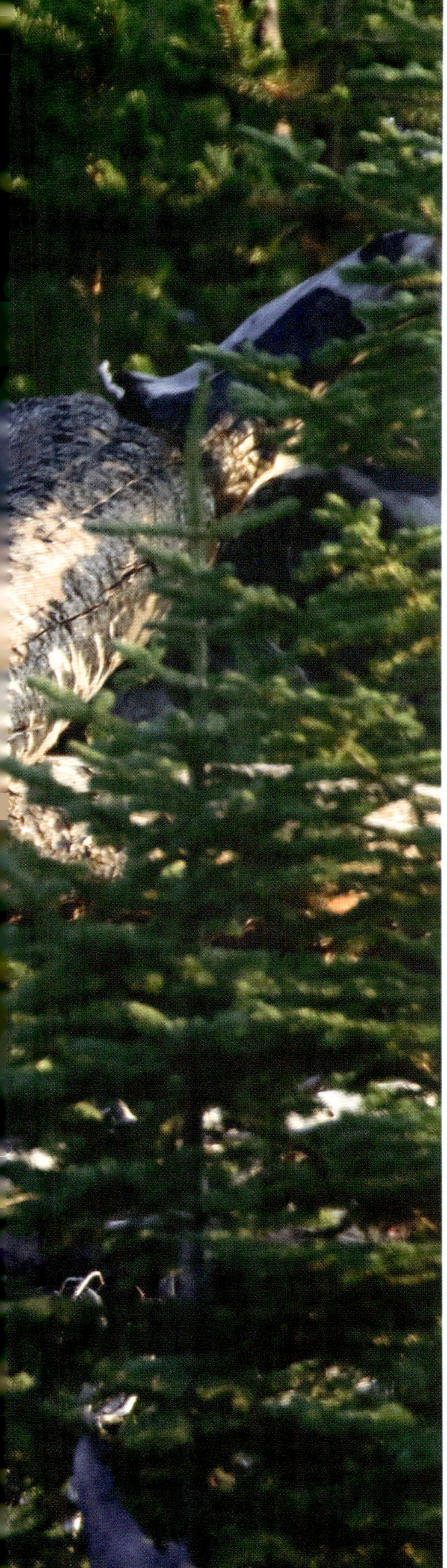

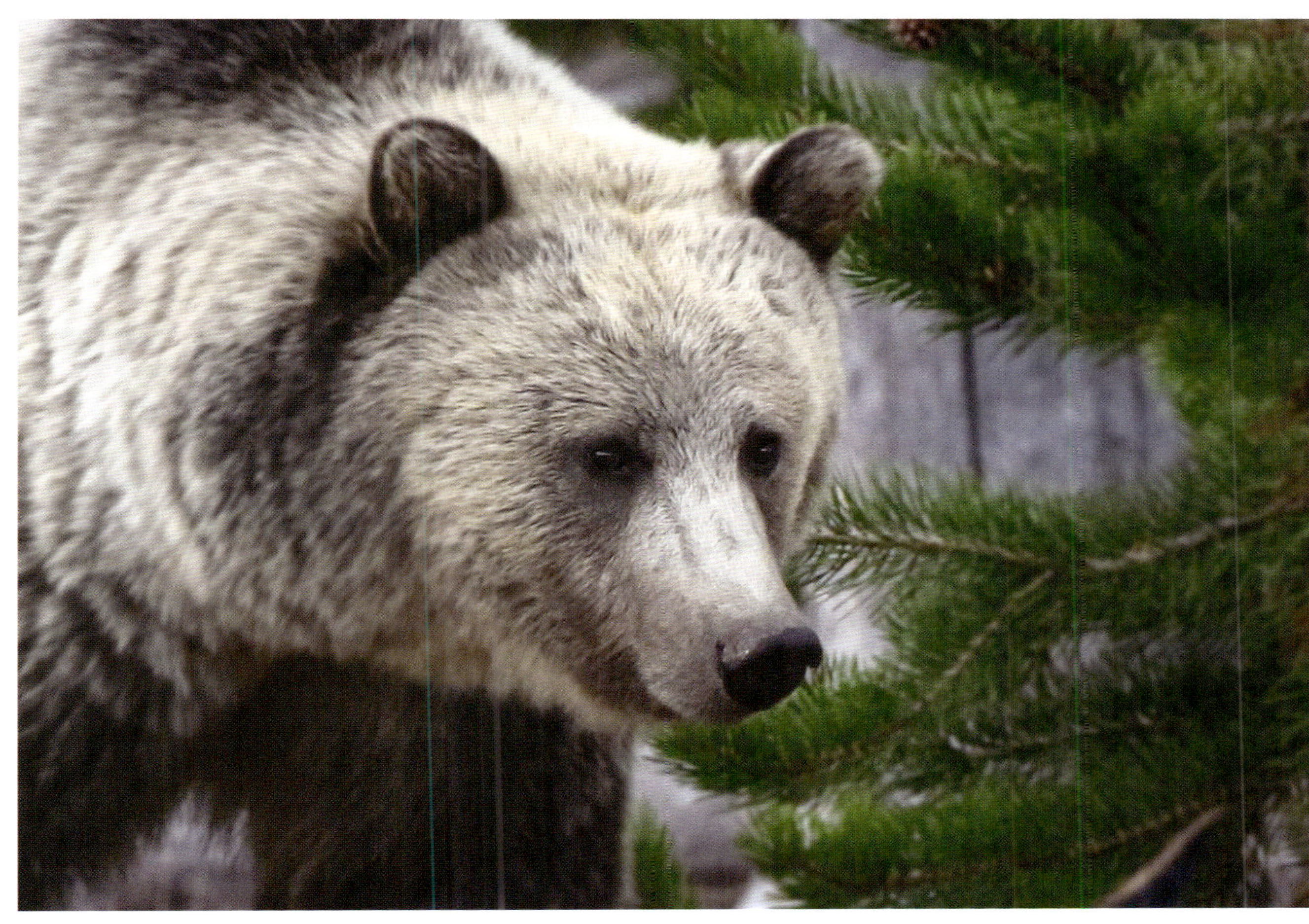

The mountains have always been here, and in them, the bears.

—*Rick Bass*

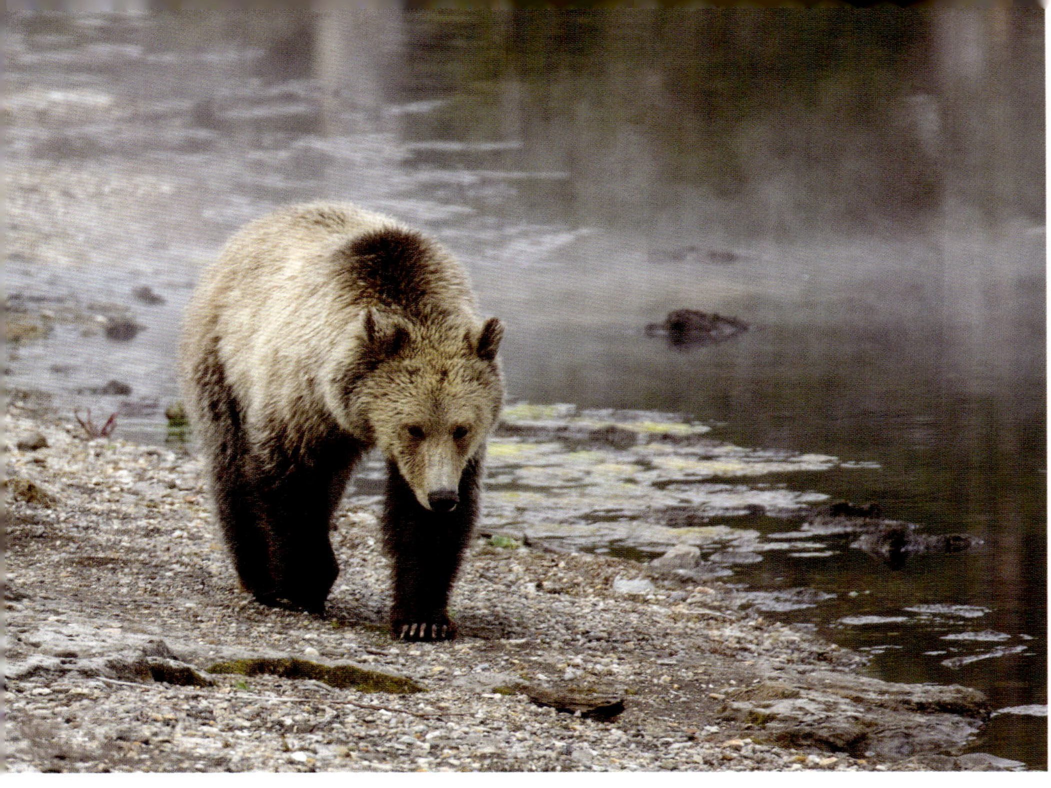
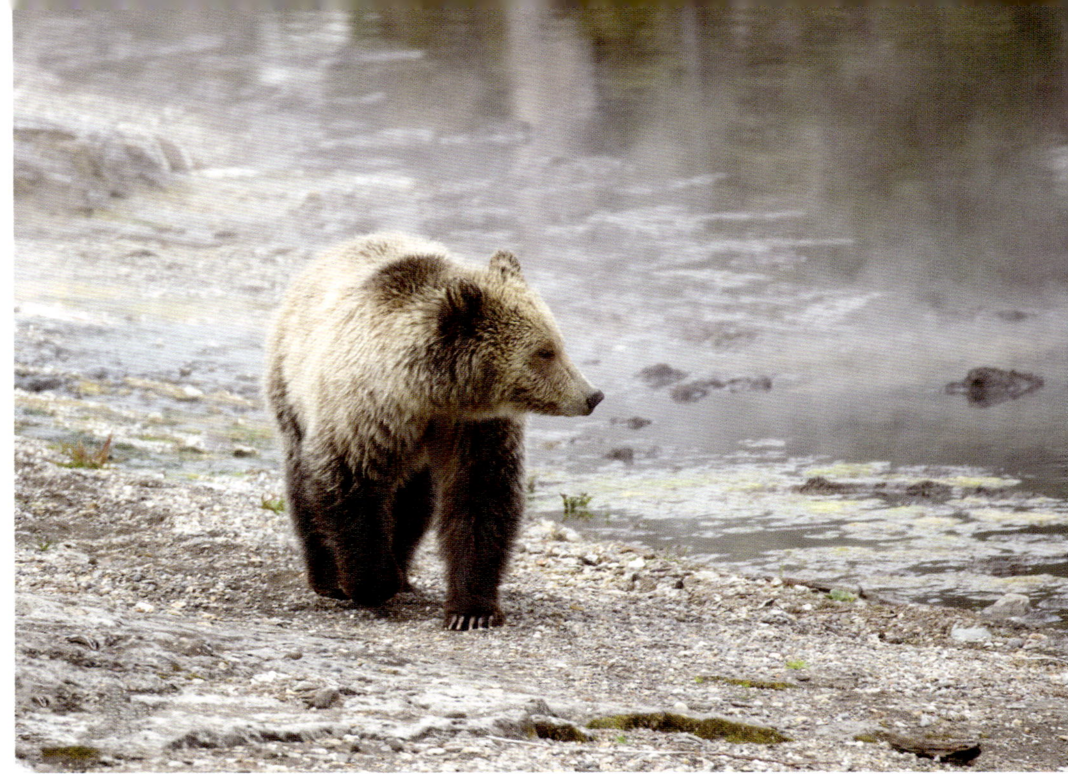
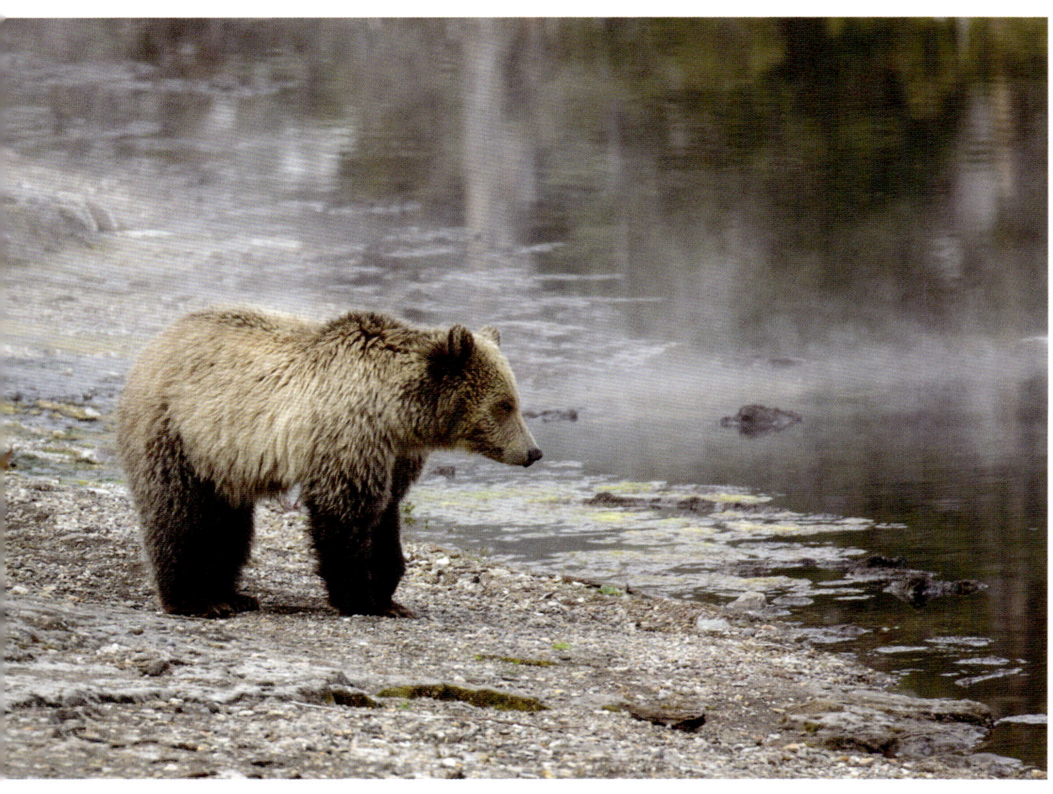
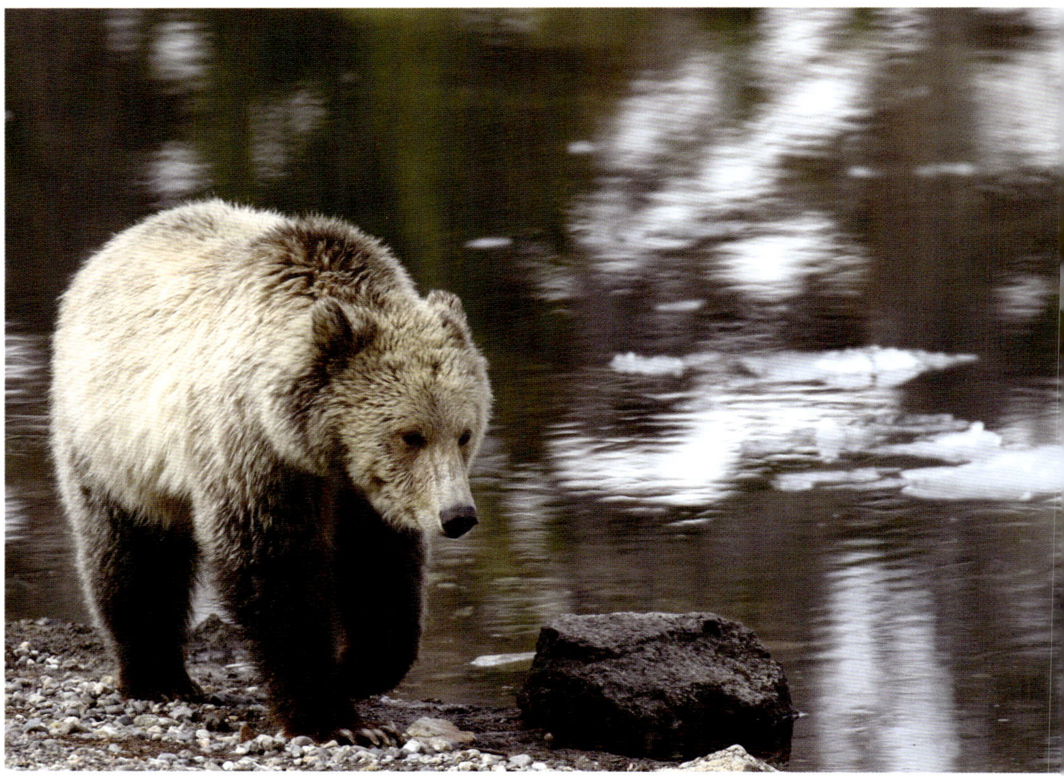

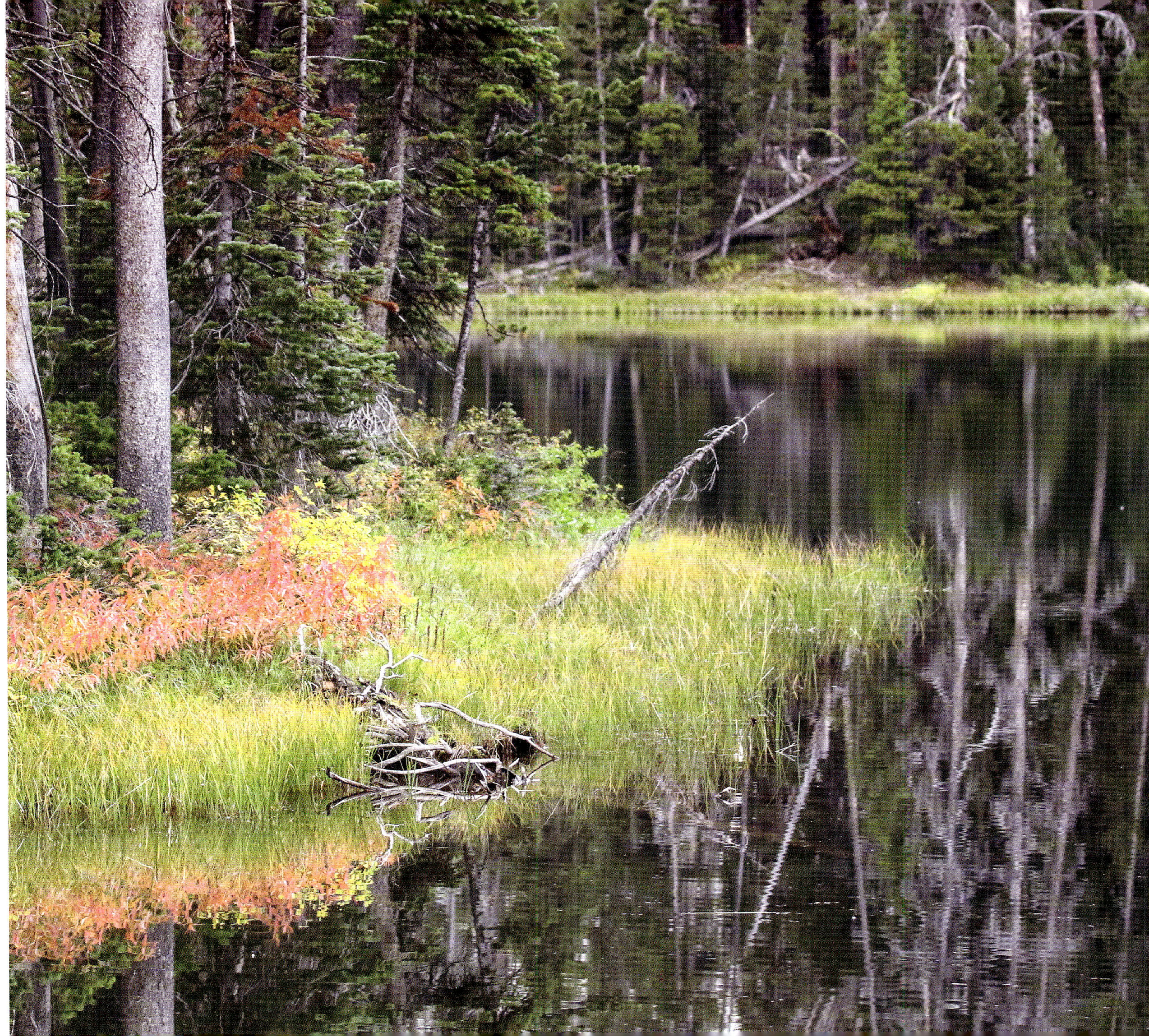

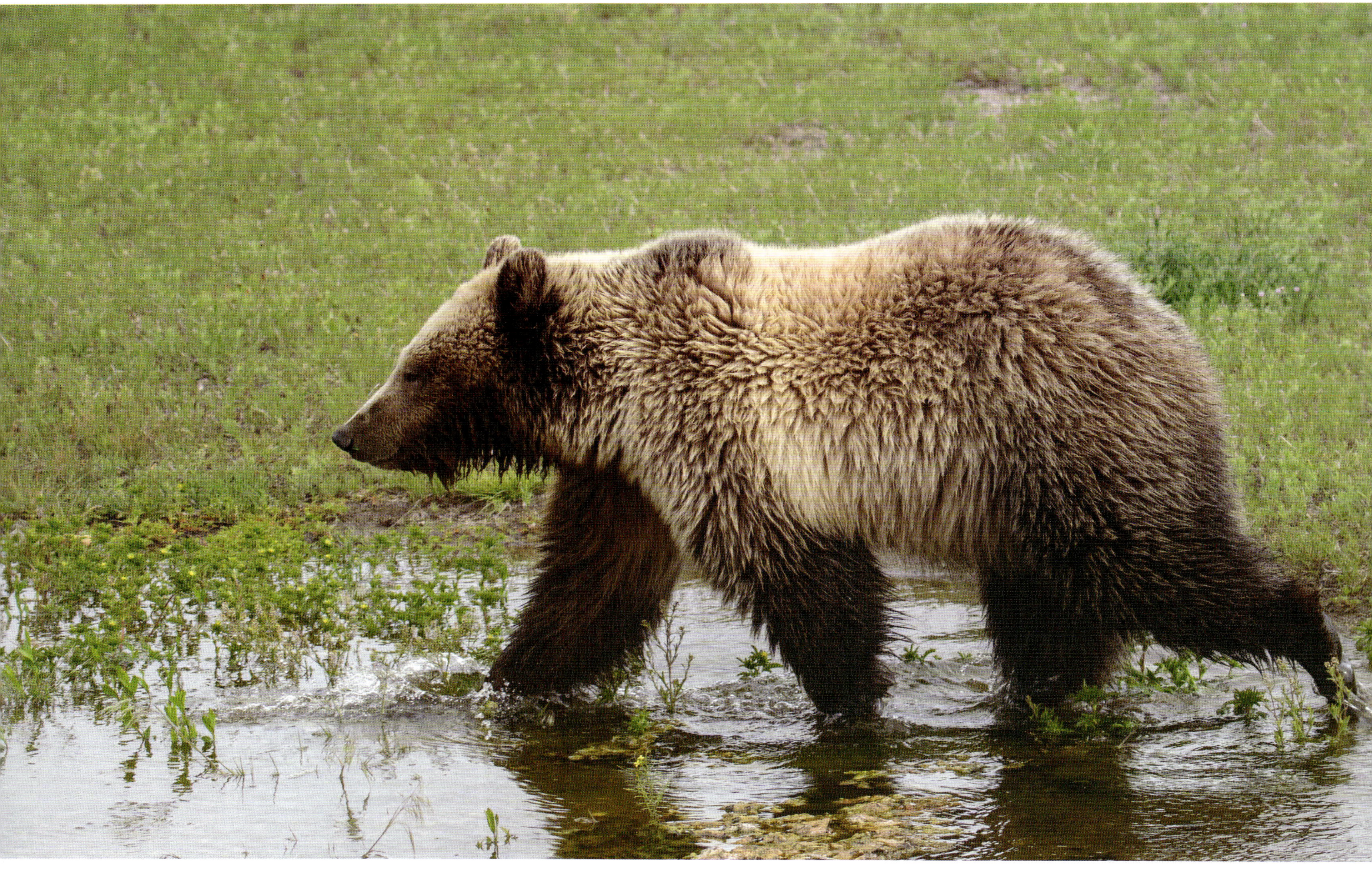

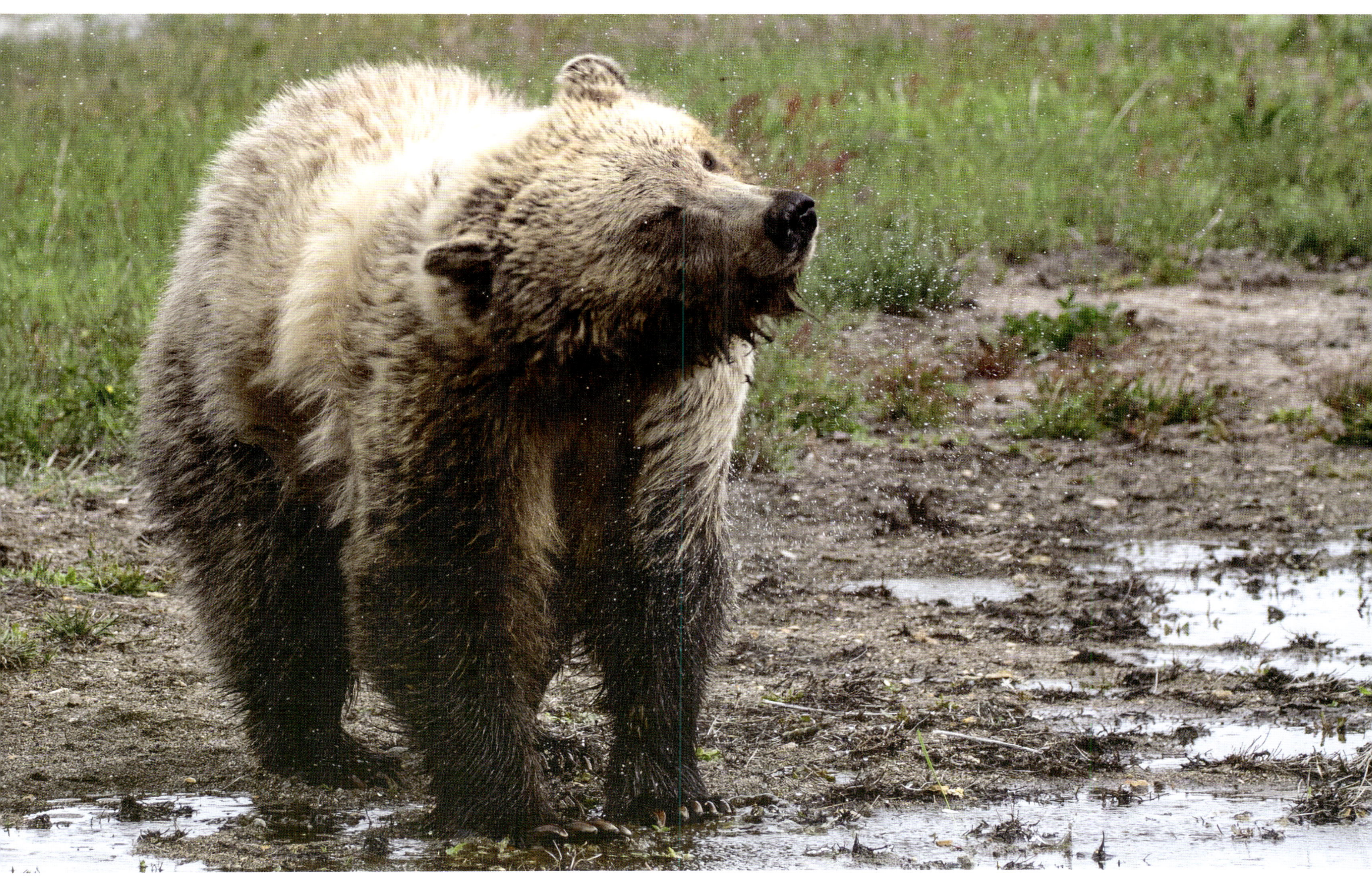

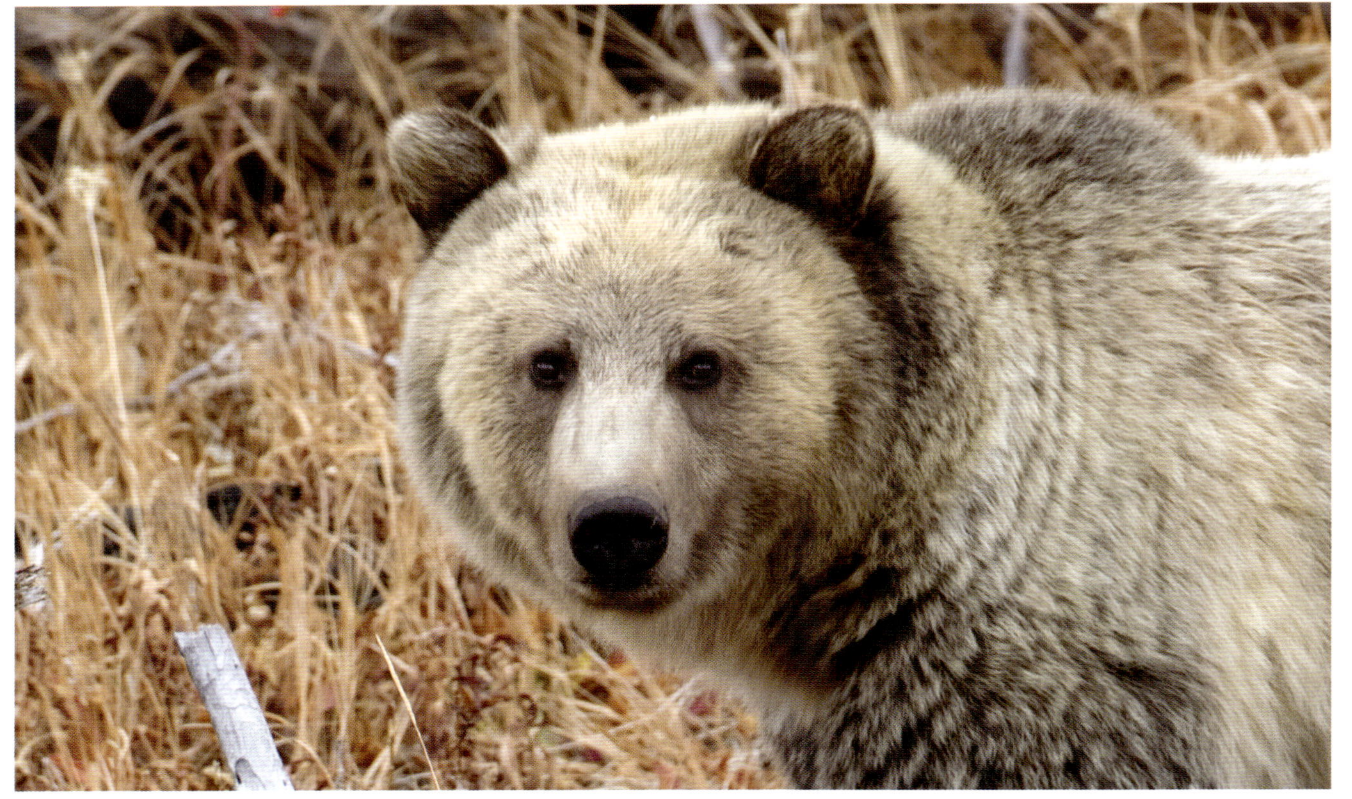

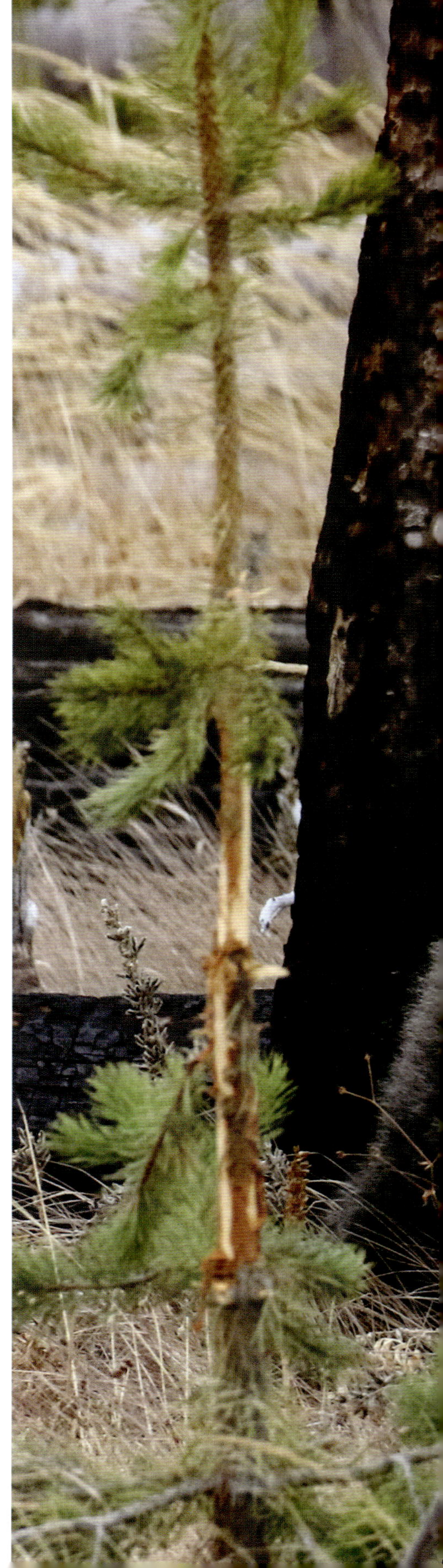

There is something about staring into a bear's eyes: an eerie innocence; a fiery drive for survival. But most of all when I look into a bear's eyes, there is this overwhelming feeling of respect. That is, for a creature that lives in heavily managed landscape, predominantly occupied by humans, but still has the capacity and intelligence to maneuver grounds so intricately to survive and coexist alongside people.

—*Tyler Brasington, Bear Biologist*

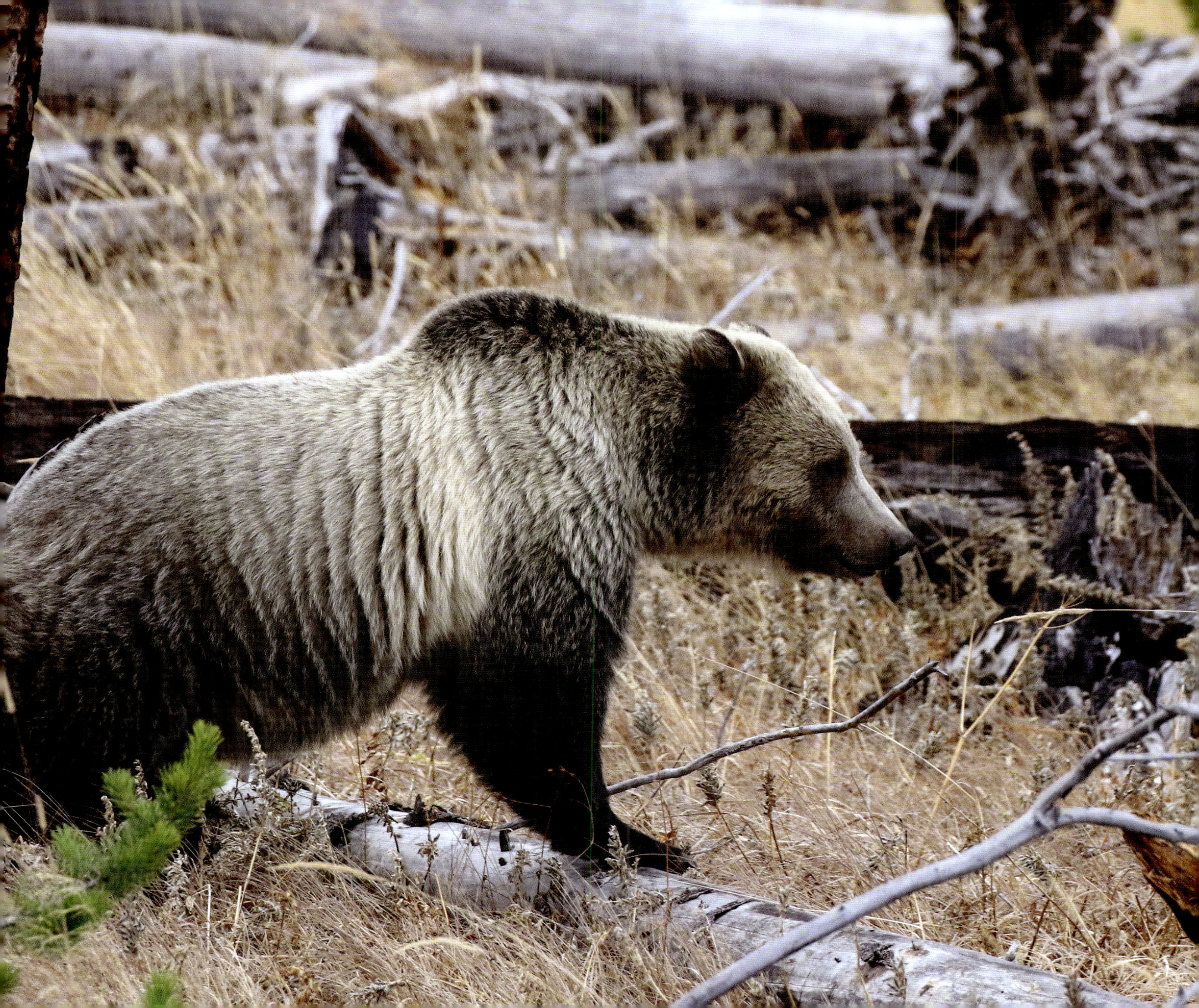

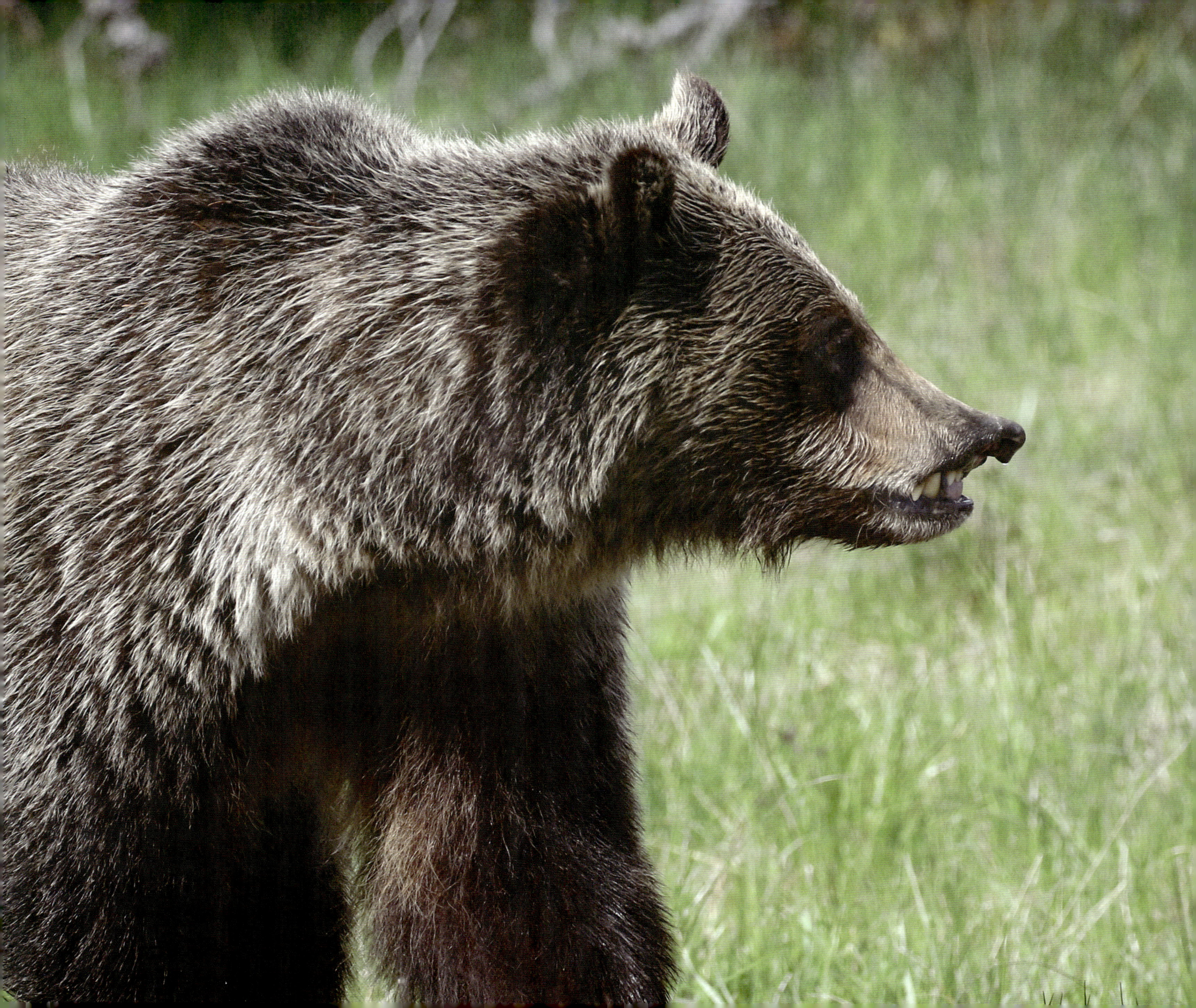

SNAGGLETOOTH

ONCE IN A WHILE YOU COME ACROSS AN ANIMAL WITH A VERY UNIQUE SITUATION. It may be a different color than normal, maybe it's missing an ear, or maybe it only has three legs. Some of these situations are the result of something that happened in life, but some of them aren't. In this particular situation, this grizzly has a rare genetic defect.

As you can see, he is missing part of his face, or his lip, near the right side of his mouth. When I first saw him, I thought he had probably been in a fight with another bear, but after asking other people about him, I learned he has a genetic defect. This defect does not seem to bother him at all, and regardless of how he might look, he is still beautiful in his own unique way.

Snaggletooth, along with his mother and sibling, originated from the area near Crandall, Wyoming. From what I was told, his sibling had the exact same facial deformity, only on the opposite side of its face.

Before coming into Yellowstone, his sibling got into trouble for chasing cattle, and was euthanized. I'm not sure about his mother, but I'm guessing it was time for Snaggletooth to be out on his own, so they probably went their separate ways.

When I photographed him, he was on the east side of Yellowstone Lake. Considering his original location, it really is amazing to think about how far a bear will actually travel. I'll be interested to see if he makes Yellowstone his home, or if he'll continue on his journey to somewhere else.

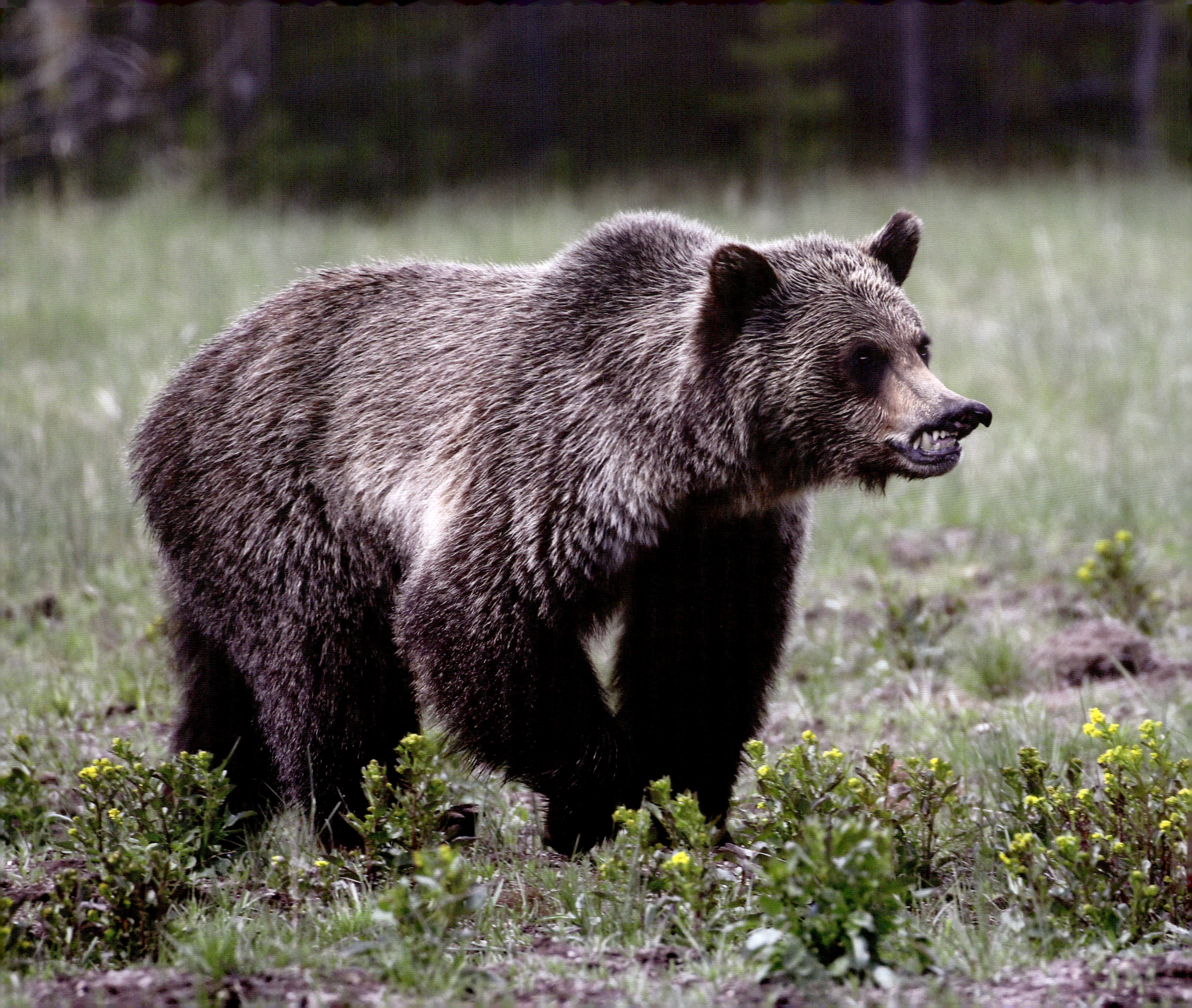

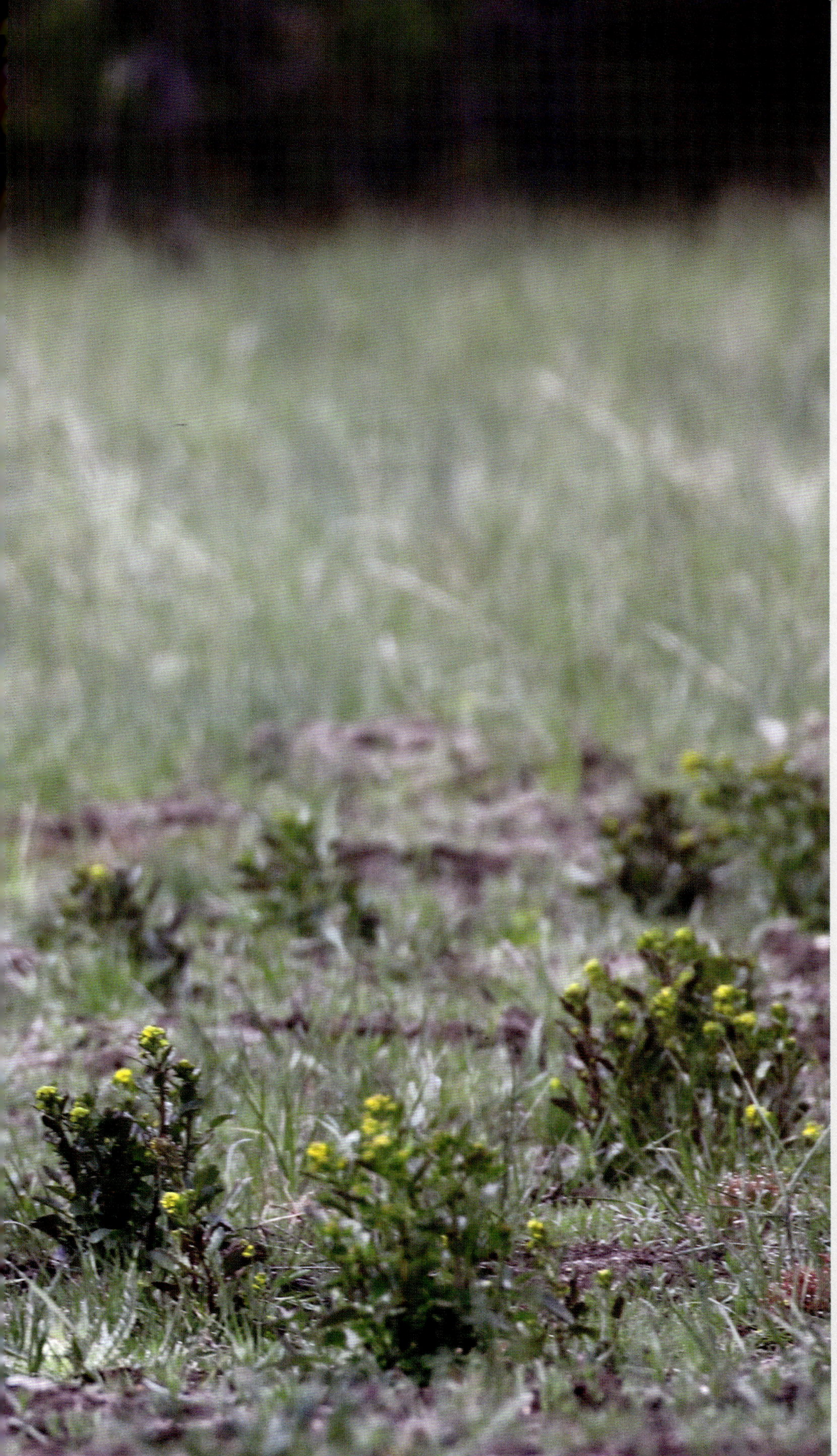

Today, you can find grizzlies in only half of their historical range. They are continuing to be displaced by human settlement, new roadways, logging, mining, increased back country travel, and a reduction of food sources. Because of this, bear-and-human conflicts continue to rise. Human-influenced death is the greatest mortality cause for grizzlies, averaging about eighty-five percent of the deaths, while natural-caused deaths average about fifteen percent.

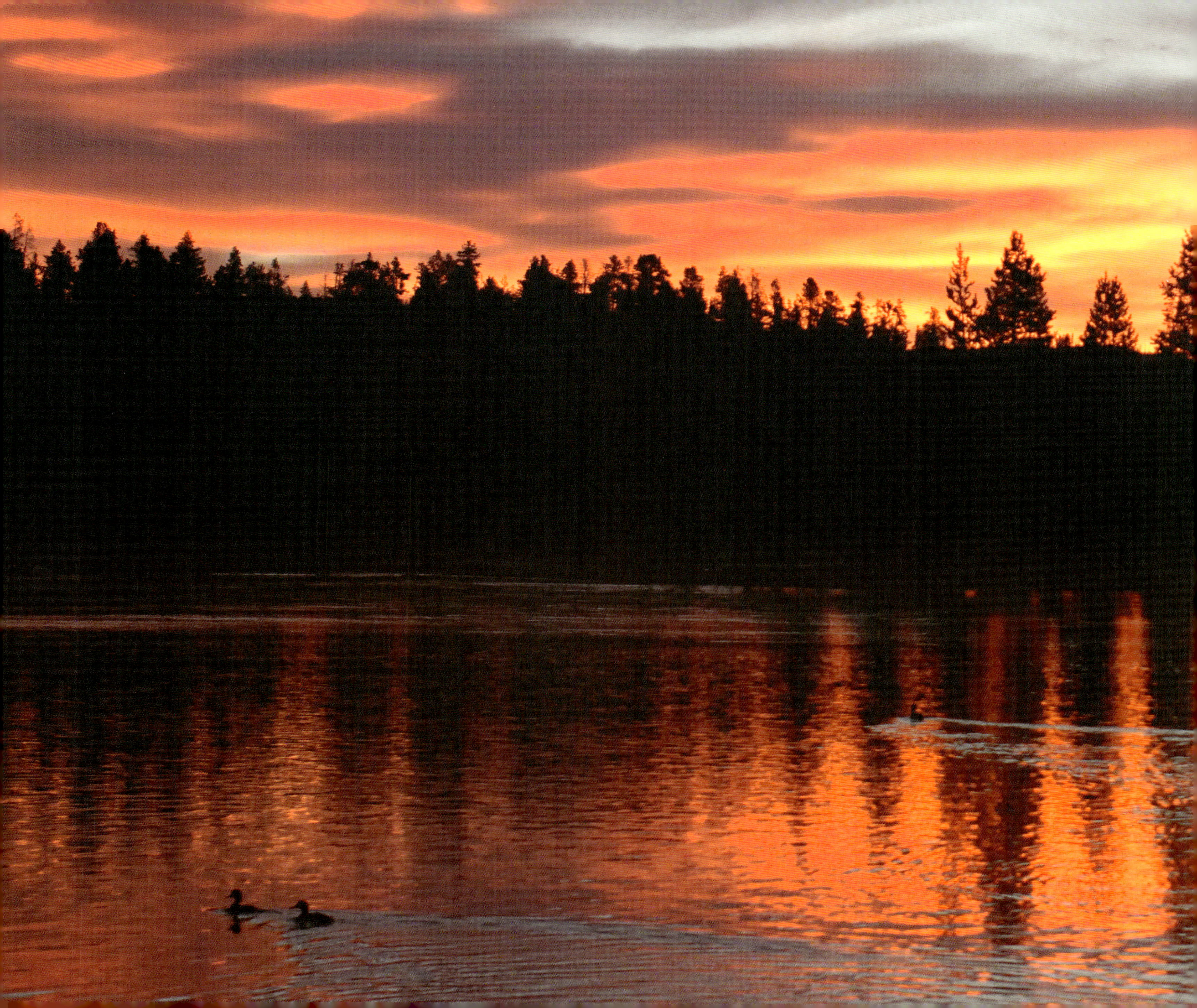

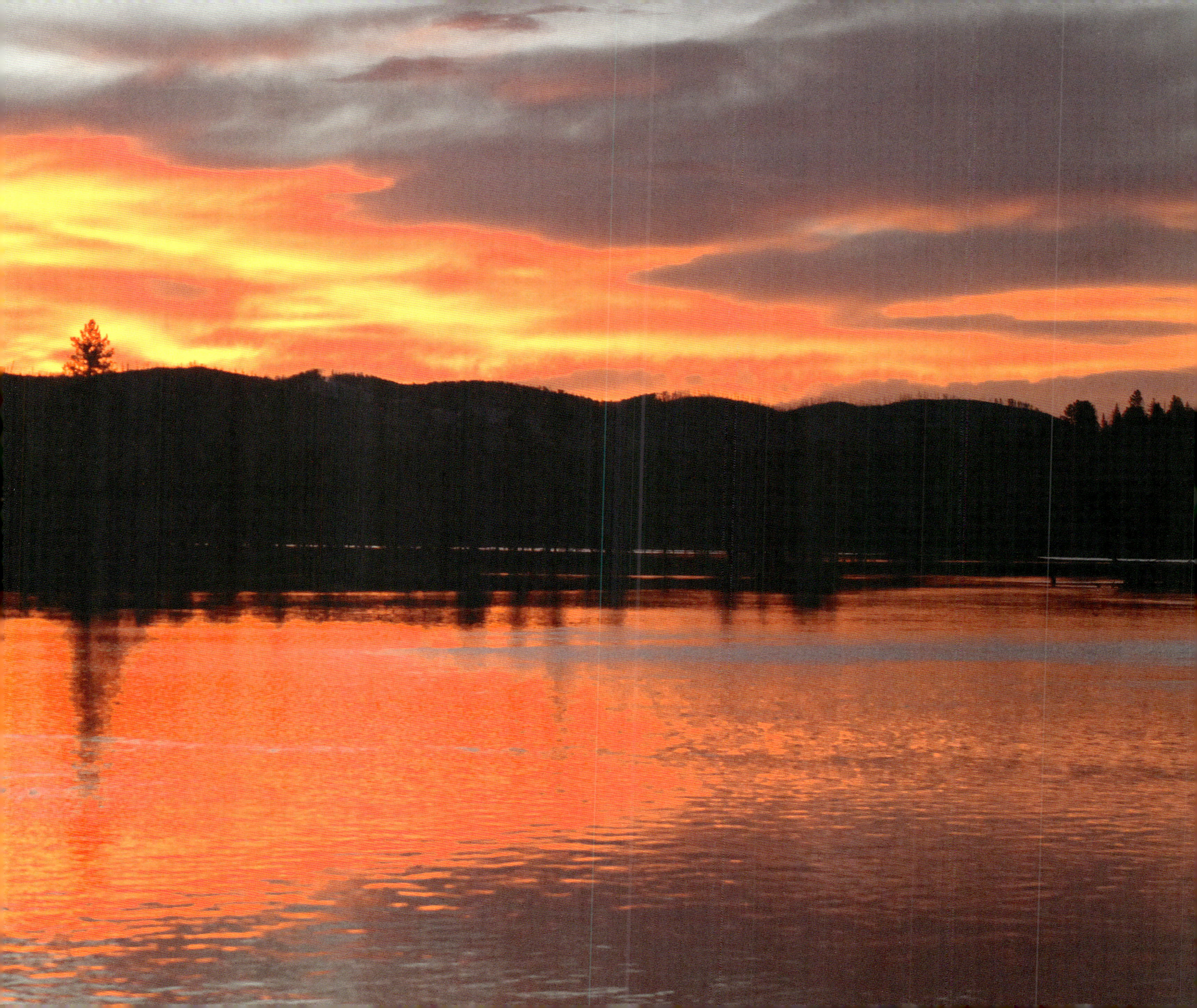

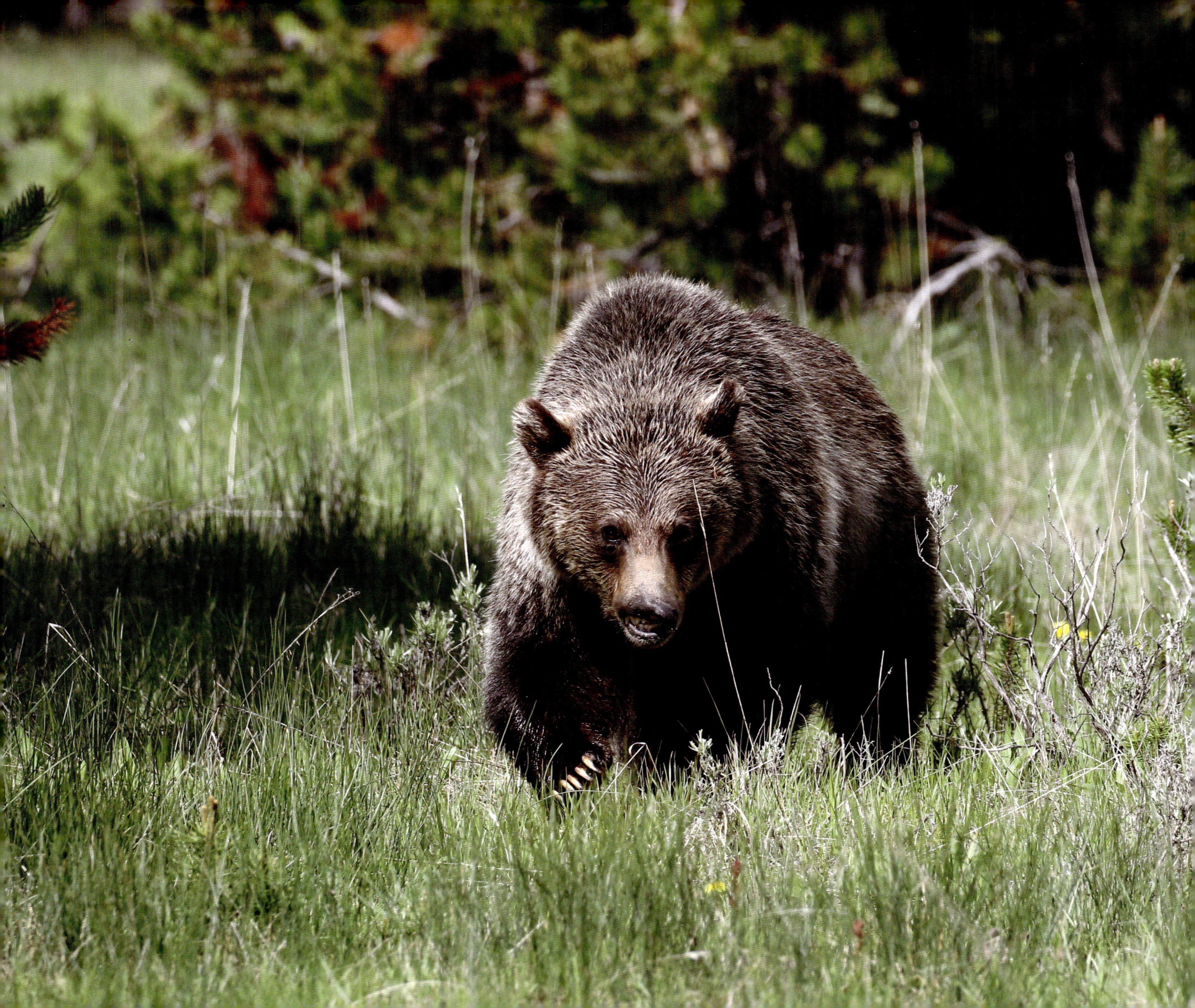

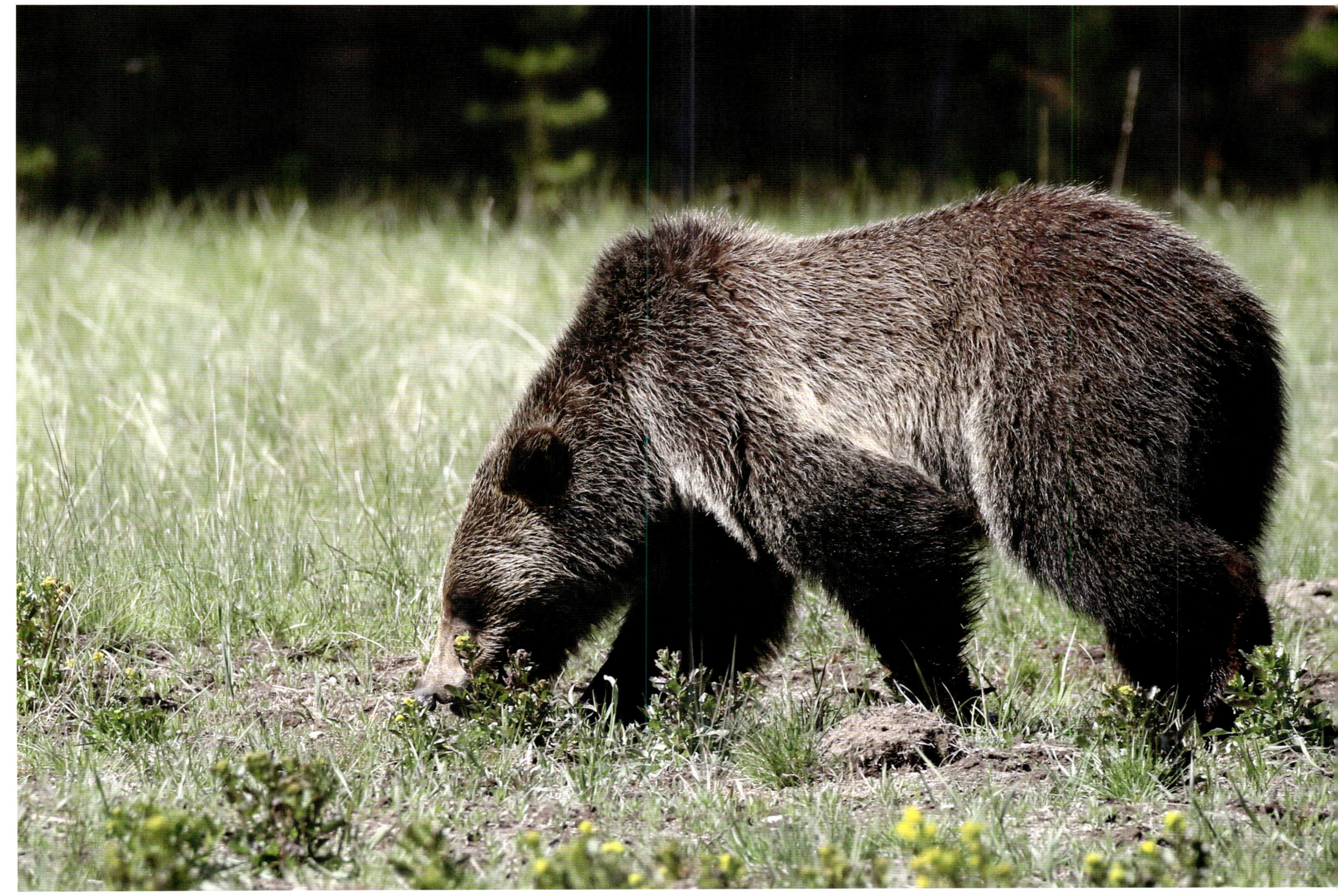

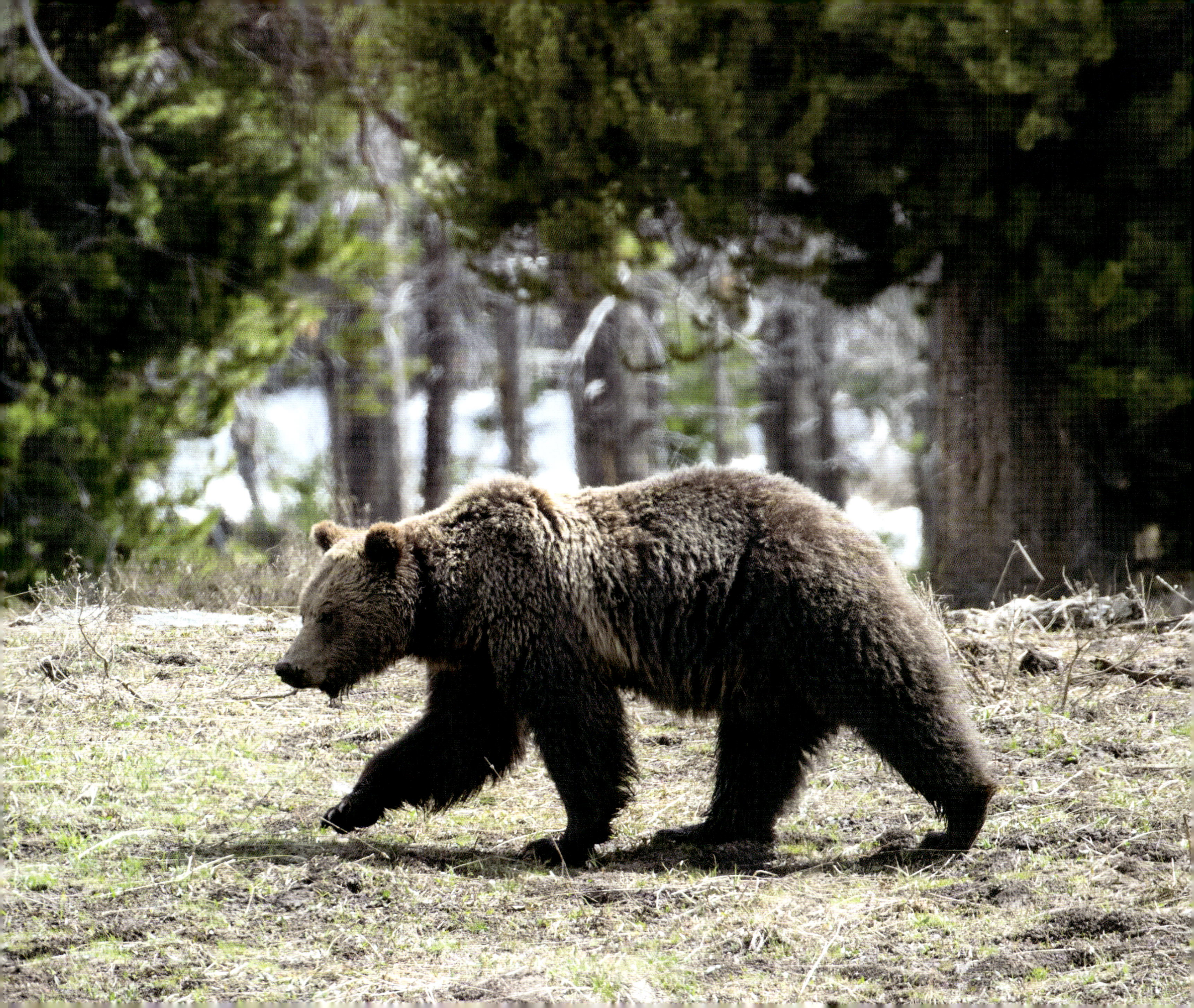

THE OBSIDIAN SOW

THE OBSIDIAN SOW IS A FAIRLY SMALL FEMALE GRIZZLY. The average weight of female grizzlies in the Greater Yellowstone Ecosystem is between 200 and 400 pounds. She is definitely on the lower end of that scale.

The Obsidian was born in 2007, and in 2015 she was collared as bear 815. She has since dropped her collar. She has had a lot of cubs, but sadly many of them have not survived. I have often wondered if that is because of her size.

In the spring of 2019, the Obsidian appeared with three cubs-of-the-year. I had the pleasure of watching this family of four shortly after they emerged from the den. What a sight that was! It was the first time I had ever seen a grizzly with three COY. Boy did she have her work cut out for her.

Two of those little cubs were very brave and would quite often venture out of the trees with their mother. The other cub was very shy, and when mom and the other two siblings would head out, it would hang back in the cover of the forest and start screaming. (At least that is what I would call it.) I remember sitting in my car with tears in my eyes as I rolled the window up whenever it would happen. I can definitely say that is a sound I do not like.

Keeping those little ones safe is a tough job, and when you have one that is screaming like that, it can quickly draw the attention of a boar that could come in and wipe the cubs out in a matter of seconds. Thankfully, once the screaming started, the Obsidian would make her way back to the cub and reassure it and then it would stop.

Eventually all three little ones were comfortable coming out of the trees. They would play together, climb over logs, and do all the things bear cubs like to do. Mom was very attentive and managed to keep all three safe for a while. Unfortunately one of the cubs disappeared and this family of four became a family of three.

These bears were one of the main attractions in Yellowstone in 2019. Sadly many visitors to the park did not follow the rules when viewing them, and, for the bears' protection, the park rangers had to haze them quite often to scare them away from people and to change their behavior. Many people didn't like this, but for the safety of the bears, it had to be done.

In 2020, both mom and the remaining two cubs were seen on occasion, and all were still doing well.

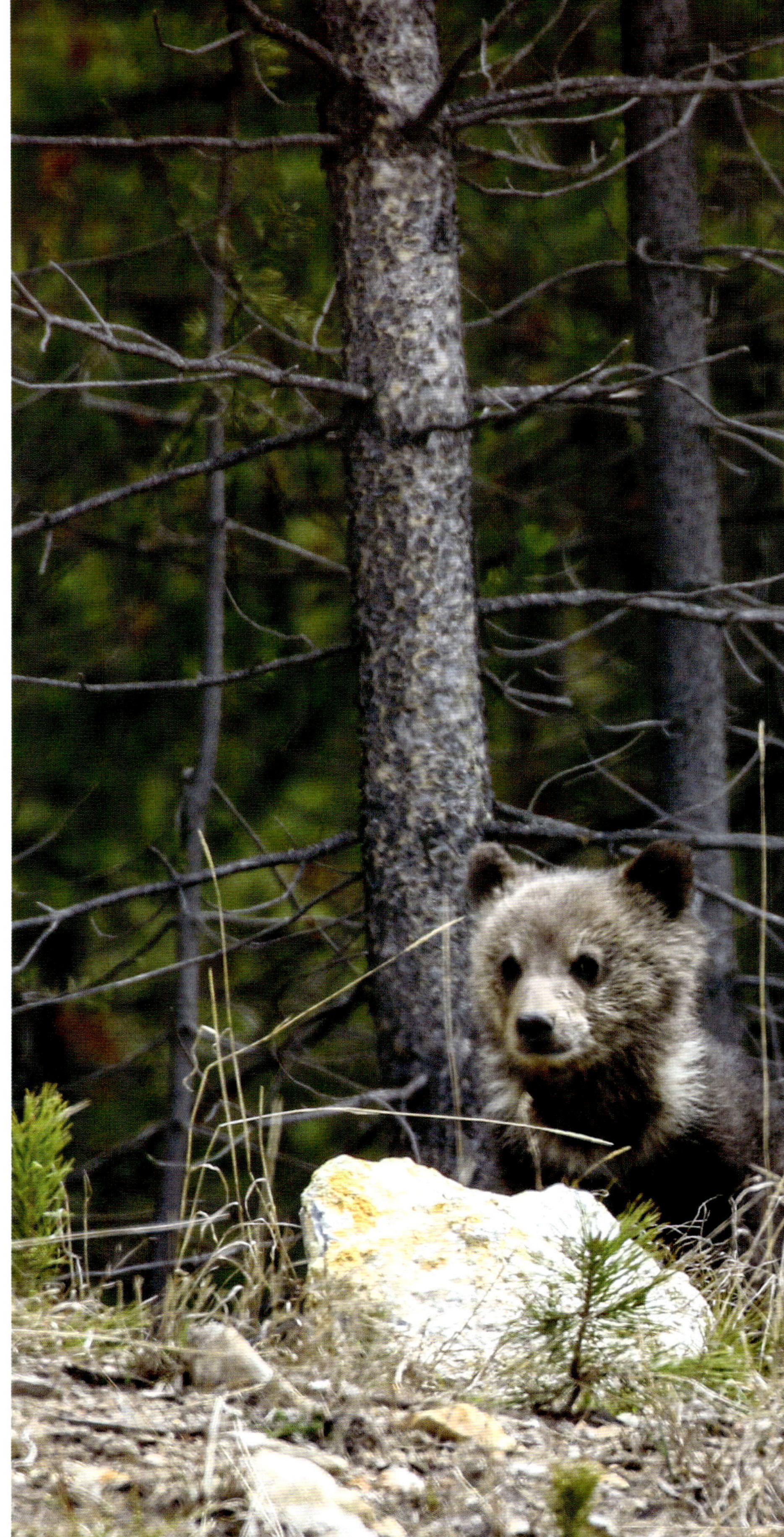

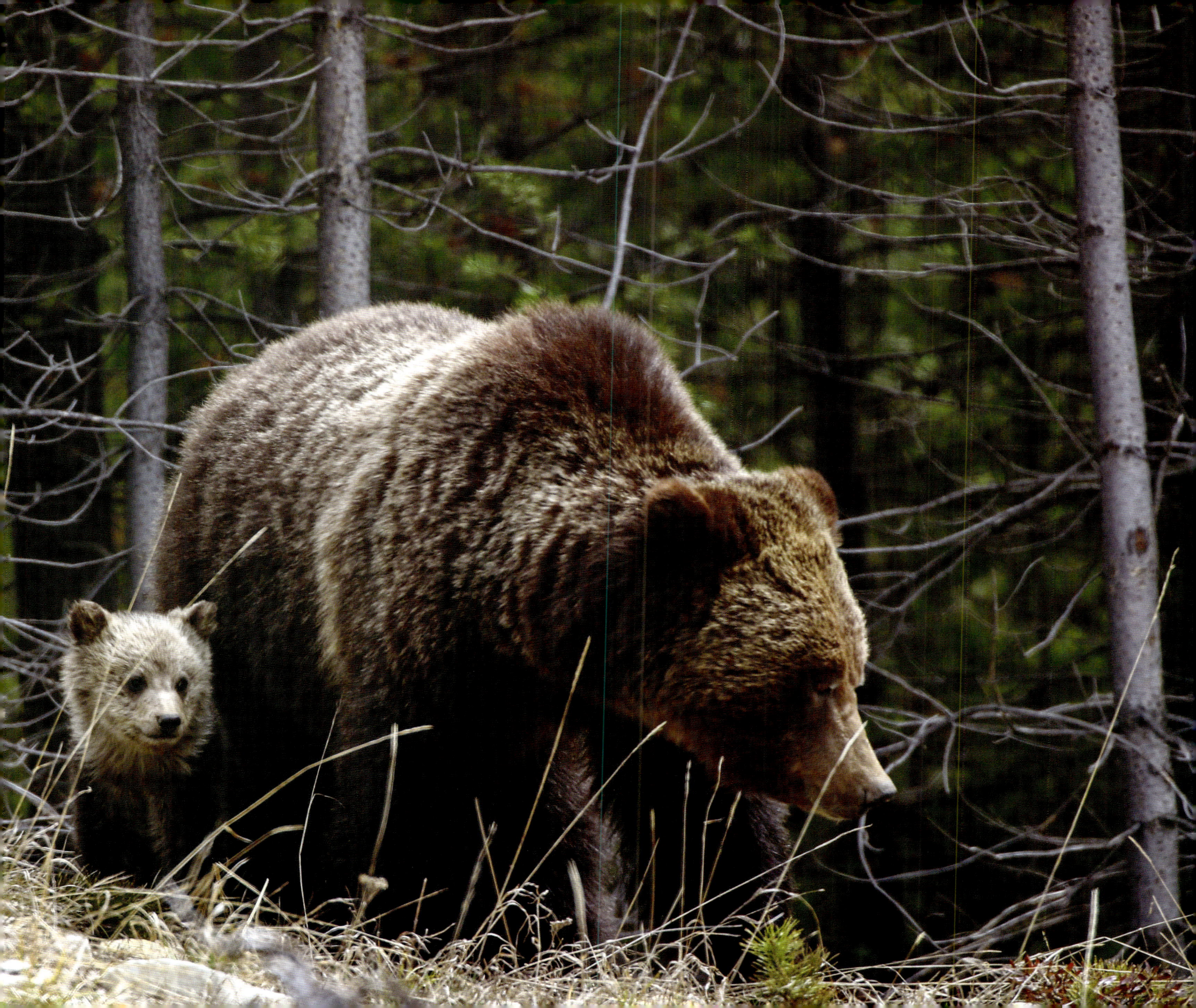

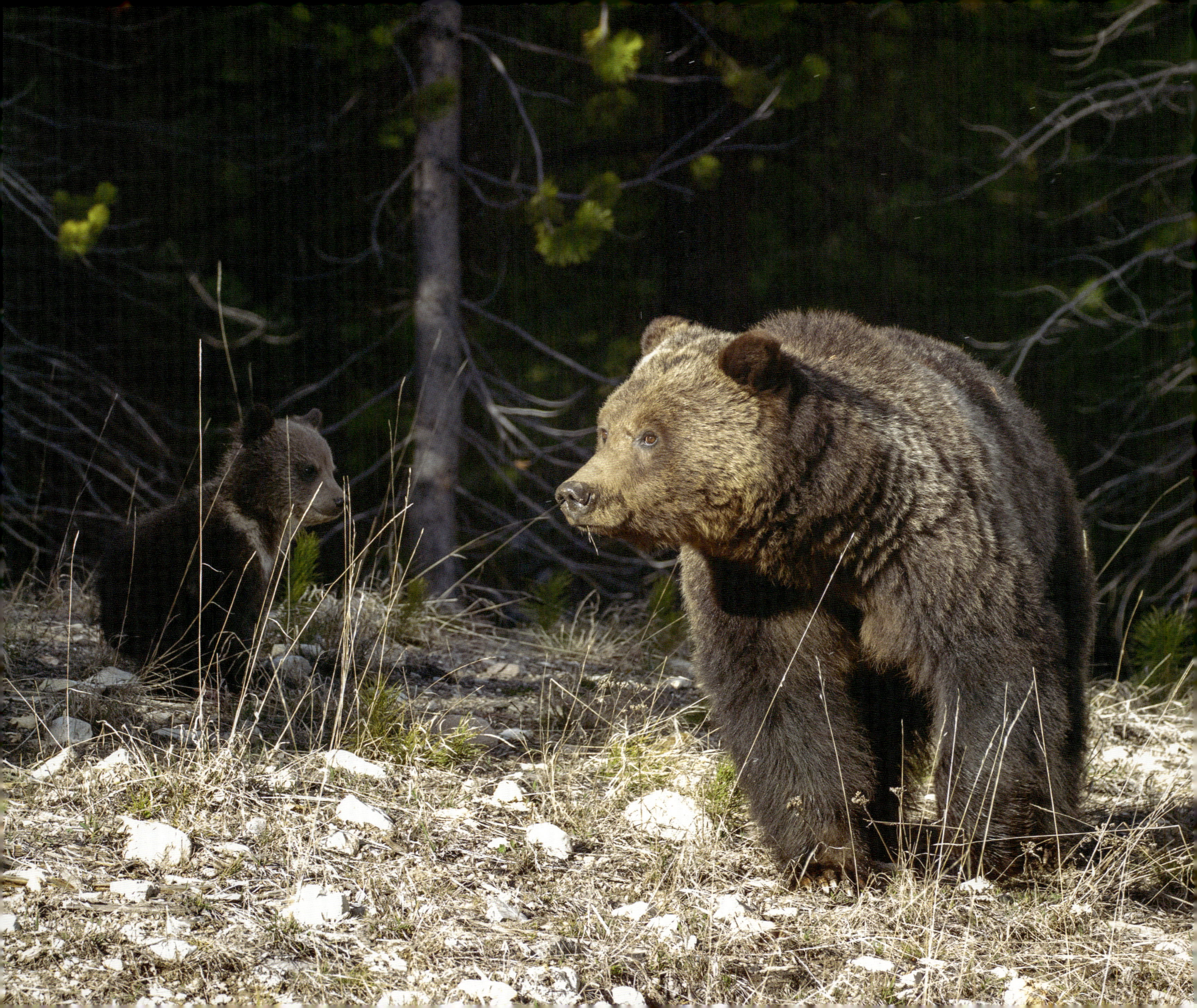

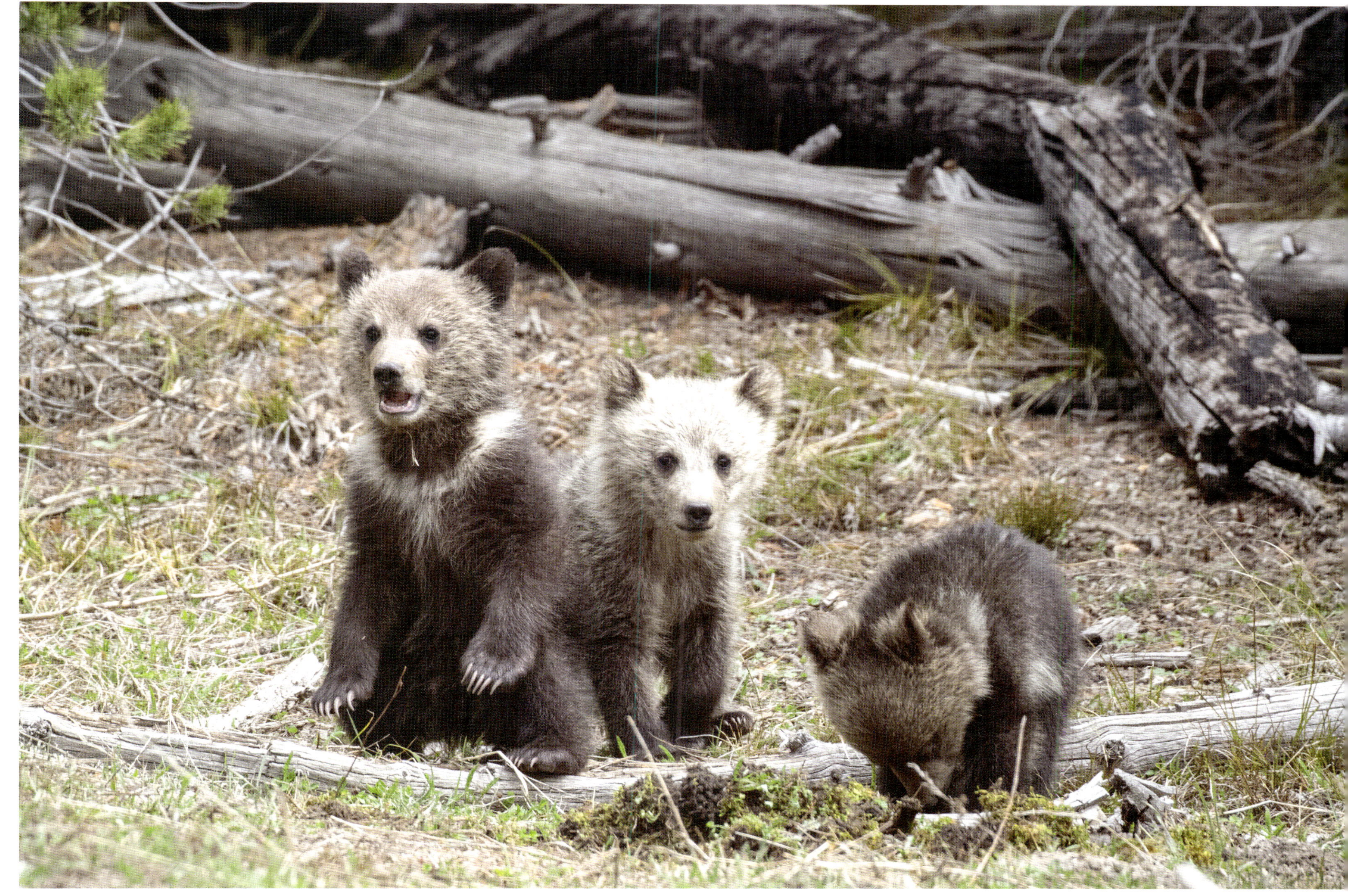

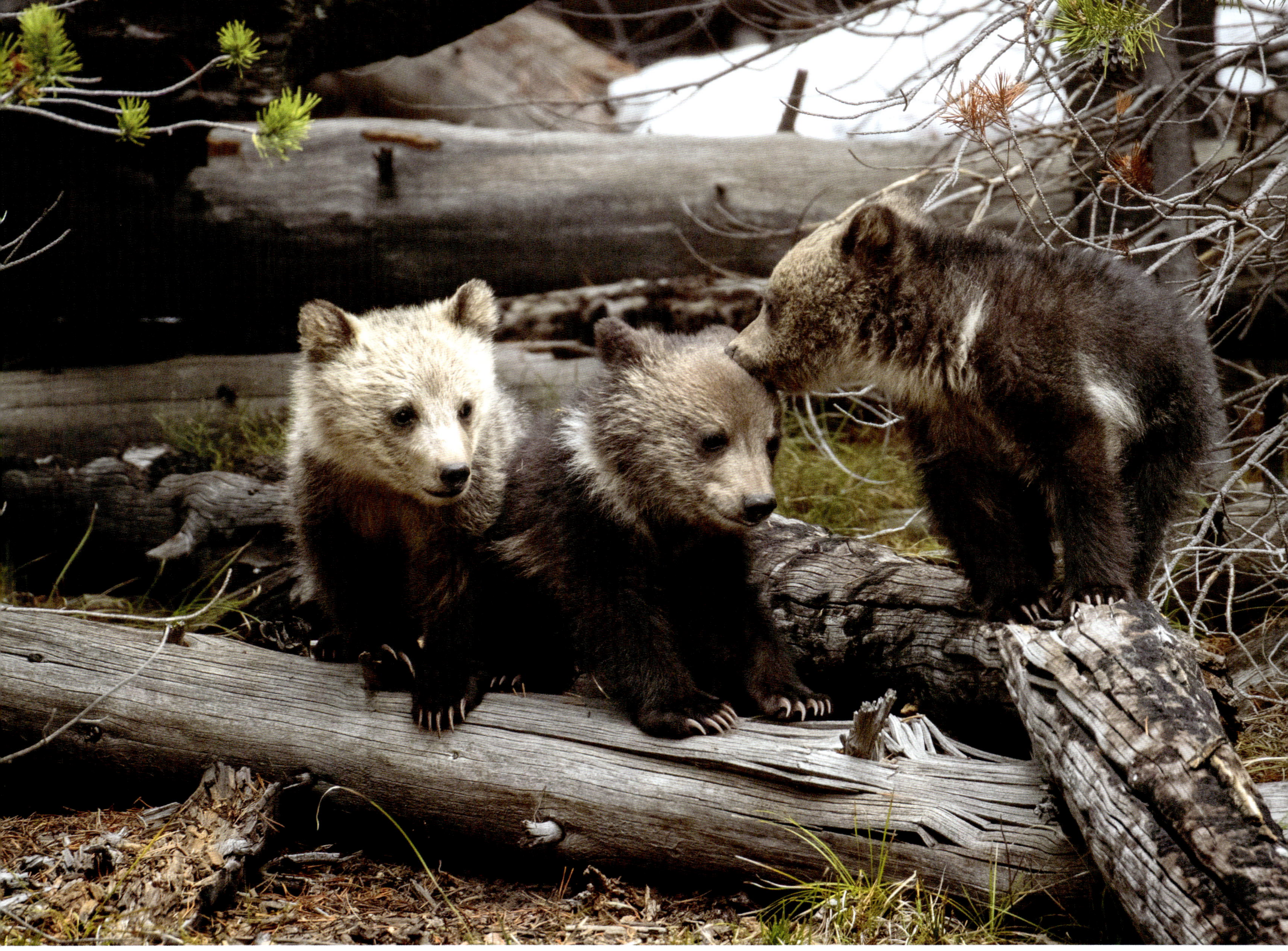

Cub survival rate is estimated based from the period when bears emerge from dens in the spring to the when the last cub observation happens prior families denning up for winter. Cub survival was sixty-four percent from 1983 to 2001, and decreased to fifty-five percent from 2002 to 2011. Yearling survival during the same time periods decreased from eighty-two percent to fifty-four percent.

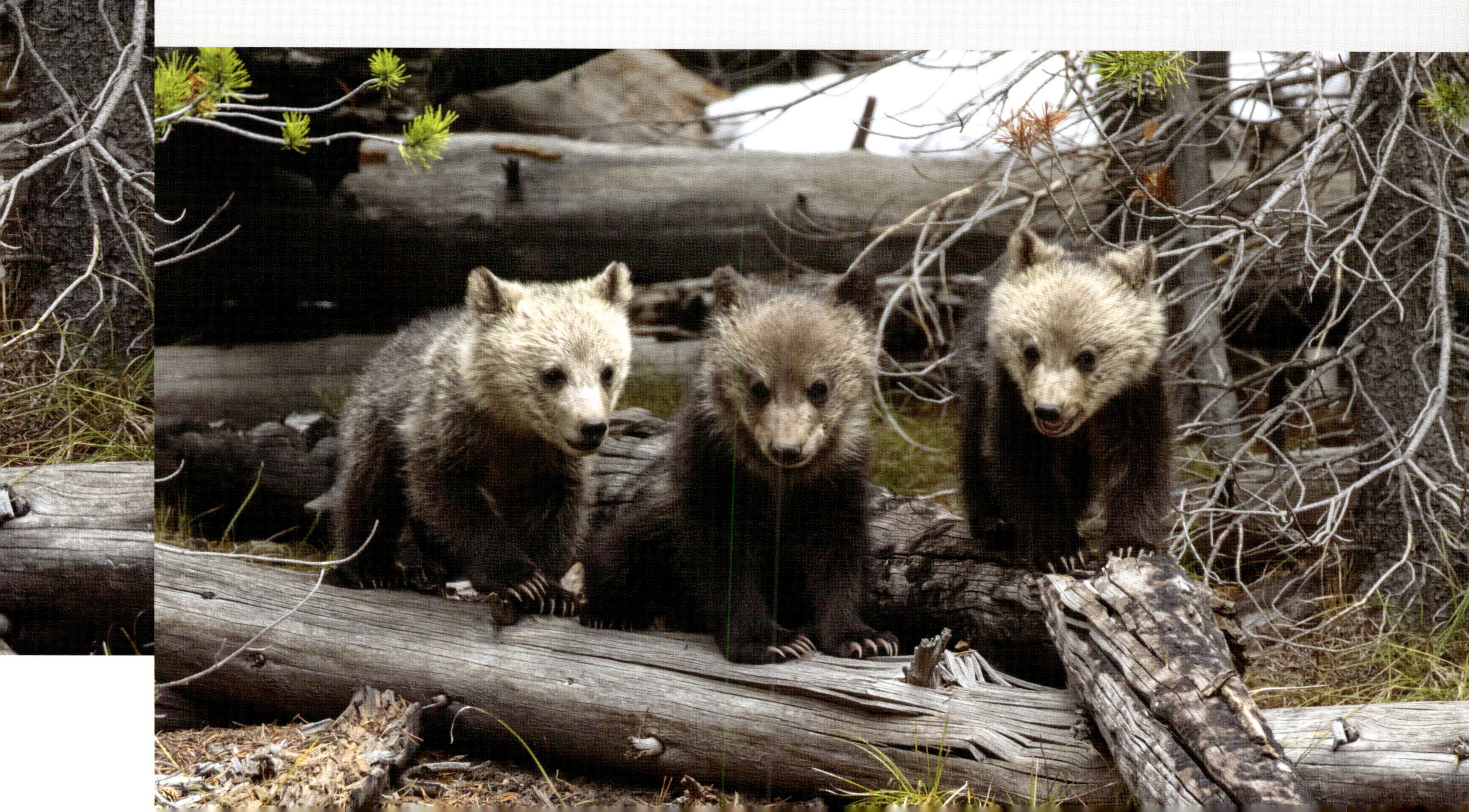

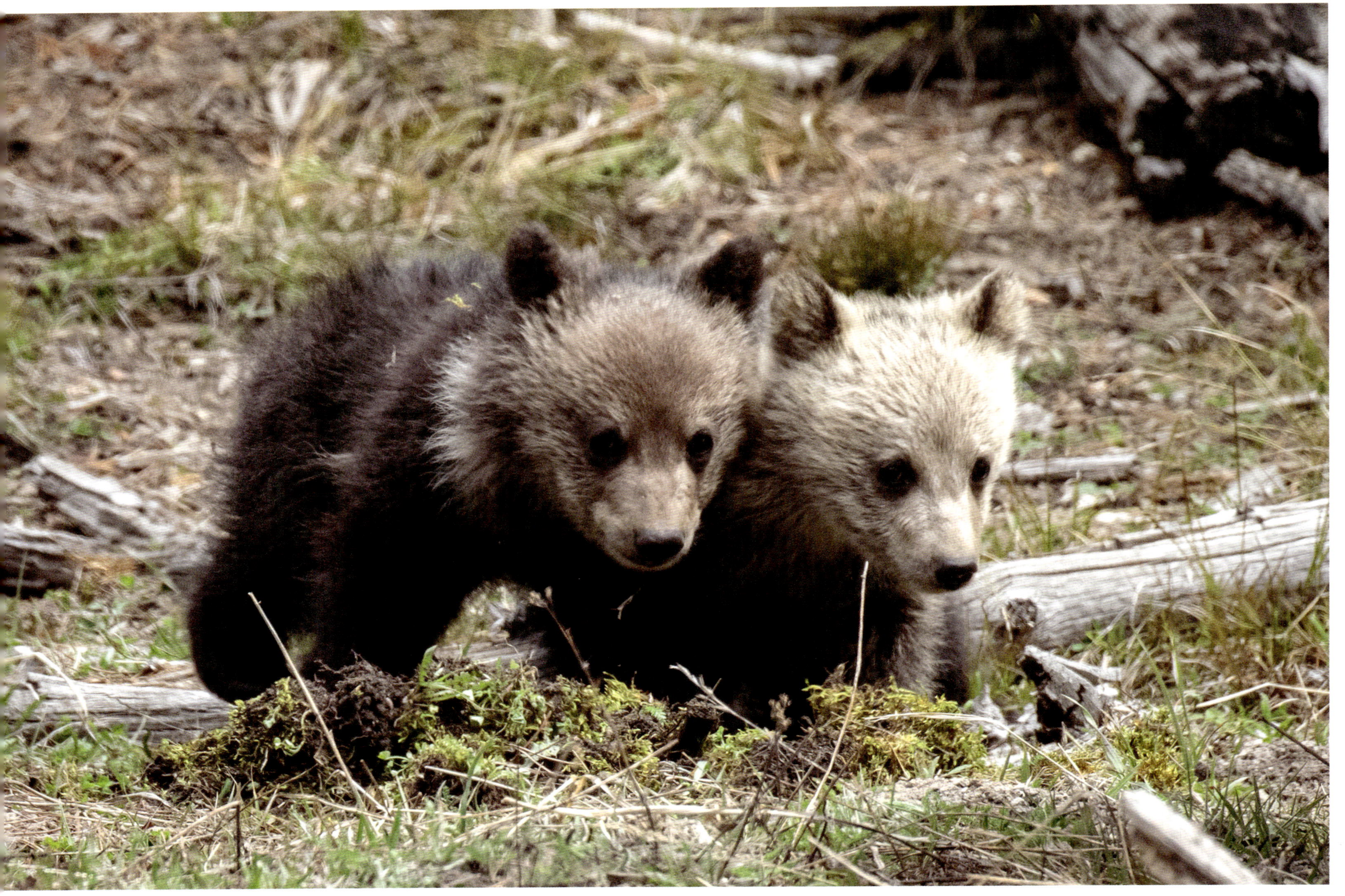

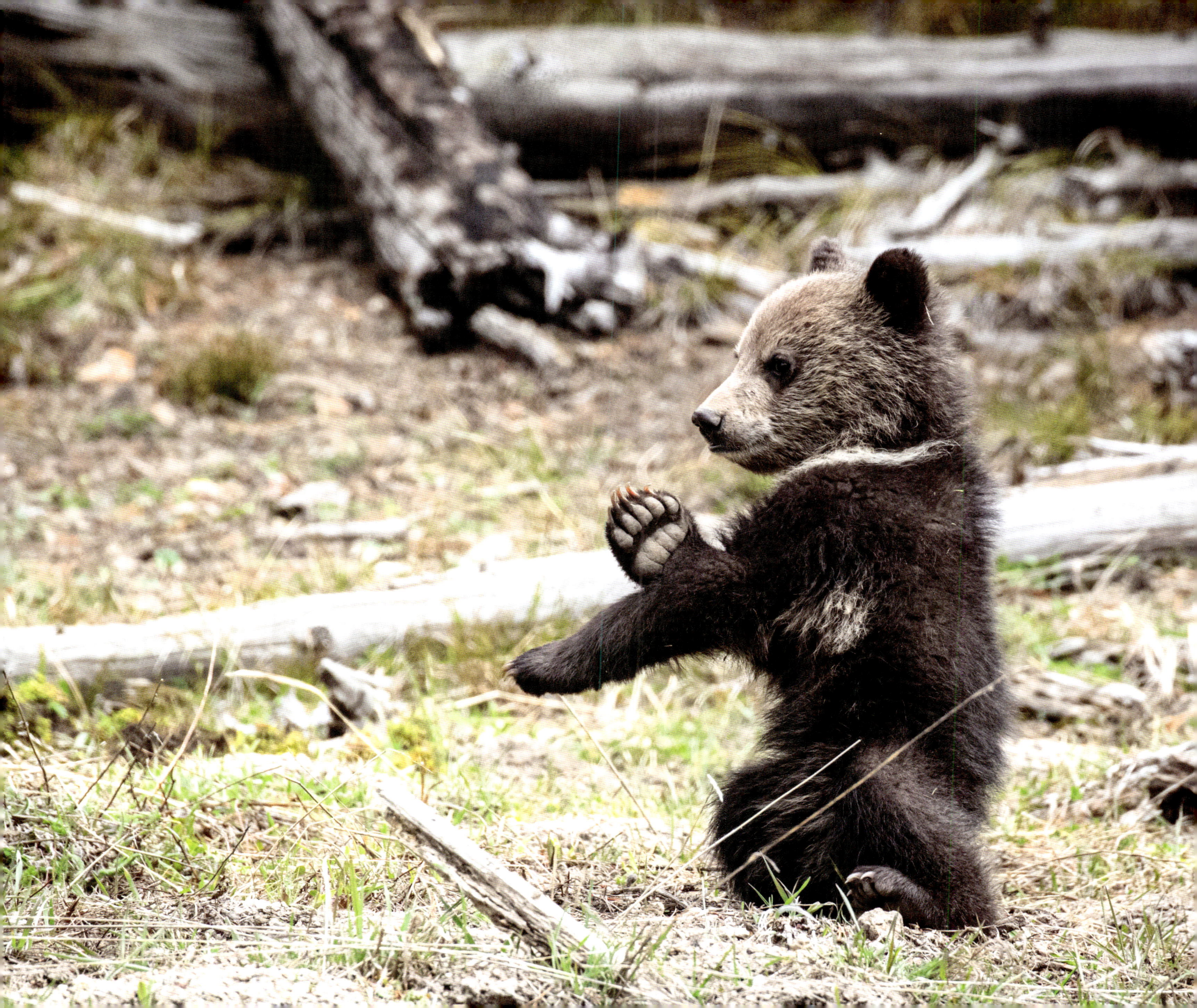

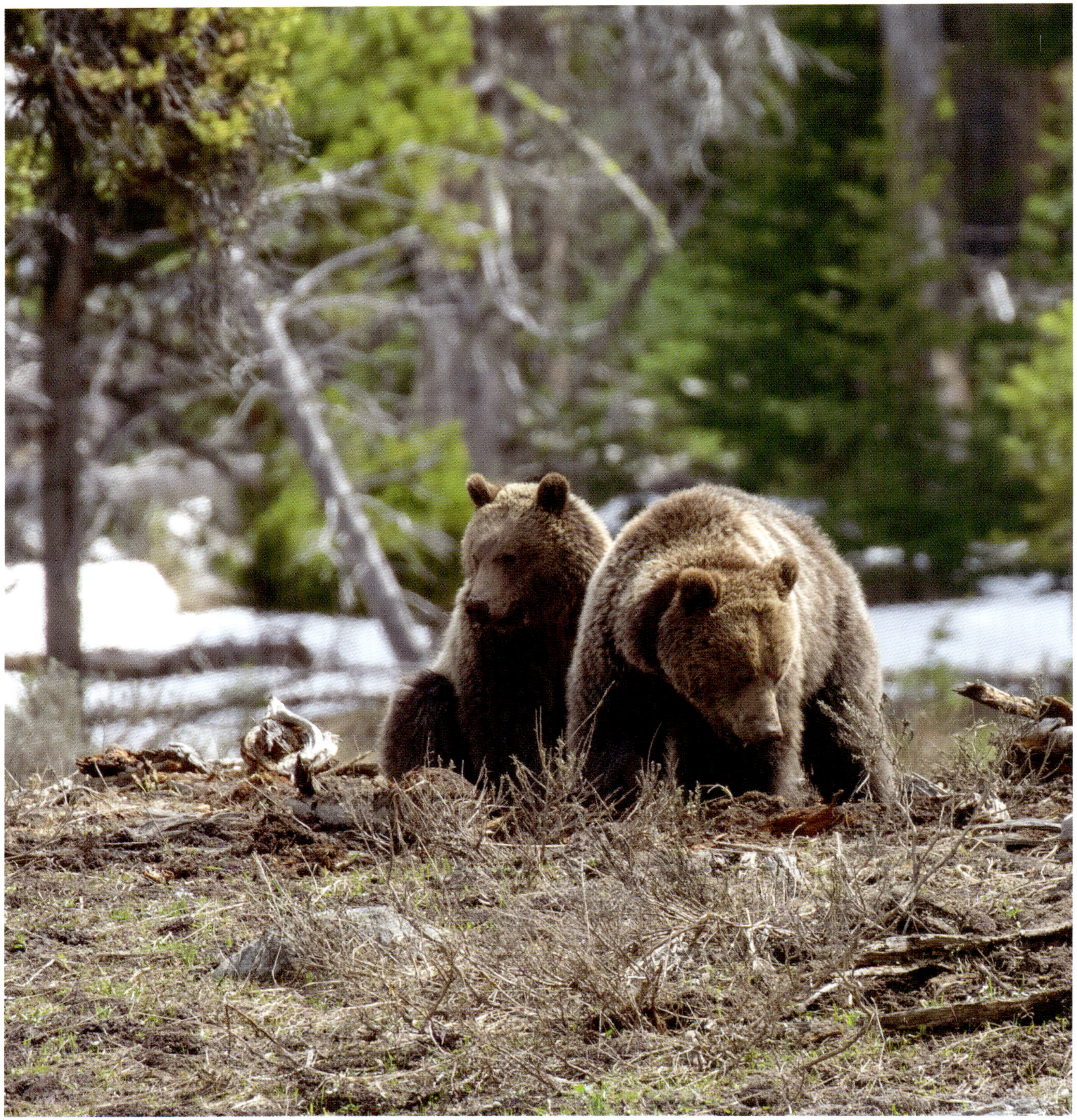

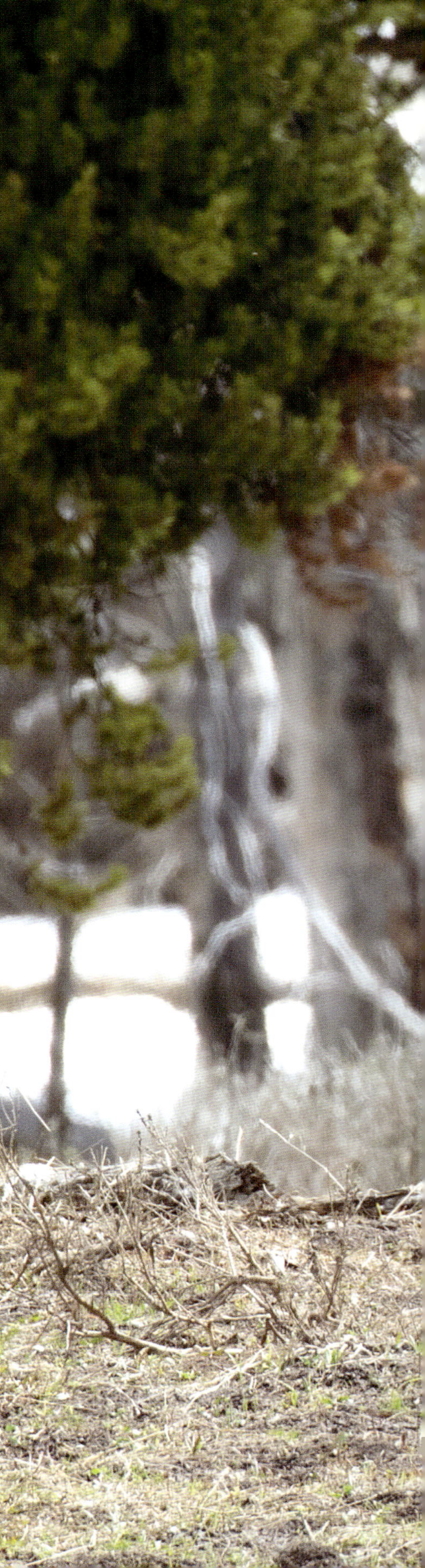

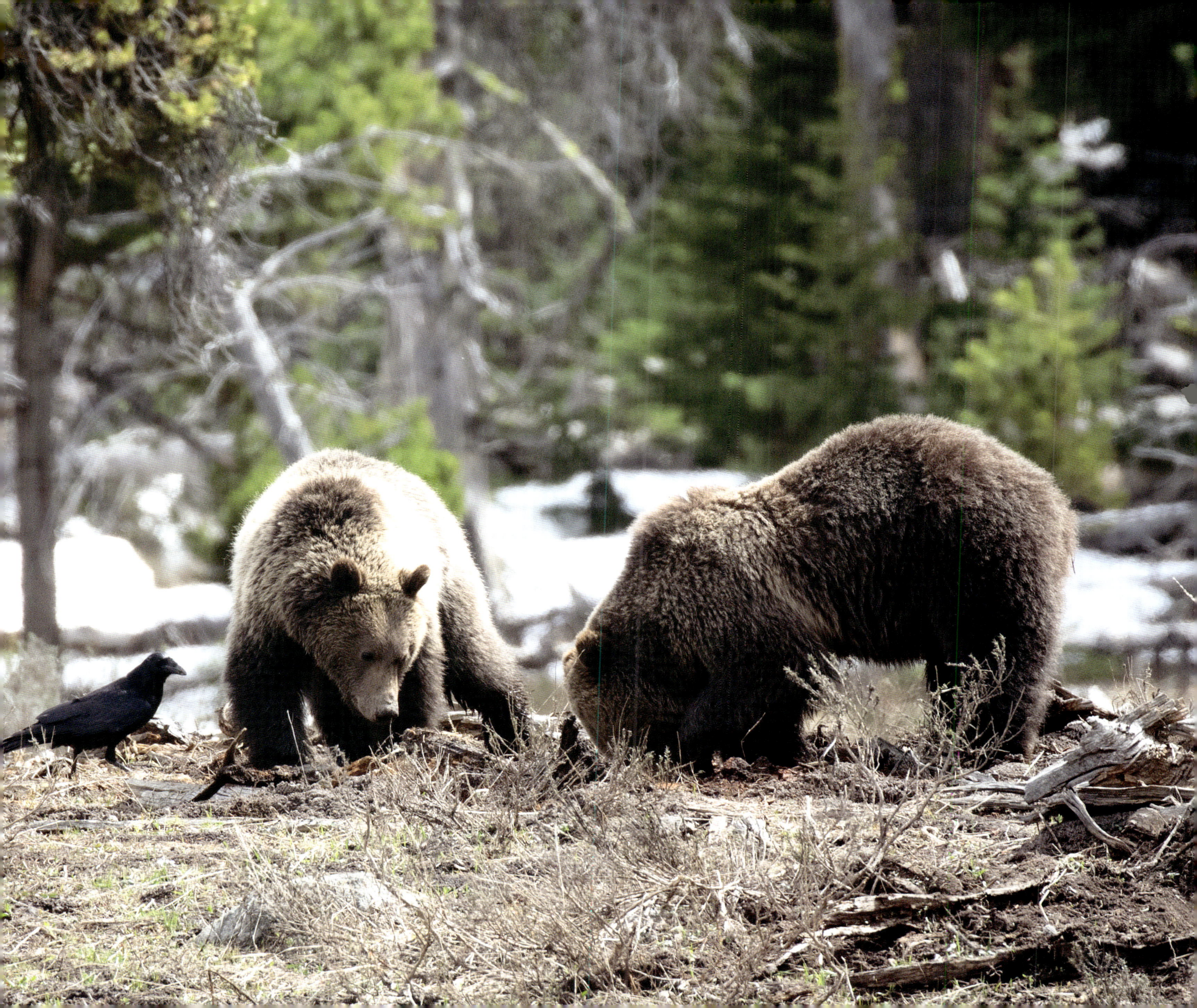

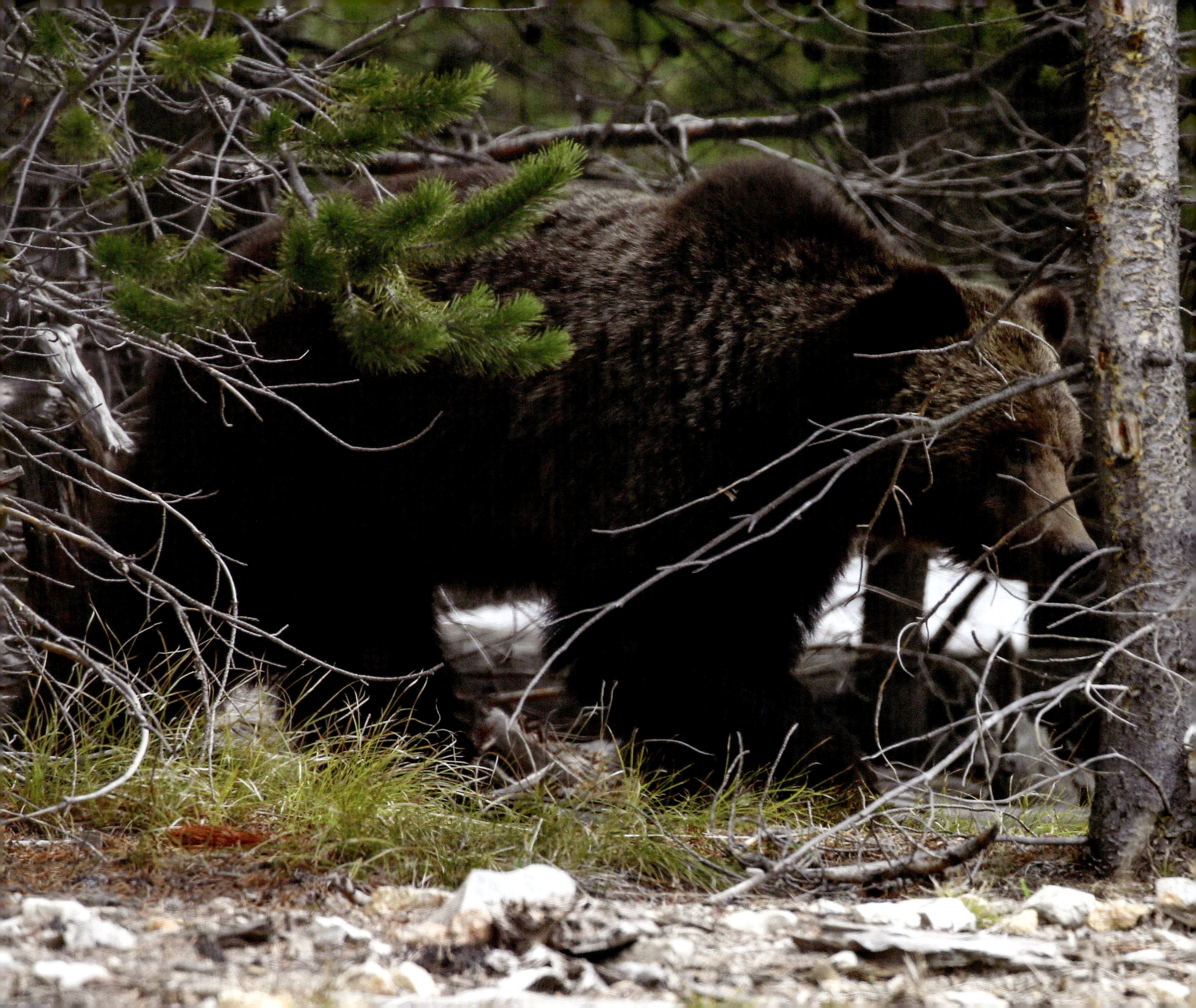

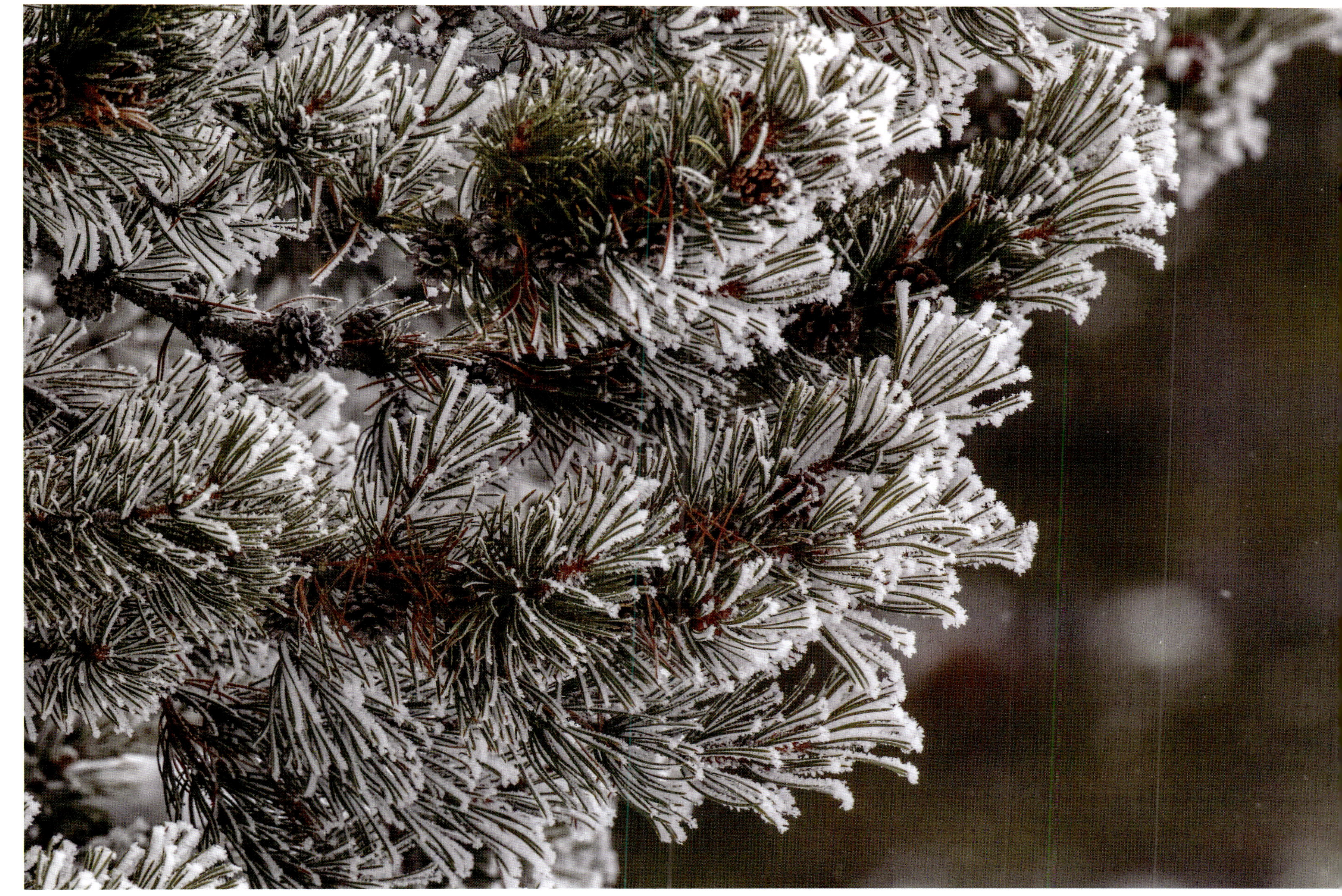

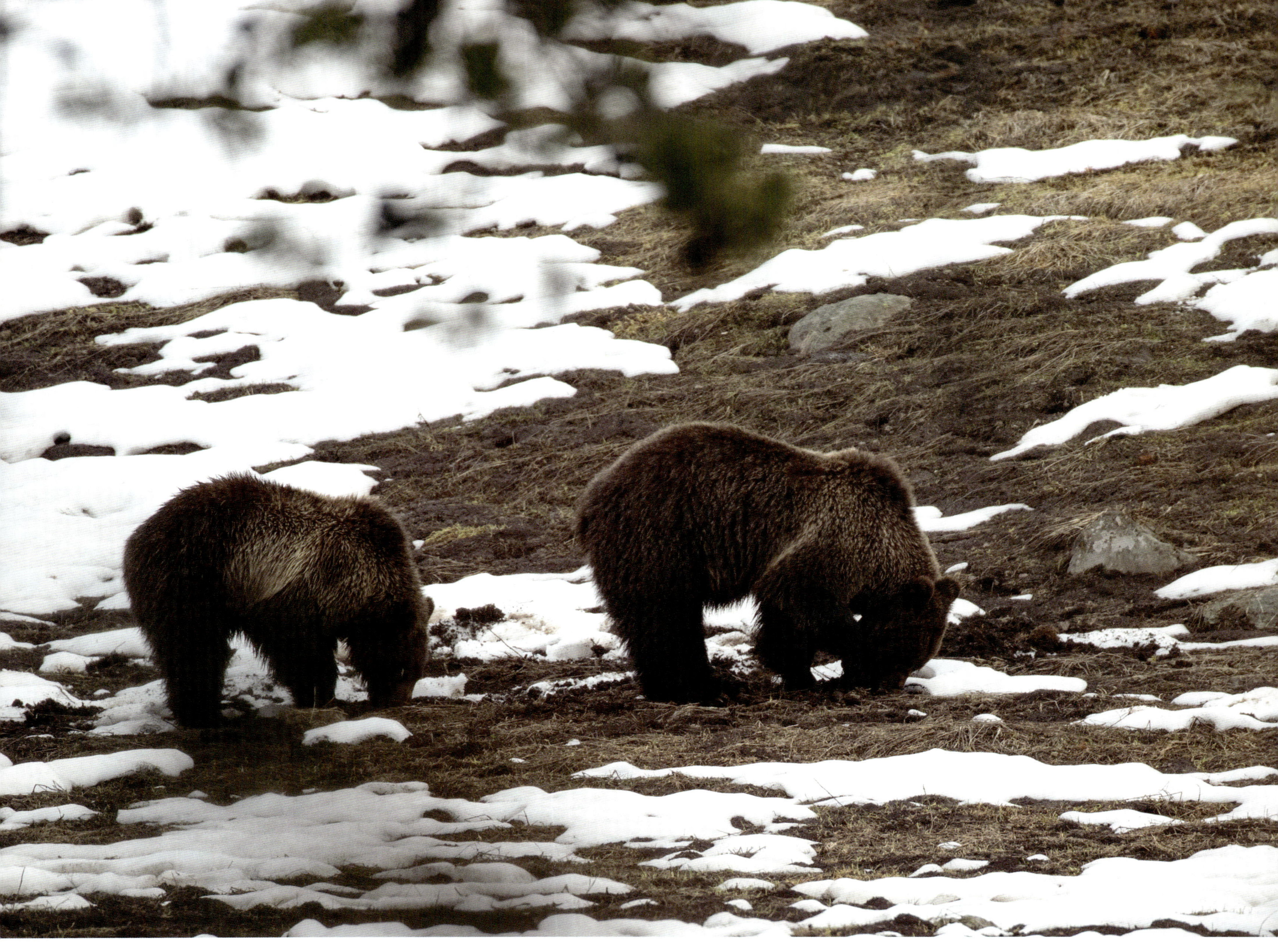

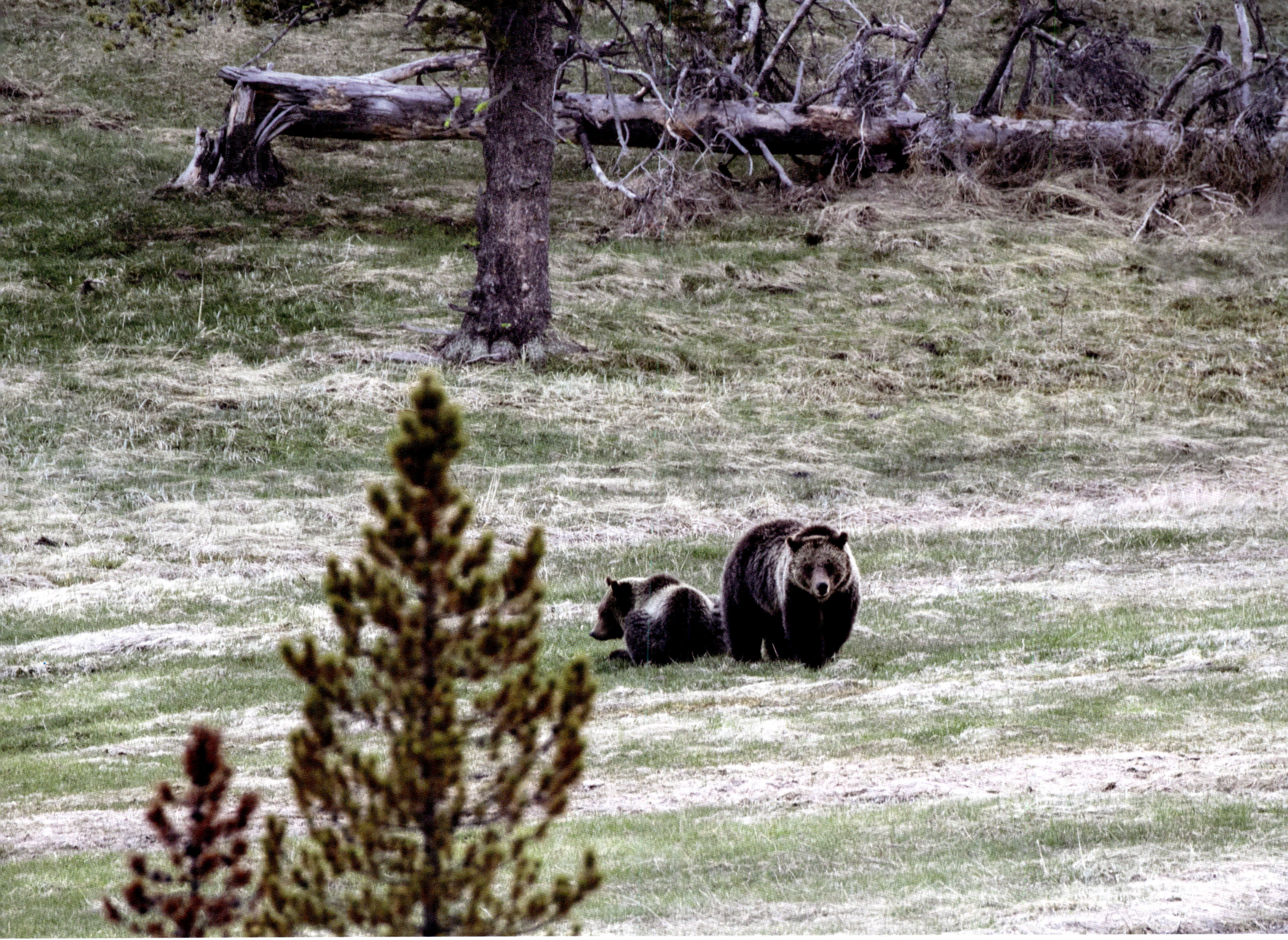

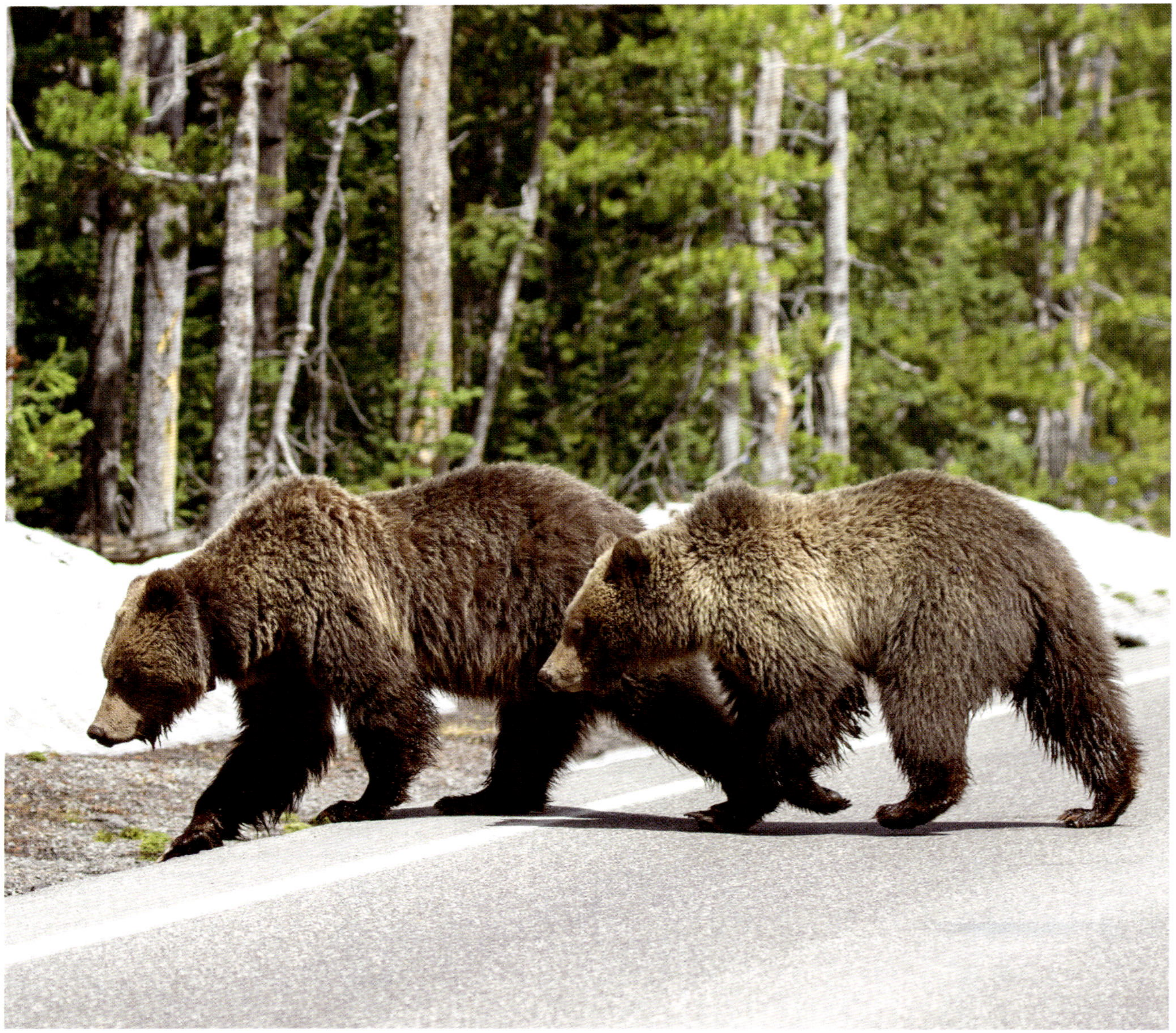

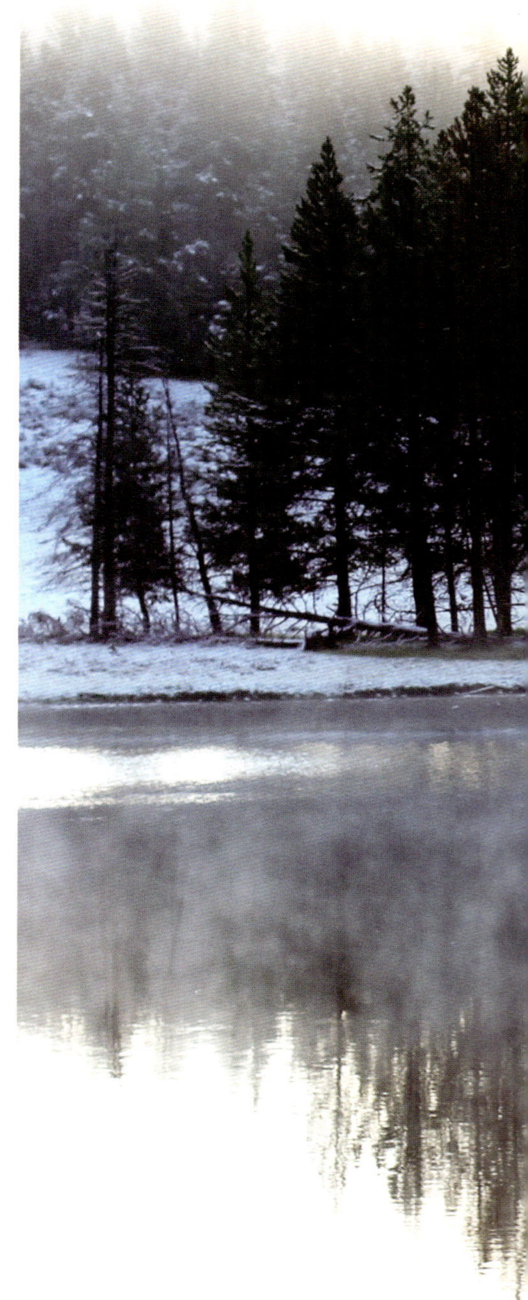

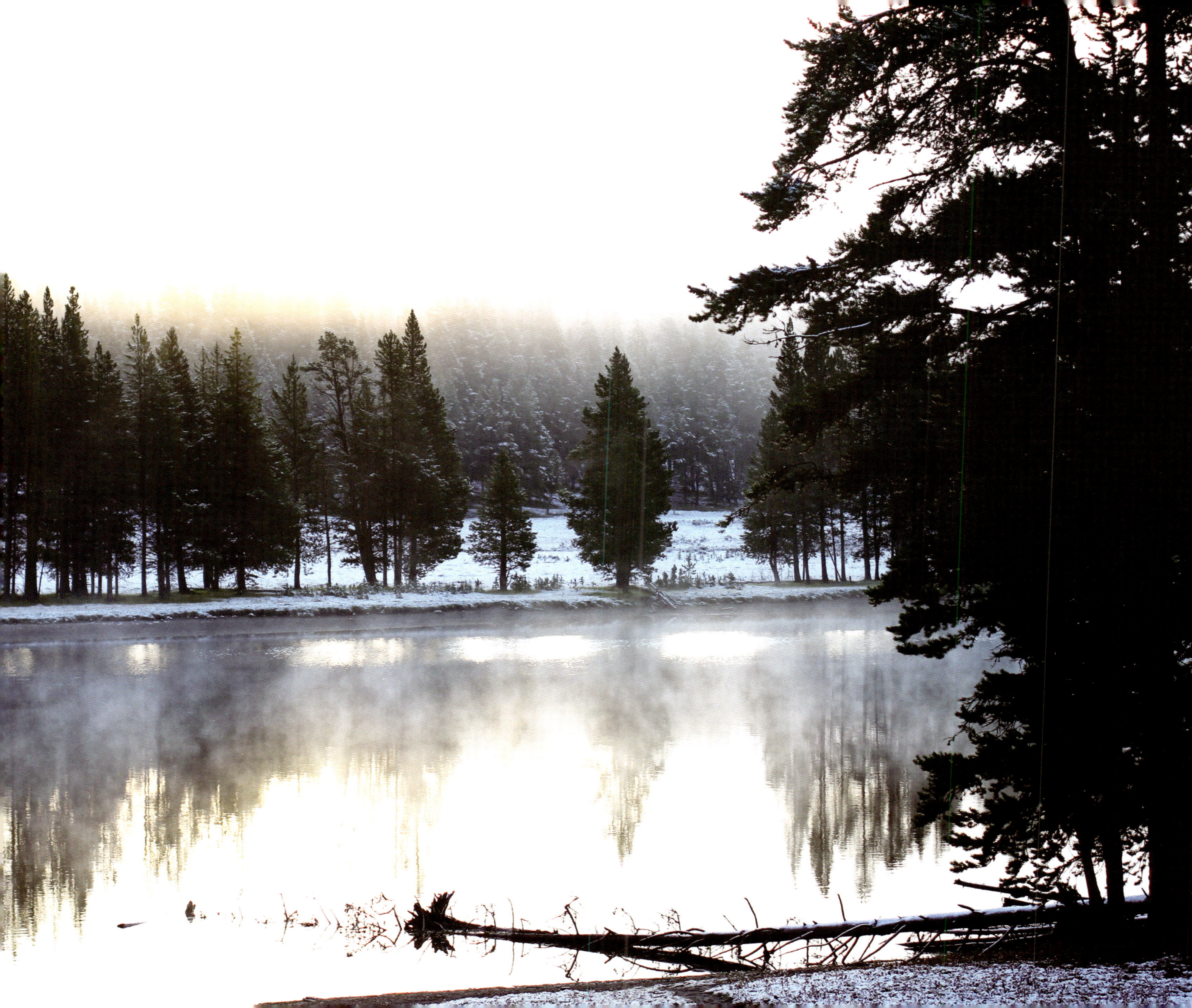

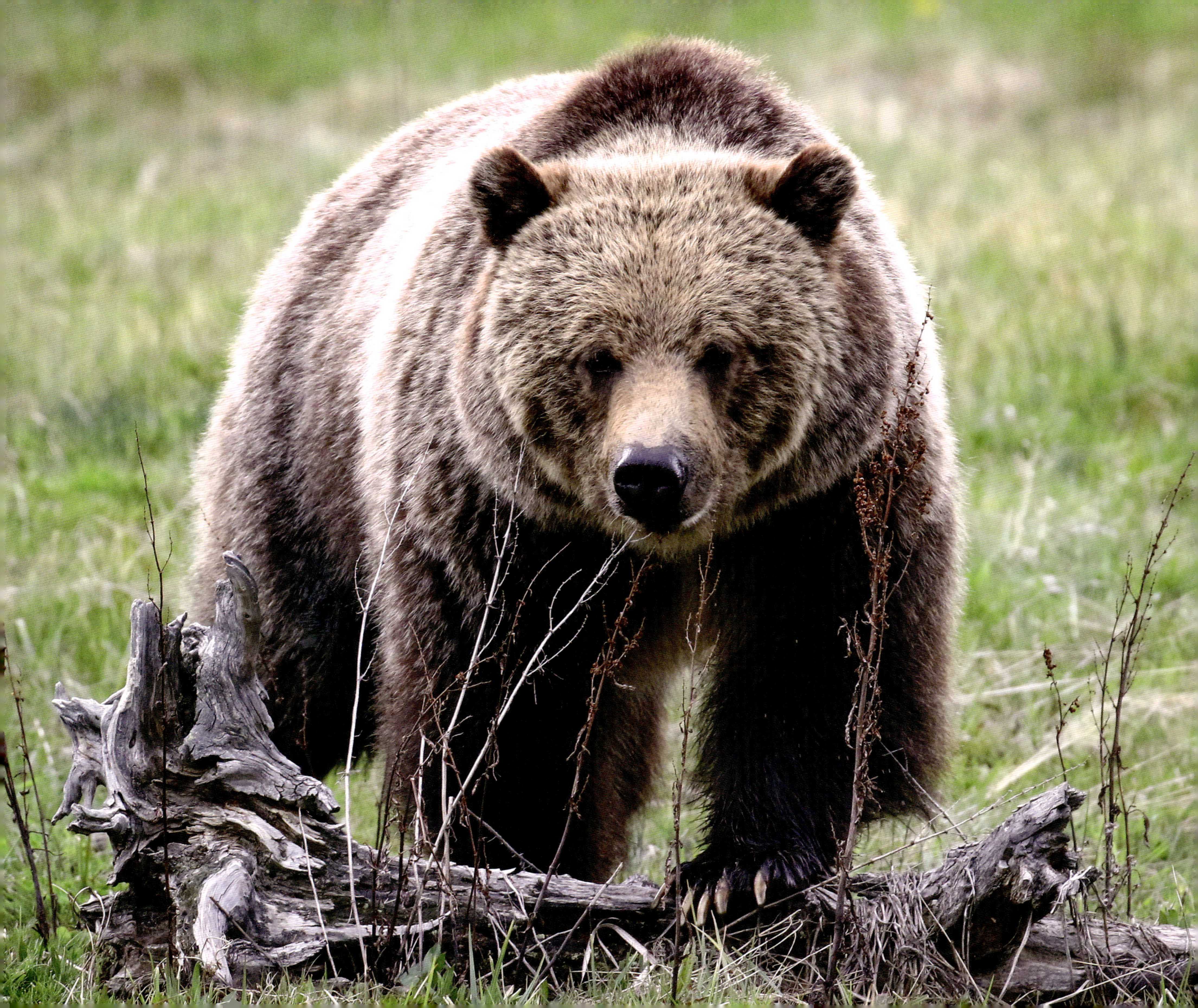

CLOSING

IT'S HARD TO IMAGINE A LANDSCAPE IN THE NORTHERN ROCKIES WITHOUT THE MIGHTY GRIZZLY. When I think about their future and some of the things that play into it such as climate change, the lack of a natural food supply, and human population growth that is compounded by people who want to hunt the grizzlies and those who illegally poach them, it makes me wonder whether they can truly survive in a world with so much uncertainty.

It is my hope that we think about that future, their future, and that we take steps to ensure their survival for many generations to come. I believe coexistence is the answer and the future for the grizzlies. Because the human population will continue to expand into traditional grizzly territory, humans will impact where the grizzlies roam. And if the mighty bears are to survive, we must find a way to coexist, acknowledging their needs and giving them the space they require to live.

RESOURCES

INFORMATION

Bear Smart
bearsmart.com

Defenders of Wildlife
defenders.org

Global Indigenous Council/Grizzly Treaty
www.globalindigenouscouncil.com

Grizzly Bear Foundation
grizzlybearfoundation.com

Hey Bear – GOAL Tribal Coalition
Facebook.com/HeyBear.GOAL

Interagency Grizzly Bear Committee
igbconline.org

Interagency Grizzly Bear Study Team
www.usgs.gov

National Wildlife Federation
www.nwf.org

Native Hope/Native American Animals
blog.nativehope.org

Save the Yellowstone Grizzly
savetheyellowstonegrizzly.org

Sierra Club/Grizzlies
www.sierraclub.org

Vital Ground Grizzly Bear Conservation
www.vitalground.org

Yellowstone National Park/Grizzly Bear
www.nps.gov/yell

VIEWING

Glacier National Park
nps.gov/glac

Grand Teton National Park
nps.gov/grte

Greater Yellowstone Ecosystem
nps.gov/yell

Yellowstone National Park
nps.gov/yell

READING

Grizzly Heart, Charlie Russell

Grizzly Years: In Search of the American Wilderness, Doug Peacock

In the Presence of Grizzlies: The Ancient Bond Between Men and Bears, Doug and Andrea Peacock

One of Us: A Biologist's Walk Among Bears, Barrie K. Gilbert

The Essential Grizzly: The Mingled Fate of Men and Bears, Doug and Andrea Peacock

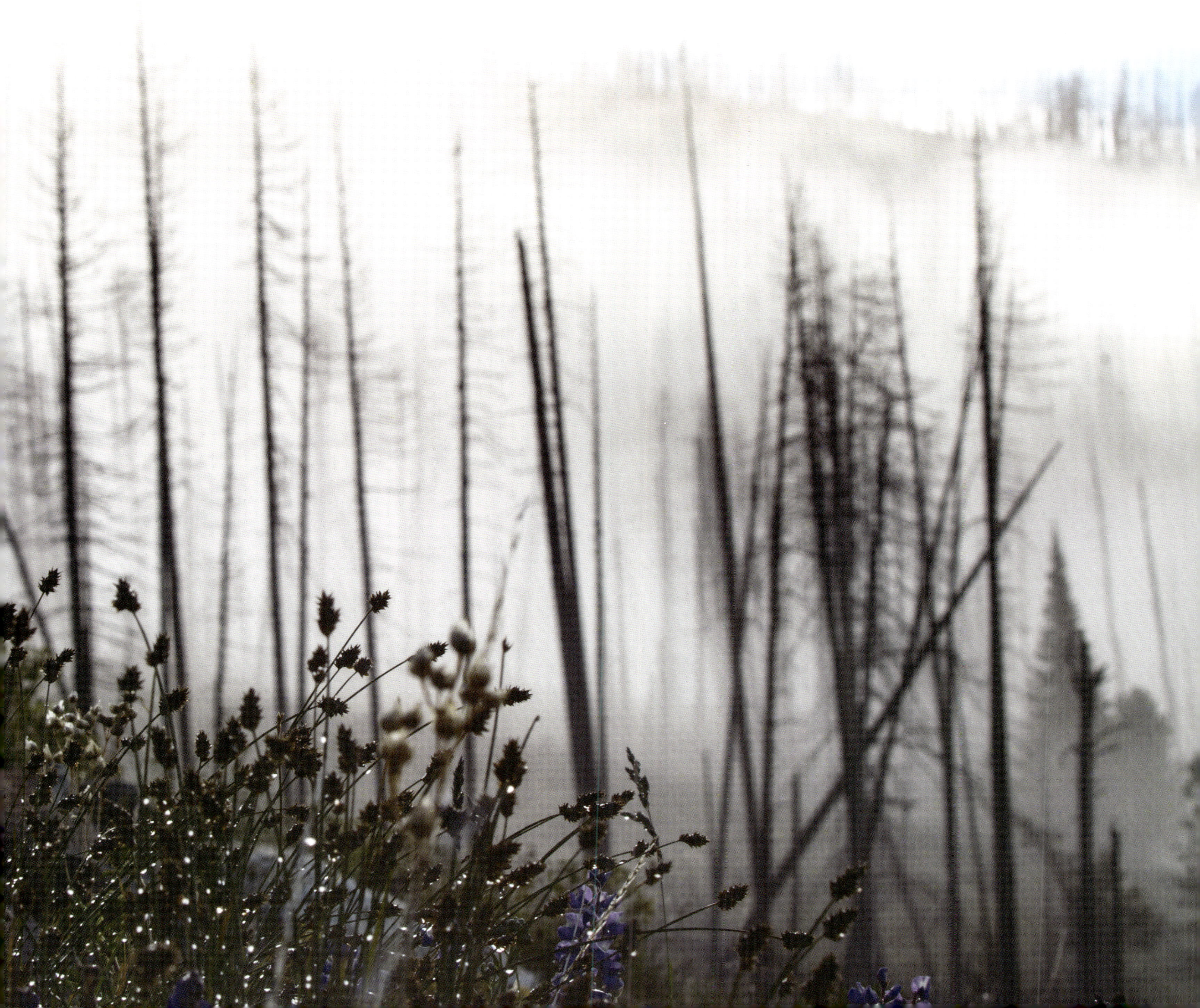

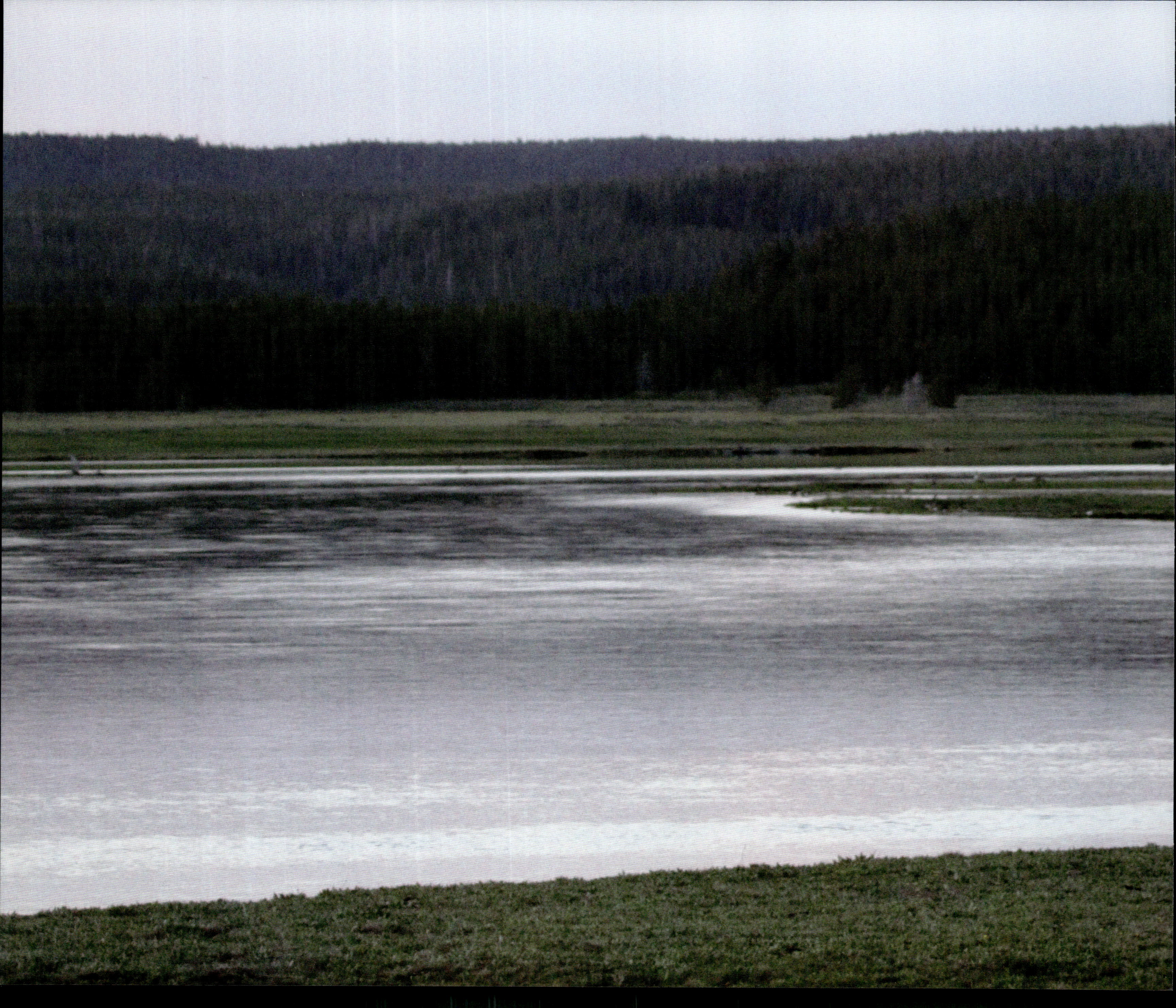

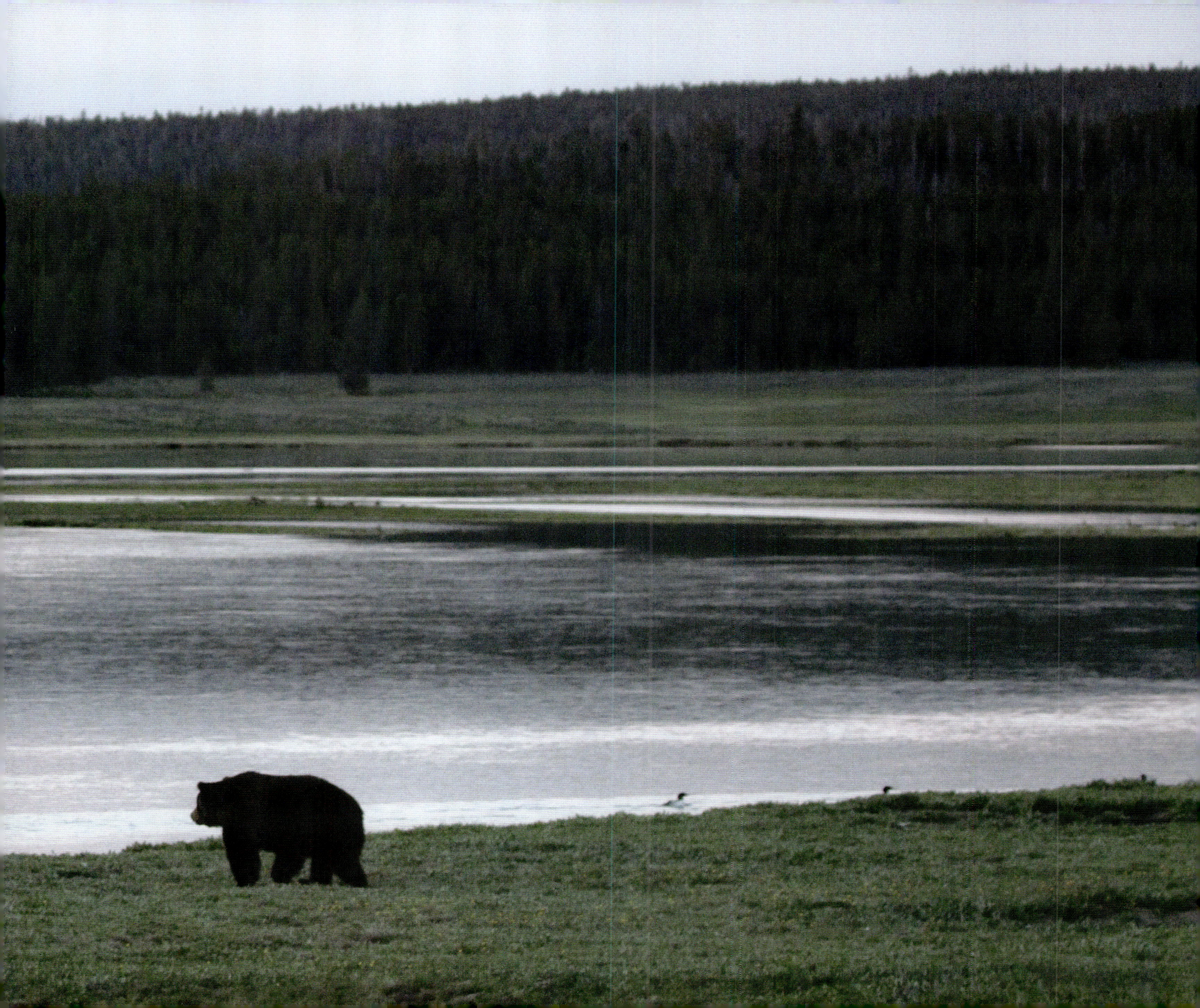

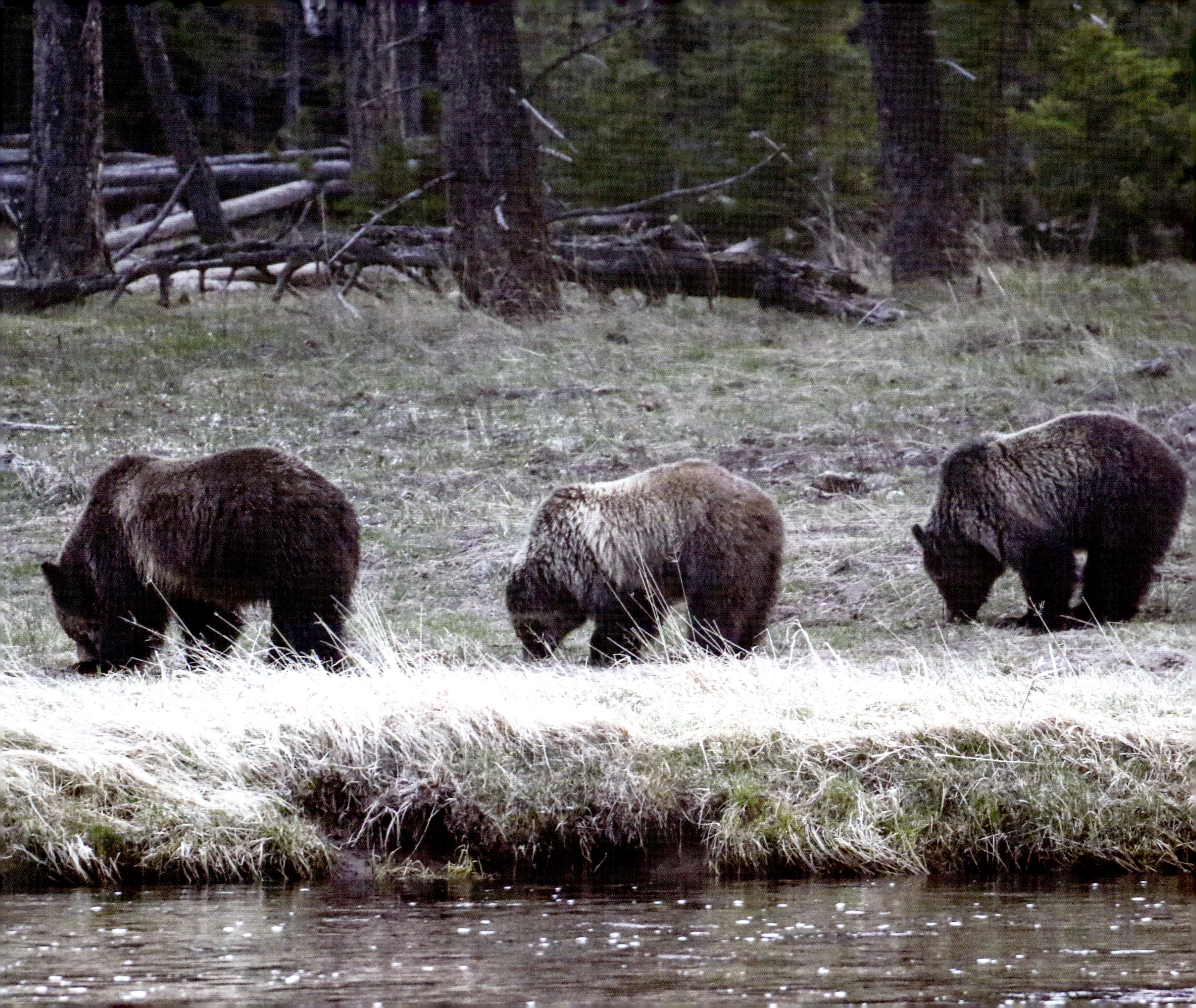

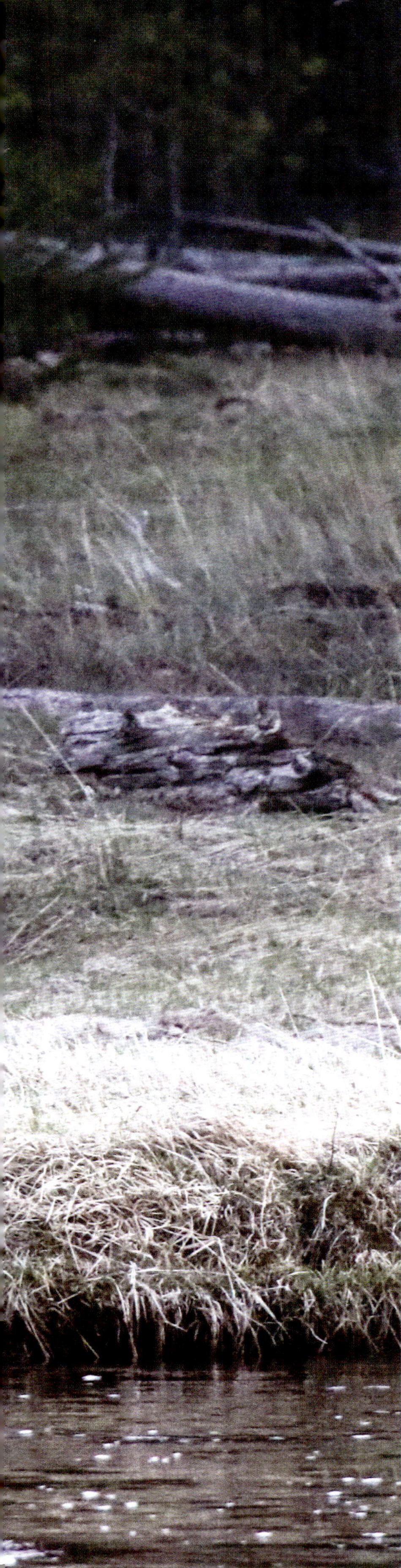

ACKNOWLEDGMENTS

I would like to thank my mom and dad for teaching me a love of nature at a very young age, and for always supporting me while I chase my dreams. I would like to thank my family, especially my kids, for allowing me to drag them out of bed at the crack of dawn and then staying out until the last light of day has disappeared. I would like to thank my Yellowstone family for all of our crazy adventures, and those who have truly inspired me with your work; Sandy Sisti, Brad Orsted, Doug Peacock, Tyler Brasington, and so many others. A special thanks to Yellowstone's Bear Management, Eric, Brady, Jessica, and all of the other rangers for watching out for these incredible animals. I would also like to thank the late Paula Ozello for her advocacy for all things wild.

First Edition
25 5 4 3

Published by
Gibbs Smith
P.O. Box 667
Layton, Utah 84041

1.800.835.4993 orders
www.gibbs-smith.com

Designed by Rita Sowins / Sowins Design
Printed and bound in China

Gibbs Smith books are printed on paper produced from sustainable PEFC-certified forest/controlled wood source. Learn more at www.pefc.org.

Library of Congress Control Number: 2021931670
ISBN: 978-1-4236-5879-5